THE SCIENCE OF WEIRD SHIT

Why Our Minds Conjure the Paranormal

CHRIS FRENCH

foreword by Richard Wiseman

The MIT Press
Cambridge, Massachusetts
London, England

© 2024 Chris French

All rights reserved. No part of this book may be used to train artificial intelligence systems or reproduced in any form by any electronic or mechanical means (including photocopying, recording, or information storage and retrieval) without permission in writing from the publisher.

The MIT Press would like to thank the anonymous peer reviewers who provided comments on drafts of this book. The generous work of academic experts is essential for establishing the authority and quality of our publications. We acknowledge with gratitude the contributions of these otherwise uncredited readers.

This book was set in Adobe Garamond Pro by New Best-set Typesetters Ltd. Printed and bound in the United States of America.

Library of Congress Cataloging-in-Publication Data

Names: French, Chris (Professor), author.
Title: The science of weird shit : why our minds conjure the paranormal / Chris French ; foreword by Richard Wiseman.
Description: Cambridge, Massachusetts : The MIT Press, [2024] | Includes bibliographical references and index.
Identifiers: LCCN 2023013925 (print) | LCCN 2023013926 (ebook) | ISBN 9780262048361 (hardcover) | ISBN 9780262375771 (epub) | ISBN 9780262375764 (pdf)
Subjects: LCSH: Psychology. | Parapsychology. | Ghosts.
Classification: LCC BF121 .F72 2024 (print) | LCC BF121 (ebook) | DDC 150—dc23/eng/20231116
LC record available at https://lccn.loc.gov/2023013925
LC ebook record available at https://lccn.loc.gov/2023013926

10 9 8 7 6 5 4 3 2 1

For Lucy, Kat, and Alice

Contents

Foreword *by Richard Wiseman* ix
Preface xiii

1 **WEIRD SCIENCE** *1*
2 **WAKING NIGHTMARES** *33*
3 **HIGH SPIRITS, PART 1: GHOSTLY ENCOUNTERS** *69*
4 **HIGH SPIRITS, PART 2: COMMUNICATING WITH THE DEAD** *111*
5 **FANTASTIC MEMORIES OF ALIEN ENCOUNTERS** *133*
6 **MANY HAPPY RETURNS?** *171*
7 **DYING TO KNOW THE TRUTH** *191*
8 **NO SUCH THING AS COINCIDENCE?** *215*
9 **TRICKS OF THE MIND** *235*
10 **SKEPTICAL INQUIRIES** *247*
11 **DREAMS OF THE FUTURE** *267*
12 **LESSONS FOR (THE) LIVING** *283*
  EPILOGUE: THE LIMITS OF SKEPTICISM *301*

Author's Note and Acknowledgments *317*
Notes *321*
Index *365*

Foreword

I have always been skeptical. Indeed, when I was seven years old, I had a sneaking suspicion that I was eight. During my teenage years I became fascinated by seemingly paranormal phenomena but always preferred the debunk to the bunk. As a result, I spent many happy hours reading books that explored the science behind the supernatural, including titles such as *UFOs Explained*, *Nostradamus Exposed*, and *All of Your Dreams Are Meaningless*. In my twenties, I obtained a degree in behavioral science, completed a doctorate on the psychology of the paranormal, and started to conduct my own research into the unexplained. Time and again I discovered that seemingly extraordinary phenomena had perfectly ordinary explanations, and it therefore came as quite a shock when, in my midthirties, I found myself regularly coming face to face with the Devil.

The encounters always happened in the middle of the night when I was fast asleep and involved me suddenly sitting upright in bed with my eyes wide open and my palms sweating profusely. Each time, I was convinced that Mr. Splitfoot had somehow found his way into my bedroom and that he was now standing ten feet away from my bed with his red eyes staring straight at me. I can easily see how the scary experience could cause a believer in demonic entities to contact their local exorcist, to carry out magical rituals in an effort to protect themselves from the Prince of Darkness, or, at the very least, to shit themselves. However, as a lifelong disbeliever, I decided to discover if there might be a more scientific explanation for my nightly encounters. A few hours searching the internet revealed that I was experiencing a strange, but surprisingly common, sleep disorder known as a night terror. The internet also offered a range of opinions about the causes of these experiences, with the smart money being split between falling asleep feeling anxious, having an overly warm bedroom, and positioning your wardrobe perilously close to a portal into the underworld.

Chris French has also spent much of his career studying weird phenomena. Like me, he has jettisoned his Ouija board and dowsing rods in favor of a more rational and scientific approach to the topic. I have known Chris for over twenty years, and I can think of no better traveling companion to guide you through the heady world of the paranormal. We first met in a coffee shop in central London. Neither of us knew that the other one was going to be there, so the meeting was a lovely coincidence. I recognized Chris as soon as he walked in and intuitively knew that he would order a large decaf latte with sprinkles. When he saw me, he levitated over, with blood pouring from stigmata on his hands and feet, and the two of us chatted about our shared skepticism for matters paranormal.

More seriously, I have always admired Chris's ability to befriend both believers and skeptics alike, to boldly go where few scientists have gone before and to have a good time going there. In this book, he will introduce you to a wide range of seemingly paranormal phenomena, including ghosts, out-of-body experiences, reincarnation, déjà vu, mediumship, hauntings, alien abductions, and déjà vu. In each instance a lesser scientist might be tempted to adopt a cynical attitude and to argue that investigating such outlandish and weird phenomena is a waste of time. However, like a modern-day Charles Fort, Chris convincingly demonstrates that there is much to be gained from spending time at the fringes of science, from studying anomalies and from taking the road less traveled. Time and again, Chris's adventures into the unknown deepen our scientific understanding of the human mind and provide important insights into the psychology of attention, perception, memory, and much more. And unlike the authors of many other books in this genre, Chris isn't an outside observer reporting the work of others. Instead, he has been there, done that, and got the bent spoon to prove it.

The benefit of Chris's sojourns into the supernatural go beyond science. A large percentage of the public is fascinated with the strange, the mysterious, and the uncanny. As a result, documentary makers and journalists frequently produce television programs and films about the topic. Alas, all too often these outpourings are aimed at those who believe in such phenomena, with skeptics struggling to make their voices heard. Chris has repeatedly fought the good fight and frequently appeared on these programs as the (often solitary) voice of rationality. This book describes his behind-the-scenes experiences and presents a fascinating and eye-opening account of how such shows are conceived, filmed, and edited.

Finally, the book also provides a rare insight into how science works. Many years ago, Sue Blackmore (who influenced the thinking and careers of both Chris and myself) wrote a wonderful autobiography describing her journey from paranormal believer to skeptic. Much of her book describes how scientists actually design and carry out their research, as opposed to the rather sanitized and sterile accounts often presented in academic papers and talks. Chris's book is full of equally valuable insights and will be of great interest to anyone wanting to know what it's really like to work as a psychologist. Often chatty, it describes the origin story behind several studies (including a lovely anecdote about how a chance observation by his daughter led to one of his most influential papers) and offers fascinating descriptions of the trials and tribulations of trying to carry out research in this tricky area.

The time has come for me to hand you over to my old pal and trusted colleague Chris French. Regardless of whether you are a firm believer in the paranormal, a staunch skeptic, or an undecided waverer, I know that you are going to have a great time. Some say that the truth is out there, but in my opinion it's in here. Enjoy.

Richard Wiseman
Professor, University of Hertfordshire

Preface

IN THE BEGINNING

The paranormal has always fascinated me. When I was a child, it also terrified me—especially the idea of ghosts. Indeed, it was not until fairly late in my childhood, maybe at the age of ten or eleven, that I was able to sleep on my own without a nightlight. Despite the terror that thoughts of spooks and bogeymen could often ignite in my childish imagination, my fascination for the paranormal would tempt me to read scary short stories or watch spooky TV programs—activities that I sincerely regretted later as I lay awake in my bed trying to get to sleep, listening out for any noise that might indicate an unwelcome nocturnal visitor.

In my teenage years, my interest in the paranormal widened and my fear of ghosts subsided somewhat. I became interested in many other aspects of the paranormal, particularly the books of Erich von Däniken,[1] who argued that evidence strongly supported the idea that our planet had been repeatedly visited in the past by advanced alien civilizations. As a curious and somewhat nerdy teenager, I was totally taken in by von Däniken's arguments and "evidence." I naively believed that it must be true because it was written in a book. I was somewhat surprised that many people did not find von Däniken's ideas as fascinating and mind-blowing as I did. Having said that, clearly enough people shared my perspective to ensure that his books sold in the millions. It was not until many years later that I came across critiques such as Ronald Story's *The Space-Gods Revealed: A Close Look at the Theories of Erich von Däniken* and fully appreciated how shoddy von Däniken's arguments were and how frankly dishonest his presentation of evidence was.[2]

Another memorable moment in my gullible teenage years was the appearance on British TV screens in 1973 of a certain Uri Geller.[3] Geller appeared to have amazing psychic abilities. He could apparently read minds, divine the contents of sealed envelopes, fix broken watches with the power of his mind, and, most notably of all, bend metal objects such as forks and spoons merely by gently stroking them (see figure P.1). From the outset, some scientists, including British physicists John Taylor and John Hasted, endorsed Geller as having genuine paranormal ability (although Taylor subsequently changed his mind).[4] At the time, I too was totally convinced of Geller's powers. I so wanted him to be genuine.[5]

In 1974, I went to Manchester University to study for a BSc in psychology. The topics covered in this book were rarely, if ever, covered as part of my undergraduate degree. After graduating in 1977 and working for a short period as a research assistant at the University of Bangor in North Wales, I embarked on a period of postgraduate research at the University of Leicester under the supervision of Dr. J. Graham Beaumont. My PhD was not in the area of anomalistic psychology. Indeed, I had never heard of anomalistic psychology at that time, not least because the term had yet to be invented. Instead, my research used electroencephalography (EEG) in a largely unsuccessful attempt to study the function and anatomy of the cerebral hemispheres of the human brain.

THE TURNING POINT

I can remember fairly precisely when I stopped believing in the paranormal—and what caused me to do so. After I completed my postgraduate research, I took up a one-year post as a lecturer at Coventry Polytechnic (now Coventry University). As I recall, it was during this period that a friend recommended to me a newly published book called *Parapsychology: Science or Magic?* by Canadian social psychologist James Alcock.[6] It was the first detailed academic critique of parapsychology that I had ever come across. It also put forward plausible nonparanormal explanations for a range of ostensibly paranormal experiences. As my friend had suspected, I loved the book—but I

Figure P.1
When Uri Geller appeared on TV in the UK in the 1970s, millions of people (including the author) were convinced that he possessed genuine psychic powers, including the ability to bend metal by gently stroking it.

had no idea how massive a role it would play in my life. It opened the door to a whole new world of skepticism that I had never suspected existed.

Back then, just like today, there were vastly more uncritical pro-paranormal books, articles, and TV and radio programs than skeptical treatments of the paranormal. If anything, the situation was even more biased in favor of paranormal claims than it is today. However, it turned out that there was some skeptical literature out there if only you knew where to find it. I noticed lots of references in Alcock's book to a publication called the *Skeptical Inquirer*. I cannot now remember how I did so, but I managed to track down Mike Hutchinson, the UK distributor for this American magazine, and take out a subscription. When an issue arrived, I would eagerly read it from cover to cover.

SOME EARLY INFLUENCES

Alcock's book also made me aware of the work of James ("The Amazing") Randi. Many readers will no doubt already be aware of this larger-than-life individual, but I confess that back then, in the early 1980s, I was not. I now know that if skeptics were allowed to have patron saints—which of course we're not—James Randi would probably occupy that role.[7]

Randi truly lived up to the word *amazing* in his stage name. During his long career, he had, for example, performed as a conjurer in a carnival roadshow and subsequently in many countries around the word, written an astrology column under the pseudonym of Zo-Ran for a Canadian tabloid (he did this by simply shuffling astrology items from other newspapers and randomly pasting them into his column), escaped from safes and prison cells on numerous occasions, remained submerged in a sealed metal coffin in a swimming pool for 104 minutes, appeared in and hosted numerous radio and TV programs, performed as both a mad dentist and executioner on Alice Cooper's *Billion Dollar Babies* tour, and escaped from a straitjacket while suspended by his feet above Niagara Falls. Anyone who would not describe such a career as "amazing" is in serious need of a good dictionary. But it is not for achievements such as these that Randi is held in such high esteem within the world of skepticism.

Born in 1928, Randi retired as a performing conjurer at the age of sixty having achieved international recognition for his skills in that domain. But by this time he had also achieved recognition as a skeptical investigator of paranormal claims. He first came to prominence in this role back in the early 1970s, when a certain Uri Geller was the focus of the world's media with his sensational claims of amazing paranormal abilities. Randi challenged these claims, accusing Geller of being a fraud and a charlatan who used nothing more than standard conjuring techniques to achieve his effects. I have a vague memory (which may or may not be a false memory) of seeing Randi on television at this time, demonstrating Geller's most famous feat of apparently bending a piece of cutlery, often to the point where the bowl of the spoon would become detached from the stem, just by gently stroking it. While not revealing how it was done, Randi insisted it was simply a magic trick. At that tender age, I saw Randi's performance as entirely irrelevant. Okay, so you can do something as a magic trick that *looked the same* as what Geller was doing. But Uri wasn't doing a magic trick, was he? He was using his psychic powers to achieve the effect. Dumb or what?[8]

I now have quite a few friends who are professional conjurers and, as they point out, if Uri really is using psychic powers to achieve the effect, he's doing it the hard way—it really does look exactly the same when done as a magic trick![9] It should be noted that Geller has never managed to successfully perform this allegedly psychic feat under properly controlled conditions that would rule out the possibility of achieving the effect through the use of conjuring techniques.

As I look back some four decades or so, I can now more fully appreciate the influence that a number of other skeptics had on my own thinking even though I was not so aware of it at the time. I can no longer remember where or when I met most of them, or even the order in which I met them, but each made an impression in their own way. Prominent among them would be Dr. Susan Blackmore. Sue was certainly the most recognized and well-informed skeptic on the UK scene when I first became aware of her back in the 1980s. Extroverted and articulate, she often appeared on TV presenting the skeptical perspective on a range of ostensibly paranormal phenomena, sometimes with her hair dyed three different colors.

Like me, Sue had been a strong believer in the paranormal for many years before embarking on her own parapsychological research, in her case at the University of Surrey. Her belief in the paranormal had been reinforced by a profound out-of-body experience (OBE) she had had as a student in Oxford in 1970, brought on by a combination of tiredness and mind-altering substances.[10] She began her PhD research at Surrey convinced that she could produce strong scientific evidence to prove the existence of extrasensory perception (ESP) despite the general skepticism of the wider scientific community. Several years later, after expending much time and effort on this quest, she ended up as a vocal skeptic. The fact that she arrived at this skeptical position after years of research in the area was pretty compelling evidence to me that maybe ESP did not exist after all.

It was also around this time that I met Richard Wiseman. In contrast to Sue and myself, Richard had never been a believer in the paranormal but had always been fascinated by paranormal claims. His other obsession growing up was conjuring (which may go some way toward explaining why he was always skeptical about the paranormal). He is now a member of the Inner Magic Circle, although he decided that a career as a professional conjurer was not for him and instead registered for a degree in psychology at University College London. If my memory is correct, I met Richard shortly after he had completed his PhD at the University of Edinburgh, under the supervision of the first holder of the Koestler Chair in Parapsychology, the late Professor Robert Morris.

Richard is currently the holder of the UK's only chair in the public understanding of psychology (at the University of Hertfordshire) and is one of the most recognizable British psychologists today. Not only is he an extremely gifted communicator of science, often appearing in the media, but he is also very, very funny. He has over 130,000 followers on Twitter (I just checked), his books have sold over three million copies, and he has been described by Elizabeth Loftus (Past President, Association for Psychological Science) as "one of the world's most creative psychologists."[11] Not surprisingly, I hate him (joking—he's actually one of my favorite people to hang out with).

As I began attending skeptics' conferences, I was also lucky enough to meet Professor Ray Hyman, an American psychologist whose writings I was already familiar with. Ray is respected by those on both sides of the debate regarding the existence of paranormal abilities because his criticisms of parapsychology, like Jim Alcock's, are based on a detailed knowledge of the field in contrast to the superficial and ill-informed dismissal that unfortunately is all too common among some skeptics. I confess I was somewhat in awe of Ray, but he turned out to be one of the friendliest people you could hope to meet.

Ray was a founding member of a group called the Committee for the Scientific Investigation of Claims of the Paranormal back in 1976. CSICOP is usually pronounced "psi cop," which seems appropriate given that *psi* is often used as a general term to collectively refer to all types of paranormal ability. The group was founded in response to what was perceived to be an upsurge in interest in the paranormal in the United States at that time. Other founding members included Paul Kurtz, Martin Gardner, James Randi, Carl Sagan, and Isaac Asimov. As well as publishing the *Skeptical Inquirer* and organizing skeptics' conferences, CSICOP members often provided skeptical commentary on paranormal claims in the media.

In 2006, CSICOP changed its name to the Committee for Skeptical Inquiry (CSI; not to be confused with the TV program of the same name). This name change came about partly because it made clear that "the paranormal" was, even in those early days, only one area of interest to CSICOP. In the words of Kendrick Frazier, editor of the *Skeptical Inquirer*:

> our underlying interest has never been the paranormal per se, but larger topics and issues such as how our beliefs in such things arise, how our minds work to deceive us, how we think, how our critical thinking capabilities can be improved, what are the answers to certain uninvestigated mysteries, what damage is caused by uncritical acceptance of untested claims, how critical attitudes and scientific thinking can be better taught, how good science can be encouraged and bad science exposed, and so on and on.[12]

MOVING ON

After I completed my contract at Coventry Polytechnic, I was fortunate enough to take up a two-year post back at Leicester University, working on a project on automated assessment for my former supervisor, Dr. Beaumont. Next, I worked for a year as a research fellow in the psychometric research unit at Hatfield Polytechnic (now the University of Hertfordshire). Although I believe that psychological tests can be useful, the idea of spending the rest of my academic career as a psychometrician did not set my heart pounding.[13] For me, it is not one of the most exciting areas of psychology. And so it was with great relief and gratitude that, in 1985, I accepted the offer of a permanent lectureship at Goldsmiths College, University of London.

At this point, my interest in skepticism was still more a hobby than a serious research interest, but I did feel I knew enough to contribute a two-hour lecture critically assessing parapsychology to a module on theoretical issues as part of our BSc (Hons) psychology program. The students appeared to enjoy the lecture, and I continued to give it once a year for the following few years. In 1990, I presented a paper on cognitive biases associated with paranormal belief at the British Psychological Society's (BPS) annual conference in Swansea. Two years later, I published a version of that paper in *The Psychologist*, the BPS's monthly magazine.[14] This was the same year that I published my first empirical paper on paranormal belief. You may have missed it? It was in the *Australian Journal of Psychology*.[15] Thus, I took my first tentative steps as a researcher in anomalistic psychology.[16]

In 1994, I realized that I knew enough to put on a whole module on anomalistic psychology as an option for our final-year BSc students. Originally, I gave my module the title "Psychology, Parapsychology and Pseudoscience" (much later, I revised the syllabus and gave it the title "Anomalistic Psychology"). It proved to be very popular with the students, and I thoroughly enjoyed teaching it. As you might guess from the title, although many of the lectures did cover paranormal topics, I also wanted to be able to cover additional topics such as homeopathy, cold fusion, and N-rays.

Over the next few years, I published a few more papers in the field of anomalistic psychology, but I was aware of the fact that my interest in "the weird stuff" was only tolerated by my then head of department, not actively encouraged.[17] I was pretty much explicitly told that I could carry on publishing the occasional paper on paranormal belief and related topics so long as I mainly published in more academically "respectable" areas. I meekly did what I was told and continued to publish on more conventional topics (such as the relationship between cognition and emotion and the functions of the cerebral hemispheres) as well as in the area of anomalistic psychology.[18]

With the benefit of hindsight, I wish I had been more courageous and just concentrated on the area that interested me the most. Although the more conventional topics that I was researching were genuinely interesting and important, they did not hold the same fascination as the paranormal. Furthermore, whereas many dozens of researchers all around the world were engaged in investigating these more mainstream topics, there were only a few active researchers in anomalistic psychology. The idea of being a big fish in a small pool, as opposed to an average-size fish in an ocean, certainly had an appeal, and I was already more widely known for my weird research than for my more conventional stuff despite, at that stage, having published relatively little in the area. Furthermore, people like Susan Blackmore and Richard Wiseman were living proof that it was possible to have a (more or less) respectable academic career in the area.

Back in those days, it was quite common for academic colleagues to view research in anomalistic psychology as being pretty much a waste of time. After all, they would say, we all know that ghosts don't exist. We all know that ESP isn't real. We all know that people aren't really being abducted by aliens. So why are you interested in this stuff?

This always seemed to me to be rather missing the point. As discussed further in chapter 1, most nonscientists *do* believe in the paranormal, and a sizable minority of the population claim to have personally had paranormal experiences. Furthermore, such beliefs and experiences have always been reported throughout history in all known societies. Clearly, they are part of what it means to be human. Their ubiquity might indicate that paranormal phenomena really do exist. If so, the wider scientific community should study them in the same way that scientists investigate other natural phenomena. But if, in fact, psi does not exist, then we can learn a great deal about human psychology by investigating the factors underlying experiences that appear to be paranormal even though they are not.

(AT LEAST) TWO TYPES OF SKEPTICISM

In most situations, humans prefer simplicity over complexity. We all have a natural tendency to perceive and think about the world around us in simple binary terms such as good versus bad, right versus wrong, us versus them.[19] This is understandable. It makes decision-making much easier if we can immediately see the correct course of action in any given situation. The only snag is that the world around us is in reality complex and nuanced. As Algernon Moncrieff, a character in Oscar Wilde's play *The Importance of Being Earnest* famously said, "The truth is rarely pure and never simple."[20]

Looking back now at some of the views I held about the paranormal when I first discovered the joys of skepticism, I am very much aware that I was a victim of such black-and-white thinking. For example, I would pretty much have agreed with all of the following statements.

- Most (if not all) people who claim to be psychic (as well as astrologers, tarot card readers, practitioners of alternative medicine, and so on) are deliberate frauds, knowingly deceiving their clients for monetary gain, or else suffering from some sort of psychopathology.
- Most (if not all) people who claim to have personally experienced the paranormal are either stupid, lying, or suffering from some sort of psychopathology.
- Most (if not all) experimental parapsychologists are either incompetent when it comes to experimental design and statistical analysis or knowingly fraudulent.

I would not endorse any of those views now. Having said that, there is certainly a grain of truth in each. The history of psychical research is indeed littered with examples of frauds claiming to possess paranormal abilities, and there are many notable examples on the modern scene (although one must be careful when it comes to naming names, as being on the receiving end of a libel case can be very expensive even if one ultimately wins the case).

Every year, many con artists stand trial for defrauding their victims, having based their deception on a claim that the victim was cursed—and that the fraudster could lift the curse in return, of course, for a substantial sum of money. But my personal opinion is that the vast majority of those who claim psychic powers genuinely believe that they possess them. They are fooling themselves as much as, if not more than, they are fooling others.

When it comes to those reporting paranormal experiences, the evidence is pretty clear that those making the claims are typically not stupid, lying, or mad.[21] True, some studies investigating the correlation between intelligence quotient (IQ) and paranormal belief have reported small but significant negative correlations, but some find no such correlation. Equally, there is little doubt that sometimes people *do* knowingly lie about having had a paranormal experience. But we now know enough about anomalistic psychology to be certain that many ostensibly paranormal experiences can be caused by a range of psychological factors that lead the claimant to sincerely believe that they have had a genuinely paranormal encounter. Finally, studies do indeed consistently show significant correlations between paranormal belief and reports of paranormal experiences, on the one hand, and a range of measures of psychopathological tendencies, on the other. But such correlations are small and account for only a fraction of the variance associated with levels of paranormal belief. Besides, given the high levels of paranormal belief within society (see chapter 1), it would seem somewhat perverse to insist that paranormal belief is necessarily an indication of psychopathology.

My belief that parapsychologists are typically incompetent scientists was revised simply as a result of actually meeting and getting to know several of them (and, to a lesser extent, actually reading reports in parapsychology journals). When it comes to fraud on the part of parapsychologists, there are indeed several notable cases in the history of the field—but the same can be said for all other branches of science, including psychology.[22]

One notable example of a parapsychologist who influenced me to revise my negative view of parapsychology was the late Professor Bob Morris. Bob was, as stated, the first holder of the Koestler Chair of Parapsychology in the Koestler Parapsychology Unit (KPU) at the University of Edinburgh, taking up the post in December 1985, just a couple of months after I started at Goldsmiths.

If my memory is correct, I first met Bob at the Fifth European Skeptics Conference at the University of Keele in 1993. I was impressed and, I must admit, rather surprised that an actual professor of parapsychology would turn up at a skeptics' conference, given most skeptics' generally antagonistic attitude toward parapsychology. However, Bob went one better than that. When it became apparent that one of the keynote speakers was not going to be able to attend after all, Bob stepped into the breach and delivered an excellent overview of his approach to parapsychological research. It was clear that he was well aware of skeptics' concerns about such research and was committed to trying his hardest to meet those concerns.

The more I got to know this likable American over the years, the more I appreciated his open-minded approach. Bob and his team at the KPU devoted much time and effort to directly testing paranormal claims. However, they devoted almost as much time and effort to investigating psychological factors that may lead people to believe they have experienced the paranormal when, in fact, they haven't. In other words, they were as interested in (and as knowledgeable about) anomalistic psychology as they were about parapsychology.[23] Bob actively encouraged skeptics such as myself to visit his lab and to point out any weaknesses we could detect in their experimental methodologies so that they could be eliminated. I realized from my encounters with Bob that parapsychologists were not, in general, the incompetents that many skeptics assume them to be. In some respects, such as the need for double-blind methodologies, parapsychologists were often ahead of more mainstream sciences (including psychology).

For those who knew him, both believers and skeptics, Bob's sudden and unexpected death in August 2004 came as a real shock. Only a couple of weeks before this tragic news broke, I had been hanging out with Bob and enjoying his wicked sense of humor at a parapsychology conference in Vienna, blissfully unaware that his days were numbered. His legacy lives on, though. Bob supervised many postgraduate students, including, as mentioned, Richard Wiseman, many of whom went on to successful academic careers. Professor Caroline Watt, the current holder of the Koestler Chair of Parapsychology was another of Bob's postgraduate students. Like him, she is a great example of a truly open-minded approach to the paranormal and, as a result she is also respected by those on both sides of the psi controversy.

Looking back, I can see that when I first rejected my pro-paranormal beliefs, I went from being a believer to what I now view as being a rather dogmatic and dismissive skeptic. Tribal thinking is natural for humans, and in those early days of skeptical discovery I rather enjoyed viewing all self-proclaimed psychics as con artists, all parapsychologists as either incompetent or dishonest, and all believers as either stupid or crazy. I had found my tribe, and we were the good guys—honest, rational, and, most importantly, right. I was certain of it and did not much like those occasional skeptics I had come across who seemed to believe that things were a bit more complicated. Of course, now I have turned into one of those people.[24]

THE ANOMALISTIC PSYCHOLOGY RESEARCH UNIT

From 1997 to 2000, I did a stint as head of department. At Goldsmiths, it is standard for different staff members to each serve a three-year term as head before passing the responsibility to the next victim. This system is sometimes referred to as a "rotating head" system, which always brings to mind that famous scene from *The Exorcist.*

As any head of department will tell you, the administrative responsibilities are very time-consuming, leaving little time for research if you happen to be less than superhuman. In recognition of this, the college provides the head of the psychology department with a research assistant for three years. I employed Kate Holden in that role, and it was in discussions with Kate that the idea of establishing a unit at Goldsmiths dedicated to the study of anomalistic psychology first arose. In 2000, the Anomalistic Psychology Research Unit (APRU) was founded.

Among the stated aims of the APRU was to raise the academic profile of anomalistic psychology. I think that we can reasonably claim to have had some success in meeting this aim. As mentioned previously, when I was first interested in this field, it was quite common for colleagues to view it as not "academically respectable." Although I still come across such attitudes occasionally, it is much less often. One way we attempted to achieve this aim was to primarily publish our findings in mainstream psychology journals rather than parapsychology journals. Although I have no objections to publishing in the latter, the sad truth is that mainstream science journalists pay very little, if any, attention to such journals regardless of the quality of their contents.

THE SKEPTIC MAGAZINE

By the year 2000, there were several regular magazines for skeptics, including the aforementioned *Skeptical Inquirer* (founded in 1976 and published by CSI, formerly known as CSICOP) and *Skeptic* (first published in 1992 by Michael Shermer's Skeptics Society). In the United Kingdom, Wendy Grossman founded *The Skeptic* magazine in 1987, just around the time that I was first discovering the delights of skepticism. I was an early subscriber. I don't recall exactly where and when I first met Wendy, and neither does she, but it was inevitable that we would become friends given our shared interests.

Wendy edited the magazine for the first few years before handing the baton on to Toby Howard and Steve Donnelly, who edited it for the next ten years or so. The British magazine always depended on the enthusiasm and goodwill of the editors and a small army of volunteers to help in its production. It took a lot of time and effort to produce each new issue, and so, perhaps inevitably, busy people like Toby and Steve eventually decided that they had done their bit and wanted someone else to take the reins. Somewhat reluctantly, Wendy agreed to a second stint as editor.

In 2001, I coedited (with Kate Holden) a special issue of the magazine devoted to the work of the newly founded APRU.[25] Not long after that, I finally gave in to Wendy's persuasion and took over the role of editing the magazine on a longer-term basis, along with a series of student coeditors.[26] I did so for the next decade or so. Like those who had gone before me, I found the job both a blessing and a burden. On the positive side, it raised the profile of the APRU and put us right at the center of the UK skeptic scene. I have no doubt that it raised my own profile somewhat, given that I could now be presented in the media not only as a professor of psychology at a London University but also the editor of the UK's longest-running skeptical publication. On the negative side, it was always difficult to find the time, given my other commitments as a full-time academic, to put each issue together, even with the help of numerous volunteers.

Each time a new editor took over, they would typically give the mag a makeover and improve its appearance somewhat. In our case, to mark the magazine's twenty-first anniversary, we introduced full-color covers, typically featuring fantastic caricatures, by the talented artist Neil Davies, of someone interviewed for that edition.

We also established an editorial advisory board that read like a who's who of modern skepticism, including academics (such as Richard Dawkins and Brian Cox), writers (such as Simon Singh and Ben Radford), comedians (such as Stephen Fry, Tim Minchin, and Robin Ince), and conjurers (such as James Randi and Derren Brown). We knew that in reality these busy people would not have much time to devote to the magazine, but we were grateful that they were willing to have their names listed in each issue to indicate that they shared the values of the magazine, summarized in our tag line: "pursuing truth through reason and evidence."

Eventually, I decided that I too had done enough and passed the job on to Deborah Hyde. Deborah has a much better talent for design than me, and she introduced further innovations after she took over in 2011. For example, she introduced full color throughout the magazine. She would also often include as a centerfold work by Crispian Jago, such as his wonderfully satirical (yet informative) illustrations of the Venn Diagram of Irrational Nonsense and the Periodic Table of Irrational Nonsense.

In collaboration with Neil Davies, Crispian produced a wonderful set of Skeptic Trumps cards based on the popular children's card game Top Trumps (see plate 1).[27] Each card featured a caricature and brief description of an individual active in the world of skepticism and listed their special power, weapon of choice, key output, and archnemesis. I was naturally chuffed to be included and find myself in the company of so many of my personal heroes. It took a little while for me to realize that I seemed to have just about the lowest scores of everyone featured. For example, my special power of "Hitting paranormal claims with a large science and reason stick" earned me a mere 63 percent—whereas Simon Singh was awarded a massive 88 percent for his special power of "Casting the shadow of a pineapple with his head"! Oh well, at least I'd been included.

The Skeptic is currently available for free, online only, ably edited by Michael Marshall, project director of the Good Thinking Society, supported by an editorial team from the Merseyside Skeptics Society.[28] I continue to write articles for the magazine.

1 WEIRD SCIENCE

When I first started working in this area, I would sometimes be asked, "What kind of a psychologist are you?" My standard reply was, "I'm interested in the psychology of paranormal belief and ostensibly paranormal experiences." Now, that hardly rolls off the tongue, does it? I was slightly envious of colleagues who could give a clear and concise reply, such as, "I'm a neuropsychologist" or "I'm a developmental psychologist." I had no such ready label to offer. The solution? Come up with one!

In fact, it turned out a suitable label did exist. In 1982, Leonard Zusne and Warren H. Jones had published their ground-breaking textbook *Anomalistic Psychology*.[1] That will do for me, I thought, and from then on if anyone asked, "What sort of psychologist are you?" I could reply instantly and concisely, "I'm an anomalistic psychologist." Unfortunately, there was just one small snag with this cunning plan. Even today, most people have never heard of anomalistic psychology, but back then the number was even smaller. Unless the inquirer had previously had the good fortune to have read Zusne and Jones's excellent tome, the next question would inevitably be, "And what the heck is that?" To which I would reply, "It's the psychology of paranormal belief and ostensibly paranormal experiences." Still, we're making progress, and maybe more people will get to know what anomalistic psychology is by reading this book.

There is another problem with the term that I had not anticipated. Although "anomalistic psychology" is indeed fairly concise, it turns out many people struggle to pronounce it properly. I have lost count of the number of times that I have been introduced at public lectures as being an "animalistic" psychologist.

At the time of writing, Wikipedia lists over ninety subdisciplines of psychology, a few of which, I confess, are new to me despite having been a professional psychologist for about four decades. Some of the main subdisciplines are defined in terms of their fundamental level of theoretical analysis. Thus, to give but a few examples, neuropsychologists focus on trying to understand the neural substrates of mind and behavior. Behavioral geneticists are interested in the interaction between genetics and environment in influencing aspects of behavior, such as intelligence, aptitude, and personality. Social psychologists concentrate on how our interactions with other individuals or groups influence our behavior and thinking. Other, more applied subdisciplines, such as clinical psychology, forensic psychology, and educational psychology, adopt insights from such fields and attempt to apply them in real-world contexts.

Wikipedia defines anomalistic psychology as "the study of human behaviour and experience connected with what is often called the paranormal, with the assumption that there is nothing paranormal involved."[2] That is a pretty good definition. Wikipedia also classifies anomalistic psychology as an example of applied psychology, and that seems fair too. Anomalistic psychologists typically take insights from a wide range of other subdisciplines within psychology and consider what light they may shine on topics of interest. This is the approach taken by Anna Stone and myself in our textbook *Anomalistic Psychology: Exploring Paranormal Belief and Experience*.[3] Most of this chapter will outline in broad terms what insights the various subdisciplines have to offer, and later chapters will explore them in more depth.

But before we go any further, we should take a step back. There is one crucially important word that has been used a lot so far without any attempt at proper definition. That word is *paranormal*.

WHAT DOES PARANORMAL MEAN?

When you come across the word *paranormal*, what comes into your mind? Maybe you recall the opening title sequence and spooky music of *The X-Files*?

Maybe you think of films like *Paranormal Activity*, *Close Encounters of the Third Kind*, or even *Ghostbusters*? Given that you probably already have an interest in the area (otherwise, why are you reading this book?), you might think of books you've read on the topic or magazines like *The Fortean Times*. But what exactly does *paranormal* mean?

To aid you with your reflection on this question, look through the following list of topics and decide in each case whether the label *paranormal* can be applied to them. More importantly, see if you can come up with a definition that covers everything that you feel is paranormal and excludes everything that is not. (Some of the topics may be a bit obscure, and so I have provided brief definitions.)

- Alien abductions
- Angels
- Astrology
- Auras (alleged energy fields surrounding all living things that are visible only to "gifted" individuals)
- Bermuda triangle (an area in the Atlantic Ocean where planes and ships are said to mysteriously disappear without trace; also, a hit single from Barry Manilow)
- Biblical miracles
- Bigfoot
- Clairvoyance (the alleged ability to obtain information from remote locations without the use of the known sensory channels)
- Climate change denial
- Complementary and alternative medicine (CAM)
- Consciousness
- Crop circles
- Crystal balls
- Crystal power
- Curses
- Déjà vu (the feeling that you have experienced something before even though you know you haven't)
- Déjà vu (the feeling that you have experienced something before even though you know you haven't)

- Demons
- Dermo-optical perception (the alleged ability to see with sensory receptors in the skin)
- Dowsing (the alleged ability to locate substances such as water or oil via the movement of held objects such as dowsing rods or pendulums)
- Electronic voice phenomenon (EVP; the alleged recording of spirit voices electronically)
- Exorcism
- Extrasensory perception (ESP; includes clairvoyance, precognition, and telepathy)
- Extraterrestrials
- Fairies
- Faked moon landings
- Father Christmas
- Feng shui (the practice of arranging objects around us in such a way that the flow of mysterious "energies" such as *chi* are not blocked or disrupted, allowing us to live harmoniously)
- Flat earth theories
- Ghosts and poltergeists
- Glossolalia ("speaking in tongues")
- God
- Heaven (and Hell)
- Hypnosis
- Hypnotic regression (the use of hypnosis to mentally travel back in time, e.g., to childhood, life in the womb, or even past lives)
- *I Ching* (ancient Chinese text used for divination)
- Kirlian photography (a photographic technique said to reveal the "life force" of an object)
- Ley lines (lines along which it is alleged that mysterious energy flows throughout the landscape)
- Loch Ness monster
- Lucky charms
- Lycanthropy (the alleged ability to turn into a wild creature, most often a werewolf)

- Mediumship (the alleged ability to talk to the dead—and get replies)
- Near-death experience (NDE)
- Ouija boards
- Out-of-body experience (OBE)
- Palm reading
- Possession
- Prayer
- Precognition (the alleged ability to directly obtain knowledge about events before they happen other than by use of the known sensory channels or inference)
- Psychics
- Psychic healing
- Psychokinesis (the alleged ability to influence objects in the outside world by the power of thought alone—excluding one's own body parts)
- Reading tea leaves
- Reincarnation
- Shroud of Turin (allegedly the shroud in which Christ was wrapped following his crucifixion, mysteriously imprinted with his image)
- Spirits
- Spontaneous human combustion
- Tarot cards
- Telepathy (allegedly, direct mind-to-mind contact)
- Unidentified flying objects (UFOs)
- Unlucky 13
- Vampires
- Voodoo
- Witches
- Zeus
- Zombies

I hope you were not expecting a definitive list of the "right answers" regarding which of the above really are paranormal topics and which are not because I am afraid I am going to have to disappoint you. The point of the exercise is to illustrate that it is actually quite hard to come up with a definition of *paranormal* that identifies only those concepts that we feel should be so labeled and excludes all others.

The online version of the *Cambridge Dictionary* offers the following definition: "impossible to explain by known natural forces or by science."[4] That is a pretty good stab at a working definition but not perfect by any means. Take consciousness, for example. Science has yet to adequately explain how subjective self-awareness can arise from the physical substrate of the brain, and yet few people would argue that consciousness is a paranormal phenomenon. As another example, physicists believe that 85 percent of the universe consists of *dark matter*. This form of matter is believed to exist because theoretical astrophysicists insist that empirical observations regarding gravitational effects do not make sense without it. The only problem is that dark matter itself has never actually been observed, and no one knows what it consists of. But, once again, no one would call this a paranormal phenomenon. Many similar nonparanormal examples of things that are "impossible to explain by known natural forces or by science" could be cited.

As already mentioned, parapsychologists typically limit their investigations and speculations to three main general areas, all of which would be considered by everyone to be core paranormal topics. The first of the three is extrasensory perception (ESP), which itself comes in three different flavors. The first is telepathy, the alleged ability of minds to make direct contact without the use of the known sensory channels. The second is clairvoyance, the alleged ability to pick up information from remote locations, again without the use of known sensory channels.[5] The final type is precognition, the alleged ability to obtain information about future events other than by ordinary inference.

The second main topic of interest for parapsychologists is psychokinesis (PK), the alleged ability to influence the outside world directly by the power of thought alone. Examples include the ability to levitate objects, to psychically heal ailments, or Uri Geller's spoon-bending powers. As my students will confirm, I have absolutely no qualms about using very old, very corny dad jokes in my lectures, so I will usually say to my class, "If you believe in psychokinesis, would you please raise my hand?" That one always gets a nice groan.

The final topic of interest to academic parapsychologists is evidence relating to the possibility of life after death. This covers a range of topics from our list above, including ghosts and (possibly) poltergeists, mediumship, reincarnation, and the electronic voice phenomenon (EVP).

I think we can agree that all of the topics within the remit of parapsychology mentioned above are indisputably paranormal. But even here there is a potential problem. The definition of *paranormal* offered by the *Cambridge Dictionary* implies that a phenomenon is only paranormal if it is "impossible to explain by known natural forces or by science." Let us suppose that the skeptics, including me, are wrong and that some people really do have the ability to directly read the minds of others, albeit by means currently unknown to science. Let us suppose that a method has been found to reliably demonstrate this telepathic ability under properly controlled conditions and that there was absolutely no doubt that the ability was real. This would be an incredible scientific breakthrough, and no doubt a Nobel Prize or two would be handed out in recognition of this achievement (although it should be noted that there currently is no Nobel Prize for psychology and there definitely is not one for parapsychology!). What would happen next?

Inevitably, scientists *would* then turn their attention to figuring out the mechanism by which telepathy operates. And if they were successful? Well, obviously there would be a few more Nobel Prizes to hand out. But then the prize committee need not give any more thought to introducing a Nobel Prize for parapsychology—because once telepathy could be explained by science, it would no longer be a paranormal phenomenon according to our definition. The same argument applies to all other paranormal phenomena. Parapsychology is an area of science that runs the theoretical risk of being so successful that it could annihilate itself.[6] Few would argue that such annihilation is imminent.

Of course, not everyone restricts their usage of *paranormal* to the three core concepts of parapsychology. The mass media and, as you will see, anomalistic psychologists tend to adopt a much looser conception of the paranormal that covers pretty much everything "weird and wonderful" that is generally thought of as unexplained. In fact, as I hope to demonstrate in the pages of this book, many of those allegedly unexplained phenomena do in fact have plausible, empirically supported explanations, but let's not get ahead of ourselves.

CRYPTOZOOLOGY AND UFOLOGY

Two fields that are often covered in books on the paranormal as more loosely defined are cryptozoology and ufology. *Cryptozoology* is the name given to the investigation of creatures that are believed by some to exist yet whose existence is not recognized by mainstream science, such as Bigfoot and the Loch Ness monster (figure 1.1). If it were ever proved that any of these so-called *cryptids* really did exist, this would come as quite a shock for most mainstream zoologists—but it would not require the revision or rejection of any currently accepted scientific theories.

Figure 1.1
Should the Loch Ness monster be considered a paranormal phenomenon? This famous photograph was first published by the *Daily Mail* in 1934 and is generally recognized as a hoax.

Ufology refers to the study of all aspects of UFOs. Typically, but not exclusively, ufologists tend to favor the so-called *extraterrestrial hypothesis*—that is to say, they believe that at least some of the evidence put forward is best explained in terms of aliens from other planets visiting the Earth. If it were ever proven that aliens really were visiting our planet, this would probably be a bit of a surprise for mainstream scientists but, again, it would be within the bounds of our current scientific understanding.[7] The main reason that most claims within both cryptozoology and ufology are rejected by mainstream science is simply because there is no compelling evidence to support them. Strictly speaking, they are not paranormal claims if we accept the dictionary definition of *paranormal*.

RELIGIOUS CLAIMS

There are other problematic topics on our list. For example, many religious topics fit the dictionary definition very well, and yet most people would not apply the label *paranormal* to them. Angels, demons, prayer, glossolalia ("speaking in tongues"), possession and exorcism, and so on might all be seen, by believers at least, as being beyond scientific understanding. God certainly would be. Yet we often make a distinction between religious and paranormal concepts. Perhaps the concept of the *supernatural* provides a conceptual bridge between these two realms?

There are many examples of miracles in religious texts that clearly fit our idea of paranormal phenomena. For example, there are several stories of prophetic dreams in the Bible that allegedly provide nice examples of precognition in action. There are also several examples of what appears to be amazing psychokinetic ability. Moses is said to have parted the Red Sea. Jesus allegedly raised people from the dead, fed five thousand people with just five loaves and two fishes, and, my personal favorite, turned water into wine (plate 2). The list could go on. Yet I suspect that most people do not think of these as paranormal feats.

The distinction between religious and paranormal experiences becomes very blurred indeed when we consider phenomena such as near-death experiences (NDEs). NDEs are widely considered to be paranormal in nature, and yet they are often intrinsically religious, involving spirits, angels, Jesus, and even God Himself. NDEs are discussed in more depth in chapter 7.

FAIRIES, VAMPIRES, AND ZOMBIES

You may have thought that some of the topics listed, such as fairies, vampires, and zombies, should not be considered paranormal at all because, as far as you are concerned, they are entirely fictional. This may be the view of the typical twenty-first century Westerner, but it most certainly is not a view that has been universally held across space and time. For long periods in humanity's past, and even in some parts of the world today, millions of people believed in the physical reality of such creatures—and a host of others besides. Anomalistic psychologists should not make the mistake of only considering beliefs that are widely held within our own society in our own era.

One question I was fond of asking my students early on in my course was, "How many of you believe in fairies?" Usually, no hands went up, or perhaps a single hand, but I got the strong impression that this was being done more to get a laugh from fellow students than out of a sincere declaration of belief. After all, the only people who might sincerely believe in fairies in our own society—along with Santa Claus and the Easter Bunny—are children, right? We all grow out of such childish beliefs by the time we reach puberty.

However, a century ago belief in the reality of fairies was widespread across much of Europe—including among well-educated adults.[8] After all, if we consider folklore we can see that stories of such "little people" are common in most cultures down the centuries. Surely this must indicate some basis in fact? Not only that, but there were many actual sightings of fairies, often from sane, intelligent adults who were not under the influence of alcohol or any other mind-altering substance. There was even photographic evidence.

Famously (or notoriously, depending on your point of view), Elsie Wright, aged sixteen, and her ten-year-old cousin, Frances Griffith, managed to take several photographs of fairies near a little stream known as Cottingley Beck in July and August 1917. The photographs were pronounced to be genuine by photographic experts. The case of the Cottingley Fairies caught the attention of none other than Sir Arthur Conan Doyle, the creator of arguably the most famous fictional detective ever, Sherlock Holmes.[9] Sir Arthur was convinced that it constituted strong proof of the existence of fairies, as recounted in his book *The Coming of the Fairies*.

There were, not surprisingly, those skeptical of this case from the beginning. However, right up until the mid-1970s there were still some who defended the authenticity of the photographs. Then the case began to fall apart. Computerized enhancement of the photographs revealed a string holding up one of the fairies, as well as strongly suggesting that the figures were simply paper cutouts. Further investigation continued to undermine the claim, including the discovery of a book, *Princess Mary's Gift Book*, containing identical illustrations. Eventually, Elsie and Frances confessed that the photographs had been taken as a practical joke.

Although the story of the Cottingley Fairies is, to my mind, a fascinating and still rather magical one, the most important point in the present context is simply that the considerable percentage of the population who believed in fairies one hundred years ago believed they had very good reasons to do so. There were eyewitness accounts from reputable sources. There were even photographs, pronounced as genuine by experts. On reflection, the parallels with the kind of evidence put forward by modern-day believers in flying saucers is striking.

TRADITIONAL SUPERSTITIONS AND PERSONAL SUPERSTITIONS

Lucky charms and unlucky 13 are two traditional superstitions to be found on our list but clearly it would have been easy to include dozens more (plate 3).[10] Is it really unlucky to walk under ladders or to break a mirror? Will crossing your fingers bring you good luck—or at least ward off bad luck? Will a black cat crossing your path bring you good luck (if you live in the United Kingdom or Japan) or bad luck (if you live elsewhere in Europe)?[11] And how does that work when you go on holiday?

In addition to the dozens of traditional superstitions that we are all familiar with, many people have their own personal and idiosyncratic superstitious beliefs. Such beliefs may arise when a particular object becomes associated in our minds with a particularly favorable outcome. For example, a student may do particularly well in an exam and henceforth insist on taking that "lucky pen" into all future exams. A tennis player may put in a particularly impressive performance in a tournament and from that point on insist on wearing that particular pair of shoes in future important matches.[12]

Athletes are notoriously superstitious and often feel compelled to carry out quite elaborate rituals prior to stepping out to perform. More generally, it is those professions in which success or failure is not entirely under the performer's control that are known to be associated with rich superstitious beliefs and practices. In addition to athletes, soldiers, sailors, actors, and gamblers are all known to be generally superstitious, whereas accountants rarely engage in complex rituals prior to starting work.

Now the chances are that most readers of this book are not especially superstitious and would probably reject claims based on such superstitious thinking. Yes, you may concede, *belief* in superstitions may have an indirect effect. If a performer is prevented from carrying out their preferred ritual, this may make them nervous and actually lead to a poorer performance. Conversely, being allowed to carry out the ritual may help them to focus and "get in the zone" and result in a better performance. But the idea that carrying a lucky charm or engaging in a ritual could have a direct effect, independent of belief, would probably be rejected as childish and silly.[13]

The main question to ask is, should such superstitious beliefs be considered as paranormal beliefs? The fact that you personally may not believe in them is irrelevant. If any of them did actually work, they would work by means unknown to science and therefore should rightly be thought of as paranormal. However, as with religious claims, I suspect most people would not naturally label them as such.

DIVINATION

Different forms of divination have been used for centuries, and many of them are still popular today. Our list above includes astrology, crystal balls, *I Ching*, palm reading, reading tea leaves, and tarot cards, all techniques with ancient origins that are still popular today (plate 4). In the past, fortunes were also told by looking at the patterns made by smoke, the flight of birds, or the entrails of sacrificed animals. The latter is as valid as, say, astrology, but considerably messier.

Even though these techniques may have no validity whatsoever, they can still arguably be considered to be paranormal on the grounds that, if they did work, there would be no known scientific explanation for how they did so. Arguably, they do not, however, fall within the remit of the three core concepts of parapsychology. None of the techniques are generally thought to involve the application of alleged paranormal abilities. Instead, each provides a means to interpret supposedly meaningful patterns as a way to provide guidance. These systems of interpretation could be, and indeed often have been, computerized to allow automated generation of readings. If any of them did actually work, it would be possible in theory for a believer in the paranormal to argue that they did so via clairvoyance and precognition, but this is not a widely held view. The truth is, however, they simply have no validity whatsoever.

COMPLEMENTARY AND ALTERNATIVE MEDICINE

One important aspect of the New Age movement is faith in *complementary and alternative medicines* (CAM).[14] CAM covers a very wide range of different practices, and not all of them claim that their alleged effectiveness is based on paranormal forces, but some do. The most obvious example of this is psychic healing, which involves the practitioner passing their hands over the patient's body, either lightly touching it or simply moving near it. Some psychic healers claim that they can work their magic from afar, a practice known as *distance healing*. It is claimed that psychic healing is capable, by means unknown to science, of "rebalancing subtle energies" and returning the body to health. Some CAM practitioners claim that they are able to diagnose health problems by examining the patient's aura, an alleged energy field that is said to surround all living things. Some claim that their forms of treatment, such as the use of crystals, can lead to not only improved physical health but also the attainment of psychic powers, such as ESP. All forms of

CAM have this in common: they have either never been proved to work or they have been proved *not* to work. In the words of comedian Tim Minchin, "You know what they call alternative medicine that has been proved to work? Medicine."

BELIEF IN CONSPIRACIES

Many supporters of CAM believe that Big Pharma is engaged in a massive conspiracy to suppress their allegedly safe and effective forms of treatment in order to ensure that they can make massive profits by manufacturing and selling unsafe forms of conventional treatment. Some go even further and claim that scientists deliberately manufacture deadly diseases in order to then sell us the treatments for those very diseases. While there is no doubt that drug companies are far from being paragons of virtue, and also that conventional medicine is far from perfect, most of these claims are without foundation.[15]

The idea that Big Pharma, in collaboration with medical scientists around the world, is engaged in a massive conspiracy is as unlikely as the idea that the international community of climate scientists are similarly involved in promoting a hoax and that climate change is not really happening or, if it is, it is not due to our use of fossil fuels. Neither of these particular conspiracies involve any paranormal aspects, but there are certainly some that do. For example, many ufologists believe that governments and intelligence services around the world are fully aware of extraterrestrials visiting our planet and abducting humans. The aliens are often reported as communicating with humans telepathically.

Interestingly, belief in conspiracies reliably correlates with belief in the paranormal, and many of the psychological factors that correlate with the former also correlate with the latter. Furthermore, there is little doubt that many of the more bizarre conspiracy theories (e.g., David Icke's claim that the world is run by shape-shifting lizards) are weird and wonderful. For reasons such as these, belief in conspiracies in general has become a topic of great interest to anomalistic psychologists and indeed the wider world.[16]

THE RELATIONSHIP BETWEEN PARAPSYCHOLOGY AND ANOMALISTIC PSYCHOLOGY

The online *Cambridge Dictionary* provides the following definition of *parapsychology*: "the study of mental abilities, such as knowing the future or telepathy, that seem to go against or be outside the known laws of nature and science."[17]

Elsewhere, I have stated that anomalistic psychology

> attempts to explain paranormal and related beliefs and ostensibly paranormal experiences in terms of known (or knowable) psychological and physical factors. It is directed at understanding bizarre experiences that many people have, without assuming that there is anything paranormal involved. While psychology, neurology and other scientific disciplines are rich with explanatory models for human experiences of many kinds, these models are rarely extrapolated to attempt to explain strange and unusual experiences.[18]

There is clearly a high degree of overlap between the two subdisciplines of parapsychology and anomalistic psychology, but one important difference is that anomalistic psychologists are typically interested in a wider range of topics—that is to say, anything considered weird—compared to parapsychologists. The reason for this is that the underlying psychological explanations for many topics within the strict remit of parapsychology are often the same as those for topics that fall outside that remit.

For example, mediumship clearly falls within the strict definition of *parapsychological*. Mediums claim to be able to communicate with the spirits of people who have died. If they truly do have this ability, that would clearly constitute irrefutable evidence of postmortem survival. A typical reading from a medium will consist not only of personal messages from the deceased to the client but often also more general information and advice regarding the client's current and future life. In fact, the reading will be quite similar to a reading from an astrologer, a palm reader, or a tarot card reader. However, as already stated, these forms of divination fall outside the strict definition of parapsychological as they do not obviously relate to the core concepts of the paranormal. From the point of view of the anomalistic psychologist, however, it seems likely that readings from both mediums and users of other forms of divination are based on the same psychological processes (see chapter 4 for further details). It would therefore make little sense to focus only on readings from mediums and to ignore those of other diviners.

Another phenomenon that is sometimes presented as providing evidence in support of life after death is apparent past-life memories. Such memories can arise either spontaneously or as a result of *hypnotic regression* (for more detail, see chapter 6). Empirical evidence very strongly supports the claim that, in many of these cases, the memories concerned are actually false memories; that is, they are not based on events that really took place in objective reality. Chapter 5 will present evidence in support of the claim that many reports of alien abduction are also based on false memories. Once again, we have an example of one phenomenon, that of claimed past-life memories, that fits the strict definition of parapsychological but is probably based on the same psychological processes as one that does not. Clearly, both phenomena should be considered within the wider context of false memory research.

In general, anomalistic psychologists approach their topics of interest on the basis of the working hypothesis that paranormal forces do not exist. Wherever possible, they design experiments and investigations aimed at putting their nonparanormal explanations to the empirical test. To that extent, their approach is more skeptical than that of many parapsychologists, who are typically more focused on directly testing paranormal claims rather than considering possible nonparanormal explanations. However, it would be a mistake to see anomalistic psychology as being in opposition to parapsychology. Instead, the two disciplines should be seen as complementing each other. If ever parapsychologists are able to provide a demonstration of a paranormal phenomenon that is robust and replicable enough to convince the wider scientific community that it is genuinely beyond explanation in terms of currently accepted scientific theories, then anomalistic psychologists will have provided a valuable service in helping them to sort what is genuinely paranormal from what only appears to be so.

It should also be kept in mind that it is logically possible that some examples of an allegedly paranormal phenomenon may, in principle, be genuinely paranormal while other examples of that same phenomenon are explicable in nonparanormal terms. The two approaches are not mutually exclusive. In fact, parapsychologists have one big rhetorical advantage over their generally more skeptical anomalistic colleagues. They can quite happily concede that not all ostensibly paranormal phenomena are genuinely paranormal but still argue that some are. This may appear to be a more reasonable position to adopt than the skeptical position of arguing that *all* ostensibly paranormal phenomena can be explained in nonparanormal terms, the working hypothesis of the anomalistic psychologist. However, it may just be that the wider scientific community is correct in rejecting all paranormal claims. Until relatively recently, there have been few attempts to develop and test alternative, nonparanormal explanations for ostensibly paranormal phenomena. Only time will tell how successful this approach will be.

It is sometimes claimed that you cannot prove a negative. This is not true (and, indeed, the first four words of this sentence are collectively an example of a negative statement that I intend to prove in the next couple of sentences). Some negative statements are easy to prove. For example, if I claimed that there was always a fully grown, perfectly normal tiger in my bath, you could easily prove that there was no tiger in my bath by simply having a look (but be careful—just saying!). However, the less extreme claim that some negative statements cannot be proved *is* true—and the claim that psi does not exist is one of them.[19] No matter how many paranormal claims are conclusively debunked, and no matter how many nonparanormal explanations are supported by compelling evidence, it will always be possible to argue that proof of psi is just around the next corner. However, in the absence of any compelling evidence for the existence of psi after well over a century of systematic research, the growing evidence for plausible nonparanormal explanations of ostensibly paranormal phenomena must surely increase the probability that psi simply does not exist.

WHY STUDY ANOMALISTIC PSYCHOLOGY?

As mentioned previously, when I first started getting seriously interested in the psychology of weird stuff, I sometimes came across people who did not understand the value of such research. Because they themselves did not believe in the paranormal, they would insist that any such research was simply a waste of time and effort. I can see how someone who is 100 percent certain that all paranormal claims are without foundation might indeed hold such a view of attempts to establish psi (even though I feel that their 100 percent certainty is not a scientifically defensible position to take). However, that was not primarily what anomalistic psychologists are trying to do. Instead, the main focus is on attempting to understand the psychology of paranormal belief and experience.

Opinion polls repeatedly show that a large proportion of the general population does indeed believe in the paranormal. For example, a poll of a representative sample of 1,347 British adults in 2017 by BMG Research found that one-third responded affirmatively to the question, "Do you believe in ghosts, ghouls, spirits or other types of paranormal activity?"[20] Forty-six percent responded negatively, and 21 percent had yet to make up their minds. In 2019, a YouGov survey of 1,293 American adults found that 45 percent believe that ghosts definitely or probably exist, with a similar figure obtained for both belief in demons and belief in "other supernatural beings" (only 13 percent believed in vampires).[21] Such figures are fairly typical of surveys in Europe and the United States, although some variation is to be expected based on factors such as the precise wording of the question, what is popular in the media at the time, and so on. There is also considerable variation across cultures. For example, levels of belief in psychic healing and communication with the dead are considerably higher in Latin America than in Europe.[22]

Not surprisingly, a major factor contributing to belief in the paranormal is personal experience. After all, it would be a bit weird to report that you had had a paranormal experience but that you did not believe in the paranormal, wouldn't it? Several studies show that personal experience is indeed the single biggest reason given for believing in the paranormal, although other factors, such as testimony from trusted others, can also play a part.[23]

There is no known society, either historically or geographically, where paranormal beliefs are not held and ostensibly paranormal experiences are not reported. Clearly, such beliefs and experiences are an important part of what it means to be human. On first consideration, such pervasiveness may appear to offer strong support for the existence of psi—and yet compelling scientific evidence in support of this idea still eludes us. Is it possible, then, that this pervasiveness is actually due to the fact that human brains are generally pretty similar from one culture to the next and thus prone to the same glitches that might lead us to conclude that we have had a paranormal experience when in fact we have not? If so, we can learn a great deal about human psychology by taking such experiences seriously.

WHY WEIRD STUFF MATTERS

I hope that by now I have convinced you, if you didn't believe it already, that a full understanding of the human condition must include consideration of the psychological roots of paranormal belief and experiences. But in addition to that fundamental, if rather abstract, aspect of anomalistic psychology, there are also important practical reasons for taking weird stuff seriously. Skeptics are sometimes portrayed as being miserable killjoys, out to spoil everyone else's harmless fun. Does it really matter if your best friend is convinced that knowing someone's star sign provides genuine insight into their personality? Does it matter that much if your neighbor is convinced that the moon landings were faked? So what if your aunt insists that homeopathic remedies are to be preferred to medically approved painkillers in treating her arthritis?

On first glance, none of those examples are really anything to worry about that much, are they? However, on reflection, there are real causes for concern even here. If you believe in astrology, you'd probably think it was quite reasonable for companies to sometimes take astrological data into account when selecting applicants to fill posts, as does sometimes happen. A reasonable argument can be made that this is a totally unjustified form of discrimination given the complete lack of validity of astrology as demonstrated by literally hundreds of well-conducted tests.[24] Your neighbor's belief that the moon landings were faked is not in itself harmful—but research shows that belief in any one conspiracy is usually a very good predictor of belief in others, many of which, such as those embraced by the anti-vaccination movement, do cause serious harm. If your aunt insists on taking homeopathic remedies for her arthritis, she is certainly not taking advantage of the most effective treatments available. But, more importantly, should she take the same approach toward life-threatening conditions such as cancer, the consequences could be much more serious.

The general point here is that those three claims are all indicative of a lack of critical thinking. When a particular claim is based on good empirical evidence and sound reasoning, it obviously makes sense to accept it. However, many beliefs are based on weak evidence (or even no evidence at all) and very poor reasoning. This does not prevent people from making important decisions based on those beliefs. There will be many examples of such cases throughout this book, but for a more comprehensive list, visit Tim Farley's excellent website, What's the Harm?[25]

Tim's website provides a fascinating, if somewhat depressing, catalog of documented instances in which a lack of critical thinking has led to unfortunate consequences, including embarrassment, financial loss, and even death. Thus, we can see that the danger inherent in believing ostensibly harmless paranormal and pseudoscientific claims is the potential such beliefs have for making it easier to accept more dangerous beliefs. What is required is the application of critical thinking to all of our beliefs. Anomalistic psychology provides an excellent vehicle for the teaching of critical thinking by demonstrating the limits of our cognitive systems, with respect to such aspects as perception, memory, interpretation, and reasoning. We are all susceptible to a host of cognitive biases without even realizing it.

It is, of course, impossible to come up with precise figures to tally the harm done by the failure to apply critical thinking. On the basis of the documented cases listed on his What's the Harm? website, Tim Farley has proposed the following minimum figures: "368,379 people killed, 306,096 injured and over $2,815,931,000 in economic damages."[26] These are minimum figures because they only relate to the cases documented on his website. It is certain that there are many more cases that go undocumented. Clearly, no one should dismiss such claims as "harmless."

Much of this book was written while the world was going through the (ongoing) COVID-19 pandemic that has already cost around seven million lives and will cost many more before it ends. Misinformation and disinformation regarding the pandemic are rife, including many wild and unfounded conspiracy theories regarding the coronavirus. Despite the fact that effective vaccines have been developed and deployed, there is little doubt that belief in such theories led to a substantial proportion of the population refusing to be vaccinated, thus prolonging the pandemic and causing further unnecessary suffering and death.

There is an issue of consumer protection here. It is difficult to state the total value of the "paranormal industry," as it obviously depends on exactly what we include under that heading. However, some indicative estimates are available. For example, a 2018 report from *IBISWorld* claimed that Americans spend $2.2 billion annually on "psychic services," including astrology, aura reading, palmistry, tarot card reading, and mediumship.[27] A report from the National Center for Health Statistics reported that Americans spent a staggering $30.2 billion on complementary and alternative medicine in 2012.[28] It has been estimated that the global "wellness industry," riddled as it is with pseudoscientific claims, was a mind-blowing $4.9 trillion market in 2019.[29] If, as seems to be the case, the services being provided here are actually ineffective over and above placebo effects, the public has a right to know that.

MEASURING PARANORMAL BELIEF

One of the central themes of this book is that our prior beliefs have a strong influence on how we perceive, make sense of, and recall events in the world around us. Many of the studies that I describe in later chapters compare paranormal believers and skeptics under controlled conditions in terms of such factors. It is, therefore, important to consider how exactly we measure paranormal beliefs.

One approach has already been mentioned. Opinion polls on levels of paranormal belief are regularly reported in newspapers and magazines, especially in the run-up to Halloween. Properly conducted surveys do indeed provide useful information, but it is typically at a rather crude level. Often respondents will simply be asked whether or not they believe in this or that paranormal phenomenon and given the response options of *yes*, *no*, and *don't know*. A properly conducted survey will collect data from a large sample that is intended to be representative of the population from which it came, in terms of such demographic factors as age, socioeconomic class, gender, political preference, and so on. Such data allow researchers to compare demographic groups in terms of belief levels, providing the opportunity to test some interesting hypotheses or to generate some new ones.

For example, the 2019 YouGov survey of American adults referred to previously showed that Republicans were more likely than Democrats to believe in supernatural beings.[30] This difference was most pronounced with respect to belief in demons, with 54 percent of Republicans reporting belief compared to 37 percent of Democrats. This almost certainly reflects the greater alignment of Republican views with those of fundamentalist Christians compared to Democrats, but this is an explanation that could be empirically tested in future studies.

One interesting general explanation that has been put forward in an attempt to explain differences in levels of paranormal belief between different demographic groups is the *social marginality hypothesis*. This hypothesis is based on the idea that those with relatively less influence and power within society will be more likely to adopt paranormal beliefs, presumably on the assumption that such magical thinking brings some sort of psychological compensation to believers. Thus, it would be predicted that, in general, lower socioeconomic class, female gender, ethnic minority status, younger age, and unemployment would all be associated with higher levels of paranormal belief. In fact, this hypothesis has not fared well in the face of empirical data. One possible exception is the consistent finding that females tend to report higher levels of paranormal belief than males (except for belief in aliens and cryptids, where males show higher levels of belief). Even here, there are alternative explanations for the findings. So it's a case of "nice idea, shame about the data."[31]

The astute reader may well by now have spotted a rather glaring oversimplification in some of the discussion of paranormal belief so far. It simply is not the case that paranormal belief is unidimensional. In fact, it is complex and multidimensional. For example, it is, of course, perfectly possible to believe in telepathy without believing in ghosts. Psychologists often use standardized questionnaires to measure psychological dimensions such as belief because, among other things, this allows for easier comparison of results between different studies. Furthermore, such scales have known psychometric properties, such as reliability and validity. A reliable test of a stable trait will give similar results on different occasions of testing. We can have confidence that a valid scale really is measuring what it is intended to measure.

Two of the most commonly used scales in the field reflect the different conceptions of the paranormal that were discussed earlier. The first of these, Thalbourne's Australian Sheep-Goat Scale (ASGS), is reproduced in Box 1. As can be seen, the items on the scale focus solely on belief in and experience of the three core concepts of interest to parapsychologists: ESP, psychokinesis, and life after death. It is a very easy scale to administer and score. If you complete the questionnaire yourself, award yourself zero points for every response of *False*, one point for every response of *Uncertain*, and two points for every response of *True*. The most extreme skeptic would end up with a score of zero; the most extreme believer would have a score of 36. A group of 247 psychology students tested by Michael Thalbourne had an average score of 14.90 (standard deviation = 7.61).[32] How did you do? Are you, relatively speaking, a believer or a skeptic?

You may be wondering about the somewhat unusual title of the scale. Within parapsychology, believers in the paranormal are referred to as sheep whereas skeptics are referred to as goats (based on a biblical reference).

Whereas Thalbourne's scale is based on a strict definition of paranormal, the other most commonly used scale is based on a much looser definition. Jerome J. Tobacyk's Revised Paranormal Belief Scale (RPBS) is reproduced in Box 2. As you can see, the items cover a much wider range than the ASGS. Furthermore, the scale provides both a global paranormal belief score as well as scores on different subscales. If you completed the scale, you may like to compare your scores to the mean scores (with standard deviation in parentheses) of 217 university students tested by Tobacyk. Their scores were as follows: Traditional Religious Belief: 6.3 (1.2); Psi: 3.1 (1.5); Witchcraft: 3.4 (1.7); Superstition: 1.6 (1.2); Spiritualism: 2.8 (1.4); Extraordinary Life Forms: 3.3 (1.3); Precognition: 3.0 (1.3). The students had a mean full-scale

Box 1: Australian Sheep-Goat Scale

For each item indicate your attitude using the following scale:

0 = False
1 = Uncertain
2 = True

1. I believe in the existence of ESP.
2. I believe I have had personal experience of ESP.
3. I believe I am psychic.
4. I believe that it is possible to gain information about the future before it happens, in ways that do not depend on rational prediction or normal sensory channels.
5. I have had at least one hunch that turned out to be correct and which (I believe) was not just a coincidence.
6. I have had at least one premonition about the future that came true and which (I believe) was not just a coincidence.
7. I have had a least one dream that came true and which (I believe) was not just a coincidence.
8. I have had at least one vision that was not an hallucination and from which I received information that I could not have otherwise gained at that time and place.
9. I believe in life after death.
10. I believe that some people can contact spirits of the dead.
11. I believe that it is possible to gain information about the thoughts, feelings or circumstances of another person, in a way that does not depend on rational prediction or normal sensory channels.
12. I believe that it is possible to send a 'mental message' to another person, or in some way influence them at a distance, by means other than the normal channels of communication.
13. I believe I have had at least one experience of telepathy between myself and another person.
14. I believe in the existence of psychokinesis (or 'PK'), that is, the direct influence of mind on a physical system, without the mediation of any known physical energy.
15. I believe I have personally exerted PK on at least one occasion.
16. I believe I have marked psychokinetic ability.
17. I believe that, on at least one occasion, an inexplicable (but non-recurrent) physical event of an apparently psychokinetic origin has occurred in my presence.
18. I believe that persistent inexplicable physical disturbances, of an apparently psychokinetic origin, have occurred in my presence at some time in the past (as, for example, a poltergeist).

Michael A. Thalbourne, "Further Study of the Measurement and Correlates of Belief in the Paranormal," *Journal of the American Society for Psychical Research* 89, no. 3 (1995): 233–247.

Box 2: Revised Paranormal Belief Scale

Please put a number next to each item to indicate how much you agree or disagree with that item. Use the numbers as indicated below. There are no right or wrong answers. This is just a sample of your own beliefs and attitudes. Thank you.

1 = Strongly disagree 2 = Moderately disagree 3 = Slightly disagree 4 = Uncertain 5 = Slightly agree 6 = Moderately agree 7 = Strongly agree

1. The soul continues to exist though the body may die.
2. Some individuals are able to levitate (lift) objects through mental forces.
3. Black magic really exists.
4. Black cats can bring bad luck.
5. Your mind or soul can leave your body and travel (astral projection).
6. The abominable snowman of Tibet exists.
7. Astrology is a way to accurately predict the future.
8. There is a devil.
9. Psychokinesis, the movement of objects through psychic powers, does exist.
10. Witches do exist.
11. If you break a mirror, you will have bad luck.
12. During altered states, such as sleep or trances, the spirit can leave the body.
13. The Loch Ness monster of Scotland exists.
14. The horoscope accurately tells a person's future.
15. I believe in God.
16. A person's thoughts can influence the movement of a physical object.
17. Through the use of formulas and incantations, it is possible to cast spells on persons.
18. The number "13" is unlucky.
19. Reincarnation does occur.
20. There is life on other planets.
21. Some psychics can accurately predict the future.
22. There is a heaven and hell.
23. Mind reading is not possible.
24. There are actual cases of witchcraft.
25. It is possible to communicate with the dead.
26. Some people have an unexplained ability to predict the future.

Note. Item 23 is reverse scored. Traditional Religious Belief = Mean of Items (1, 8, 15, 22); Psi = Mean of Items (2, 9, 16, 23); Witchcraft = Mean of Items (3, 10, 17, 24); Superstition = Mean of Items (4, 11, 18); Spiritualism = Mean of Items (5, 12, 19, 25); Extraordinary Life Forms = Means of Items (6, 13, 20); Precognition = Mean of Items (7, 14, 21, 26)

Jerome J. Tobacyk, "A Revised Paranormal Belief Scale," *International Journal of Transpersonal Studies* 23, no. 1 (January 2004): 94–98.

score of 89.1 (standard deviation = 21.9; note that the full-scale score is simply the grand total of the scores across all items).

This scale can be used to provide a profile of an individual's paranormal belief in addition to an overall global score. Thus, two people with the same global scores may have very different profiles, highlighting the fact that paranormal belief is multidimensional. It should be noted that although the RPBS has been the subject of various criticisms, it continues to be widely used.[33] It has proved itself to be a useful tool within the field. The fact that paranormal belief is multidimensional should always be kept in mind. It is likely that the psychological factors that lead to belief in ghosts may be very different from those that lead to belief in, say, telepathy.

INSIGHTS FROM PSYCHOLOGY

Psychology is often defined as "the study of mind and behavior." As already stated, it is divided into subdisciplines, each of which is typically presented as a separate chapter in introductory textbooks. To a large extent, this is simply a matter of presentational convenience. In practice, there are few, if any, psychological topics that are exclusively within the remit of a single subdiscipline. To give but one example, a memory researcher is likely to be interested in how information is encoded, retained, accessed, and lost (cognitive psychology) as well as the underlying neural processes involved (neuropsychology), how memory changes across the life span (developmental psychology), and the influence of others on memory performance (social psychology).

The rest of this chapter will outline in general terms some of the ways anomalistic psychology gains insights from the various major subdisciplines of psychology. Some of these have already been touched on. For example, in investigating why people's level of belief varies, we might consider individual differences in demographic and personality factors. As we've seen, the social marginality hypothesis, despite being intuitively appealing, did not stand up well in the face of empirical evidence. On the other hand, there are several personality variables, such as fantasy-proneness, absorption, dissociativity, and hypnotic susceptibility, that reliably correlate with paranormal belief and the tendency to report ostensibly paranormal experiences.

It was also pointed out previously that there exist a number of measures of psychopathological tendencies that reliably, albeit moderately, correlate with paranormal belief. Furthermore, it is widely accepted that sometimes serious psychopathology, such as certain types of schizophrenia, does involve delusions regarding paranormal beliefs (for example, that one is able to read other people's minds or that one is broadcasting one's thoughts to others). However, correlations between psychopathological tendencies and paranormal belief and experience within the general population are nowhere near strong enough to justify any simplistic claim that everyone reporting ostensibly paranormal experiences is suffering from serious psychopathology. But they do need to be taken into consideration in any comprehensive account of paranormal belief. Thus, clinical psychology may provide important insights. As we will see in later chapters, these associations with personality and psychopathological tendencies may provide a clue in explaining some accounts of paranormal experiences.

The psychobiological (or neuropsychological) perspective within psychology focuses primarily on the underlying neural substrate of mind and behavior. Most scientifically minded psychologists assume that ultimately all psychological phenomena are potentially explicable at this level of analysis. Although detailed and comprehensive accounts of psychological phenomena at this level are rare, insights from neuropsychology certainly help us to understand in broad terms such phenomena as sleep paralysis (chapter 2) and out-of-body and near-death experiences (chapter 7).

Chapters 8 and 9 consider insights that cognitive psychology can provide in understanding anomalous experiences and beliefs. The term *cognition* covers all aspects of information processing, including perception, memory, reasoning, language, and so on. The human cognitive system is amazing. During every second of your waking life, you are likely to be simultaneously engaged in several higher-order cognitive processes that, at the time of writing, even the world's most powerful and sophisticated computers could not match (it is only a matter of time, though). Nevertheless, cognitive psychologists have identified a large number of cognitive biases that can reliably lead us to misperceive, misremember, and misinterpret events under certain circumstances. As we shall see, some of these biases can lead us to conclude that we have had a paranormal experience when, in fact, we have not.

One might have expected that evolutionary pressures would have eliminated such biases in favor of cognitive systems that unerringly produce accurate mental representations of the world around us. Evolutionary psychologists have provided a plausible account of why this has not happened. Essentially, their argument boils down to this: in evolutionary terms, it makes more sense to have cognitive systems that are quick and usually accurate rather than those that are slower and slightly more accurate. This line of argument and the evidence supporting it will be discussed further in chapter 9.

Social psychologists are interested in how we might be influenced by our interactions with others. It is undoubtedly the case that many people believe in paranormal phenomena not because they believe that they themselves have directly experienced the paranormal but because trusted others, such as friends and family, have told them of their experiences. The media's role in promoting paranormal claims would also be a topic of interest to social psychologists. Techniques of persuasion would also come within the remit of social psychology. One particular example of this will be discussed in more detail in chapter 4: that is, the technique of cold reading, which you could use to convince complete strangers that you know all about them. Should you wish to convince people that you have psychic powers, this is a very useful skill to have.

Developmental psychologists are interested in changes and consistencies in our mental lives and behavior across the life span, from birth to old age.[34] It is a commonly held view that, subjectively at least, children live in a more magical world than adults. This was certainly the view of the influential Swiss psychologist Jean Piaget (1896–1980). He argued that it was not until around the age of twelve that children could reliably distinguish between reality and magic. Adults, by contrast, have outgrown childish magical thinking and instead think rationally about the world around them—or so it was once believed.

There is certainly no shortage of evidence that children do often engage in magical thinking. Most young children believe in such fantasy figures as Santa Claus, the Easter Bunny, and the Tooth Fairy, not to mention scarier entities such as ghosts, monsters, and witches. Most believe that wishes made when blowing out the candles on a birthday cake are likely to come true. Many children take superstitious beliefs very seriously, such as avoiding stepping on cracks in the pavement or wishing on the first star seen in the evening.

Psychologists, such as Jacqueline Woolley and colleagues at the University of Texas in Austin, have reviewed the evidence relating to children's magical thinking and carried out numerous investigations of children's belief in fantastical beings (such as Santa and the Easter Bunny), the power of wishing (and praying), and so on.[35] Summarizing a vast amount of empirical evidence, we can conclude that belief in magic is prevalent between the ages of three and eight years and peaks around five or six years. It is during this period that children engage in the highest levels of pretend play and are most likely to believe in imaginary playmates and other fantastical beings.

So, was Piaget right to view young children as being unable to distinguish between reality and fantasy? It turns out that the situation is a little more complicated than that. Even very young children do understand that there is a difference between the two. However, there are factors that might sometimes lead them to draw the wrong conclusion regarding the reality status of certain concepts. In addition to the fact that children have less knowledge and direct personal experience of the world around them, it turns out that most adults, especially parents, are engaged in a dastardly conspiracy to convince children that magic is indeed real. Films, books, and plays for children almost always include magic in the form of talking animals, fantastical creatures, and superhuman powers. We typically provide physical evidence for the existence of Santa, the Easter Bunny, and the Tooth Fairy. It makes sense for children to believe what their parents and other carers tell them, as they are more likely to avoid coming to harm that way—but that means they will also be likely to fall for blatant lies when adults tell them.

Having said that, even very young children do their best to establish the reality status of objects, whether they are contrasting real objects with toys, imagination, or pictures. They know that real objects have different properties to those in the latter categories and behave accordingly. Not surprisingly, children are very good at correctly judging that things that they have directly experienced, such as teachers, chocolate, and cars, are indeed real. But they sometimes struggle to make accurate judgments regarding objects that they have not directly experienced—such as dinosaurs, knights, and germs. As a result, children sometimes dismiss such real objects as being fictitious unless given specific evidence to the contrary. It is possible to be too skeptical!

It is also the case that adults sometimes attribute more magical thinking to children than is actually justified. This is nicely illustrated by an event that took place when my eldest daughter, Lucy, was little. Lucy, in common with an estimated 65 percent of children in general, had an imaginary playmate. Both my wife and I are psychologists, and we were well aware that this was perfectly normal and not a cause for concern. Lucy's imaginary friend was called Gunda, and we were quite used to Lucy telling us what Gunda was up to and what Gunda had done or said that day.[36] At the time, Lucy was an only child, and Gunda was pretty much just one of the family—albeit one that happened to be invisible and of unspecified gender. One day, when she was around six years old, Lucy told me something. I cannot now remember what it was, but I do recall that, as a joke, I replied, "I already know that. Gunda told me." Lucy looked at me as if I was crazy and declared, "Gunda's not real, you know!"

It is now thought that Piaget was somewhat overstating the case for children being unable to distinguish between reality and fantasy. So, was he at least right in claiming that adults generally live in a world of rational thought and rarely fall prey to magical thinking? Sadly, it turns out that he was mistaken in this view as well. Although adults are sometimes capable of rational thought, this book will amply demonstrate that adults also routinely engage in magical thinking. In fact, a team led by Christine Legare of the University of Texas at Austin has presented convincing evidence that supernatural explanations sometimes actually increase from childhood to adulthood.[37]

Legare and colleagues go further in that they point out that both natural and supernatural explanations are sometimes used simultaneously to explain the very same event. For example, the Azande people of Central Africa sometimes shelter from the heat of the sun in the shade beneath their grain silos. Every so often, a silo will collapse, killing the unfortunate person below. The Azande are perfectly happy to explain the collapse in terms of physical damage caused by termites. However, at the same time, they might claim that it collapsed at that precise moment in time, killing that particular individual, in terms of witchcraft. Similarly, in some South African communities, the scientific explanation of contracting AIDS (i.e., it is caused by a virus) was generally accepted. However, it was also believed that witchcraft could be involved, as witches might influence an individual to make an unwise choice of sexual partner.

Having considered in general terms the nature of anomalistic psychology, we are now ready to consider in more detail some of the specific topics of interest—beginning, in the next chapter, with the often terrifying experience of sleep paralysis.

2 WAKING NIGHTMARES

A GENTLEMAN PAYS A CALL

The following is an account of a strange experience that happened to my friend Sarah Elizabeth Cox. At the time, Sarah was a university press officer and part-time postgraduate student. She was thirty-two when this happened to her.

> To my memory this has only occurred to me once, in 2019. I was in my flat alone in Catford and sleeping. I was particularly stressed out because of the noise from my upstairs neighbour. I'd lived there about six months, and he made a huge amount of noise at night, banging on the floor, talking loudly on the phone for hours on end. Most nights I would sleep in padded ear defenders—it was that bad. I wasn't drunk as such when I went to bed but had maybe three glasses of wine in the evening, so there was definitely some alcohol in my system. It was during a period where I was probably drinking a lot more alcohol than usual and would have much weirder dreams generally but nothing particularly dark and disturbing.
>
> I remember the door to my flat being wide open (it wasn't in reality, it was locked and I always double-lock it at night, but it felt very real that it was open into the communal corridor)—the door went straight in to the main living and sleeping space in the studio flat.
>
> A figure came in, tall and hunch-backed, wearing a long black hooded cloak. He didn't have a face, but the space where the face should have been seemed to kind of glow a dark green. He seemed to be larger than life-size, perhaps around 7 or 8 feet tall, but hunched over. Even though I couldn't see his face, I felt that he probably did have a face, and would have looked similar

to those characters in an episode of *Buffy* which aired in 1999 when I was 13. I found them particularly scary at that time (especially the way they floated) and I would dream about them a lot in my early teens.

I confess that I was not that familiar with the TV series *Buffy the Vampire Slayer*, but Sarah kindly sent me a link to a YouTube clip so that I could see for myself what she was referring to.[1] The Gentlemen, as they are called, appeared in "Hush," the tenth episode of the fourth season of the popular series. This episode is widely acclaimed as one of the best ever by fans of the series. I can see why. The Gentlemen are as creepy as hell! They are pale, silent ghouls with bald heads, sunken eyes, and wide, malicious grins, revealing metallic teeth. They do not walk but instead glide along about a foot above the ground, their legs not moving. They are elegantly dressed in black suits and carry satchels containing scalpels (which are used to remove the hearts of their terrified victims). They communicate with each other using polite, graceful gestures. You would not want these guys in your bedroom. Sarah's account continues:

> The figure stopped at the side of my bed, which is a low platform bed, so he leaned right over me (possibly hovering in the style of the Gentlemen) and let out this god-awful shrieking noise which lasted for what felt like about 30 seconds, perhaps longer, in one long stream. It was a sort of deep guttural sound. I felt like it was imprinted on my brain and I could hear it for days and days afterward. I'm not sure I could replicate the noise myself but I'd recognise it if I heard it.
>
> I felt like I wanted to roll over and shriek back or to punch him in the space where his face should be, but I couldn't force myself to turn over, so I lay curled up facing the wall and tried to put my right arm over my ear to stop the noise, which seemed to go on forever. I felt like I was shaking from trying to force myself to roll over and also felt really scared that he was going to stab me with a knife in the ear.
>
> I thought it could have been a real person who had broken in and I was about to be attacked. Eventually the noise stopped, and I remember him turning around and leaving the room slowly the way he came in.
>
> Interestingly, I also had an extremely meta and perhaps quite unique experience of knowing exactly what was going on at the time it was happening. I

was thinking, "This is terrifying, I'm about to be attacked" but also, "Wow, this is sleep paralysis, isn't it? I know all about this because of Chris and Alice! Weird that this has never happened to me before but there's a first time for everything." So, there was this level of awareness about what was happening which I thought was quite interesting (and funny the next morning, perhaps not at the time).

I remember the sound of the shriek a lot more clearly than everything else.

Sarah actually realized what was happening to her as it was happening because she had already had conversations about sleep paralysis with me and another colleague and friend at Goldsmiths, Professor Alice Gregory. Alice is an expert on sleep in general and has carried out research into the topic of sleep paralysis with me, along with Dr. Brian Sharpless and Dr. Dan Denis, both leading international experts in the field. But Sarah was mistaken in thinking that her experience of recognizing what was happening to her as it happened was unique. In fact, it is quite common—and I have even had previous reports of people telling me that, during the experience itself, they have thought, "Oh, I must tell Chris French about this one!"

Sarah had sent me her account in response to an appeal I had sent out on Twitter just before I started writing this chapter. I know that such firsthand descriptions really bring to life the topic of sleep paralysis. I also know from experience that people who suffer from sleep paralysis—especially in its vivid form as exemplified above—are usually quite keen to share their experiences as they, perhaps more than anyone else, want to make sense of what can be a profoundly disturbing experience. For example, I wrote a column on the topic for the *Guardian*'s science pages back in 2009 and was overwhelmed by the number of comments from readers wanting to share their own stories.[2]

Not all episodes of sleep paralysis are as scary as Sarah's. Indeed, it is far more commonly experienced as simply a temporary period of paralysis, lasting perhaps a few seconds, occurring between sleep and wakefulness. It is typically a little disconcerting but, in its mild form, likely to be quickly (and correctly) dismissed as being nothing to worry about. As we shall see, however, it can sometimes be much more terrifying and can, in some rare cases, seriously impair a person's quality of life.

THE TERROR THAT COMES IN THE NIGHT

The title of this section is taken from folklorist David J. Hufford's classic book on the topic of sleep paralysis, subtitled *An Experience-Centered Study of Supernatural Assault Traditions*.[3] That book is required reading for anyone with a serious interest in the topic, along with the excellent volumes by Shelley R. Adler and Brian A. Sharpless and Karl Doghramji.[4] Before delving a little deeper into the science of sleep paralysis, I would like to present a few more firsthand accounts. I find reading such accounts endlessly fascinating, and I am always struck by the fact that what we are dealing with here is a common core experience overlaid with peculiar idiosyncratic elements. The main thing to bear in mind when reading these accounts is that, however we might interpret what is going on, these are real experiences happening to real people.

As stated, sleep paralysis in its mild form consists of nothing more than a temporary period of paralysis as one is drifting off to sleep or emerging from it. However, it may be accompanied by a number of other symptoms that can serve to make it altogether more terrifying. The first of these is an overwhelming sense of presence. Even though the sufferer may not be able to see or hear anything threatening in the room with them, they are convinced that someone—or *something*—is definitely there. Furthermore, whoever or whatever it is, it has extremely malign intentions toward its helpless victim. A good example of this phenomenon was sent to me by Ian Gardiner:

> My particular experiences were over a few months about 26 years ago when I was in my early 30s. I would wake up in the early hours, always lying on my right side. I knew that I was awake as I was aware of reading the time on my bedside clock radio. I was unable to move and shortly after waking I would be overcome with a sense of appalling dread which grew greater as time passed.
>
> Our bed was against the back wall of the bedroom and the door was at the bottom left corner (as viewed from the bed). So as I woke I could not see the door but had this overwhelming sensation that someone or something was in the doorway watching me. It was a horrible sensation which I can still bring to mind all these years later. I tried to shout to alert my wife but could not speak. The fear was one of impending doom or the fact that someone was "coming for me." Gradually the paralysis would ease off and I was able to

move again. The first thing I would do would be to look over at the doorway. There was never anything there.

I have had no episodes since that time and cannot put a finger on any reason why they might have started or indeed why they stopped when they did. As I recall an episode would last for between 1 and 2 minutes. Even though this was 26 years ago, the memories are still clear in my mind.

Here is another account in which the sense of presence is a dominant feature. This one was sent to me by Matt Salusbury and includes a description of a second common feature of such episodes: a sense of pressure on the chest.

I had one back in 1986. I was 18. I was travelling in the US during the summer, staying with some distant Japanese relatives in Michigan State University, Ann Arbor, Michigan. I slept on their sofa, which may have had something to do with the incident that followed. I woke up one night, I seem to remember a light had been left on in the kitchen/living room I was staying in. The first sensation I had on waking was that I couldn't move, I was being firmly pinned down onto the sofa with quite a lot of force and pressure.

Very soon afterwards I had the sensation that there was something very evil standing just out of my field of vision, immediately behind my head. I can't recall whether I actually sensed something visually in my peripheral vision, I suspect not, or was just aware of some malevolent entity that I couldn't physically see, but could somehow sense. The pressing down on my chest and my complete inability to move felt like part of that evil and malevolence, this entity was causing it.

I sensed that it was some kind of tall entity, in some way skeletal, like H. R. Giger's alien design from the *Alien* movies, or Judge Death in his helmet from *Judge Dredd*. There was some sort of gristle and muscle between the bones, particularly around the teeth and what was left of the jaw, but I couldn't see a face. I can't recall the colour, except for the white teeth with the raw flesh of the gums around it, as if the skin of the face and lips had been peeled away.

I didn't know whether it had seen me, or seen whether I was awake, or I was in denial about whether it could see me. I closed my eyes tightly, trying to pretend I was asleep, somehow feeling that if it didn't know I was awake it might not react and might leave me alone. I remember thinking what a lame and hopeless plan this was, but desperately trying to pretend I was asleep, it was for some reason very important that I did. I was absolutely terrified, perhaps the most scared I've ever been in my life.

I remember initially trying to cry out, but being unable to make a sound.

Somehow, I managed to get to sleep. I woke up the next day and the daylight came flooding in, it seemed almost impossible that the experience of the previous night had happened. So unlikely, in fact, that I didn't mention it to my hosts. I had a sneaking suspicion that there was a perfectly logical explanation.

I can't recall how long the experience seemed to last, I do remember lying there in complete terror with my eyes tightly closed for an awfully long time, perhaps hours, before I eventually drifted off to sleep.

Matt refers to a third extremely common feature of sleep paralysis episodes: intense fear. Now, you may be thinking, not unreasonably, that intense fear is a perfectly understandable reaction to the terrifying experiences described, and it would be hard to argue with that. However, it has been suggested by J. Allan Cheyne and colleagues at the University of Waterloo in Canada that the intensity of the fear experienced is a direct result of activation of the amygdala, the part of the brain that deals with the perception of threat and initiation of the fight-or-flight response.[5] Cheyne's theory will be described in more detail later in this chapter.

It is also interesting to note that in both Sarah's and Matt's accounts they felt able to describe what their unwelcome intruders looked like despite maintaining that they could not actually see them. This type of "seeing without seeing" is not uncommon in sleep paralysis episodes. However, some sufferers do actually experience vivid visual hallucinations (see figure 2.1). Garret Shanley here describes a couple of sleep paralysis episodes from his childhood. The first of these features the "Old Hag," a common nocturnal visitor during such episodes.

> Up to the age of 7, I experienced the old hag regularly. She resembled the witch from *The Wizard of Oz* or from an Irish children's television programme that was on at the time. She wore black and a pointed hat. I would see her in the corner of the room and she would gradually approach me. Rather than feeling paralysed, I remember deciding not to move. I would remain completely still so she would not know I was awake and I would only glimpse in

her direction now and again to see how close she was. When she was close, I would keep my eyes shut and feel the weight of her as she sat on my legs. I did not want her to know I was awake in case she spoke to me. Then, exhausted by the terror that would decrease over time, I would eventually fall asleep. I started to question if this was a kind of dream (it did not feel like my other dreams or nightmares at all) when my mother and siblings said it was. I once wedged several books between my bed and a wall so my bed could not be approached without knocking them over. In the morning, they were all knocked down—I suppose they just fell over by themselves, but I heard them falling, slowly, one by one during the night.

The most distressing sleep paralysis experience I had as a child was when a series of faces appeared in a mirror that hung opposite the bed I was in that night—I was not in my usual room. One deformed face of a gurning monster would be replaced by another—over and over, each more horrible than the last. My eyes were glued to the mirror and the faces were accompanied by an intense sound like feedback, but not quite. I thought the faces would become so awful that I'd eventually see one that would drive me insane. I yelled and my mother came to the room.

Garret stopped having sleep paralysis after that, only to start having episodes again in his twenties. His final episode (to date) occurred in his late thirties.

It is not just visual hallucinations that may be experienced during sleep paralysis. As Sarah's and Garret's accounts illustrate, auditory hallucinations may also occur. Indeed, it appears that the hallucinations that are experienced during such episodes can potentially affect all sensory modalities. Here is an account from Angela Keogh, a forty-three-year-old secretary for a property company in London with a lifelong history of sleep paralysis. It is an example not only of a vivid tactile hallucination but also of the strange bodily sensations that often accompany episodes and the difficulty of finding the words to properly describe the experience.

> It was January this year, the usual thing happens, I feel like something is trying to get into my chest and everything caves in in my chest, with the sensation of something running through my chest like . . . oh, sorry, I don't think I have the right words for it, but it's like a build-up of something happening, like I'm falling or going deeper into the sleep paralysis, and I'm not able to move

at all, and I really struggle with it. But this time I can see a claw at the end of my bed. The claw/hand looks like it's made out of sticks or a tree. So all brown and veiny with long brown fingers with sharp nails and I could actually feel it grabbing my foot. I can feel that sensation of it grabbing and the sharp feel on my foot. And then at this stage I literally jump out of bed and I immediately check under the bed. I think that episode happened one or two more times after that. So I started sleeping with my feet away from the end of the bed.

The following account, sent to me by Nell Aubrey, includes auditory, visual, and tactile aspects:

My first really clear memory of having a full episode, as it were, was when I was about 8 or 9. I was staying with my best friend, who lived in a very large old house. I had stayed there plenty of times before, but as her stepbrother was staying I had to sleep in one of the guest rooms, which had a four-poster bed in it. I should point out that I loved the house, and the family, it was really a magical and very happy place, and I spent as much time there as I could and would have moved in with them and never gone home again, if they had let me!

There were some parts of the house that really scared me, but not the four-poster room, and I had longed to sleep in that bed for ages. I mention this to emphasize just how scared I must have been to give up on the opportunity! I remember being very excited to go to sleep there, but becoming quite fearful and agitated once I was alone and in the dark. I had trouble getting to sleep, although that wasn't unusual. I woke up thinking that someone had called my name, I had a lot of hypno-hallucinations and still do, but I had no idea what they were at that age and they disturbed me a lot.

At first, I thought one of the cats was in the room, and had jumped on the bed, as I felt that kind of a thud on the covers. I couldn't sit up and that was when I began to feel really scared. I felt something leaning on my abdomen which was very painful and oppressive and at the same time I saw a figure at the foot of the bed, where the curtains were pulled around the bedpost. It was a thin long humanoid shape and it seemed to stretch its arm out over me, and the pain got worse as it did so. I suppose it was pain from feeling I couldn't breathe properly, but I'm not sure. It lasted quite a long time and then gradually dwindled away. When I could finally move I abandoned the four-poster and went to sleep on the floor of the room next door.

I have come across several accounts in which episodes are initially interpreted as a family pet jumping onto the bed. Indeed, I have myself had one such sleep paralysis experience some years ago. The familiar sensation of feeling the weight of our cat pressing on the mattress was slowly joined by the realization, as I emerged from sleep into full wakefulness, that the cat had died two years previously. It was not a particularly scary experience as, by this stage, I was fully aware of the nature of sleep paralysis. Carla MacKinnon, a filmmaker and friend of mine, describes a very similar experience:

> I am asleep in bed when I am woken by a cat jumping onto the bed. It must have jumped in through the window. I am on the third floor. The window is closed. I know it is a black cat although I cannot open my eyes to look. It is padding on the bed and it is not unpleasant, but it is unsettling, and when I try to sit up in bed to deal with the situation I cannot move. I cannot turn over. My whole body is lead. This time I know what is happening, I've been reading about sleep paralysis and how it manifests. I relax and as calmly as I can I examine the feelings in my body. Very soon it ends.

It has to be said, however, that this was one of Carla's least terrifying episodes. Before she became familiar with the nature of sleep paralysis, the experience was often much more frightening. For example:

> I awake in bed, next to my boyfriend. It is very dark in the room. In the corner of the room there are two men. I cannot see them but I know that they are there, and what they look like. I can hear them talking. They are talking about murder. I cannot move, my body is frozen but my senses are on high alert. One of the men comes and stands directly above me. I know he is wearing a hat, although my eyes are closed. He spits, and his spit lands in the socket of my closed eye. I can feel the impact, the wetness, the trail of slime.[6]

It is not uncommon for there to be a strong sexual aspect to these experiences. In some cases, they can even be emotionally positive experiences as the following account from William in Derbyshire illustrates:

> I could feel myself lying on the bed and knew that I was in my bedroom. I saw the part of the bedroom that I would expect to be able to see, were my eyes open, e.g. my half of the bed, the window, the wardrobe, but I could feel that my physical eyes were still closed. In addition to seeing what I would expect

to be there, there was a young woman standing by the foot of my side of the bed. She wore a fawn pullover with brown stripes and a brown skirt. She had long, thick, brown hair, partly falling down over what should have been her face, but there was just a black void where her face should have been. I recalled having had a previous sleep paralysis experience and remembered that when I'd last had such an experience, it all stopped when I opened my eyes. Sensing that this time, it had the potential to be an enjoyable experience, I deliberately kept my eyes closed, so that the experience would continue. She walked down the side of the bed (towards my head). Without verbalizing the wish, I thought that it would be nice if the woman came and got on top of me. Much to my pleasant surprise, she climbed on top of me. I therefore seemed to discover that one can control such an experience to some extent by willing the 'intruder' to do something. As she was kneeling across me, I tried to raise my arms to cuddle her, as I fancied her, but much to my frustration, I could not budge them even a millimetre—I was paralysed. So I had to give up that idea. She then lent forward and her hair fell onto my face—the feel of it was *exactly* as it would have been, had it have been a real person's hair. The sensation caused me to ejaculate. When I did open my eyes, of course there was no-one there.

Unfortunately, the sexual side of the experience, when it occurs, is sometimes far from positive, as vividly illustrated by the following anonymous account from a female long-term sufferer:

That's when the sexual side of it started. I will be brief on this part, but basically with the sleep paralysis came the "rapes." I mean, obviously they weren't real. But that is what I could feel happening to my body. The weirdest part of the sexual side was that when I was being raped (so to speak—gosh, that sounds bad, doesn't it?), the hallucinations made it a bit odd. Basically, I was being turned upside down. So pretty much like a suspended headstand. In my mind, I could feel my hair hanging down and the blood rushing to my head. But I could actually feel something entering me down below. You know I could actually feel something inside me. Weird!! I can't really remember how long that went on for. But thankfully I've been free of that now for at least 6 years or so.

I hope that the accounts provided have given you some indication of the weird and surreal nature of sleep paralysis. But, for those of you lucky enough never to have experienced the more vivid manifestations of this intriguing phenomenon, I recommend viewing Carla MacKinnnon's award-winning short film *Devil in the Room*.[7] As an animator and long-term sufferer from sleep paralysis, Carla received funding from the Wellcome Trust to support her in making her film, which is partly based on her own experiences. When she first invited me to collaborate with her on this project, she told that she was aiming to produce a film that was "half science documentary, half horror movie." I think she was 100 percent successful in achieving this.

If you're brave enough, you might also want to check out *The Nightmare*, a documentary film about sleep paralysis directed by Rodney Ascher, released in 2015. This is another film that really captures the disturbing nature of this phenomenon, in this case by a mixture of talking-heads-style interviews with eight sufferers and vivid reconstructions of their experiences. My only reservation about this film is that, toward the end of the documentary, all of those involved are asked to give their views on what they think is really happening to them. All but one of them reject the scientific explanation for sleep paralysis and instead opt, to a greater or lesser extent, for paranormal interpretations. The director cannot be held responsible for those views, of course. If that's what his contributors said, then that is what they said. But I believe that this is not a helpful message to put out to others who are experiencing this perplexing phenomenon. I would prefer to let people know that sleep paralysis, although it can be absolutely terrifying, is essentially harmless—with one possible big exception, to be discussed later in this chapter.

For people who are unaware that sleep paralysis is a scientifically and medically recognized phenomenon, actually experiencing an episode can be extremely frightening and perplexing. Depending on the nature of the episode, a single attack may be dismissed as "some weird kind of nightmare"—albeit one that feels very different from any nightmare previously experienced. Multiple episodes are more difficult to dismiss. The sufferer is likely to consider two main possibilities—either they are "going crazy" or they are experiencing something that is really happening. The latter possibility may sometimes seem marginally preferable to the former.

Sufferers are often reluctant to tell anyone else about their experience because they are afraid, with some justification, that they may indeed be judged to be "crazy" and have to endure the stigma that so often and so unfairly goes along with such labeling. On several occasions, following public talks that I have given, I have been approached by people telling me that they have had the very experience I have described but that they have never told anyone else about it before. The sense of relief that people report on first learning that what they experienced has a scientific explanation is almost palpable.

My aforementioned colleague, Alice Gregory, provides an example of this in her excellent book, *Nodding Off*.[8] I had put seventy-year-old Mrs. Sinclair in contact with Alice after Mrs. Sinclair had contacted me to discuss her sleep paralysis experiences. Living alone in the depths of the countryside, Mrs. Sinclair had eventually been convinced that her 300-year-old cottage must be haunted. For example, in one of her first experiences she had awoken one night with the feeling that someone was trying to strangle her. She had expected to see a burglar upon opening her eyes but instead was shocked to see the face of a childlike imp laughing at her. In Alice's words:

> He began pushing against her sides, tucking the bed sheets in around her. "We've nearly strangled you and now we're tucking you in," he goaded, in a way reminiscent of bullying children some 60 years previously. She tried to move but instead found herself "tarmacked into the mattress," paralysed except for the ability to move her eyes. Although non-religious, the petrified Mrs Sinclair began mentally to recite the Lord's Prayer.[9]

By the time Mrs. Sinclair contacted me, after several other equally distressing episodes, she had already realized that her cottage was not haunted and that, in fact, she was suffering from sleep paralysis. She had found out more about the topic by consulting reliable sources on the internet. She got in touch with me because she had seen me talking about the phenomenon on daytime TV and realized that I would be interested to hear about her own experiences. She told me that she had been so relieved to learn that she was not being plagued by pesky poltergeists—and that she was not losing her mind!

She had decided, not unreasonably, to visit her doctor to see if he had any advice on how she might reduce the incidence of her episodes. She was

Figure 2.1
During an episode of sleep paralysis, the sufferer may experience nightmarish hallucinations as terrifyingly real. Illustration courtesy of Hawraa Wriden.

still rather nervous at the prospect of talking to anyone about her weird adventures but plucked up the courage to describe them to her doctor, making it plain that she now realized that they were caused by sleep paralysis. Her doctor arrogantly dismissed her on the grounds that he had never heard of sleep paralysis. When I heard this, it made me very angry. Thirty seconds on the internet would have made it clear to this pompous so-and-so that sleep paralysis is actually a scientifically and medically recognized phenomenon. I sent Mrs. Sinclair some papers on sleep paralysis and jokingly suggested that, the next time she visited her doctor, she might like to hit him around the head with them. This is just one illustration of the fact that it is often not just many members of the general public who have never heard of sleep paralysis but also a large proportion of medical professionals.

WHAT SLEEP PARALYSIS IS NOT

Before discussing the defining features of sleep paralysis, we will briefly consider a few common sleep-related anomalous experiences with which it might sometimes be confused. The first and most obvious is the *nightmare*—or at least the nightmare as that term is commonly understood in modern times. Historically, the term did indeed refer to an experience that corresponded to what we would nowadays call sleep paralysis rather than just being a synonym for "a bad dream." Even in modern usage, the term should really be reserved for dreams that induce real terror rather than having a slightly negative tone. Dreaming of finding yourself naked in the supermarket or of being stuck in an interminable traffic jam on the M25 do not really fit the bill.

To really count as a nightmare, the dream might involve, for example, the sense of a real threat to one's very survival or the survival of one's loved ones. It will be extremely memorable, and the sleeper will be in a state of extreme arousal when they wake up. All of this is likely to also be true for an episode of sleep paralysis, so what are the differences? First and foremost, the latter involves an awareness of the actual surroundings of the sleeper whereas the nightmare does not. Second, paralysis is a defining feature of sleep paralysis whereas this is not the case for nightmares, although they may involve a sense of difficulty moving away from a threat. Finally, on awakening, the dreamer immediately realizes that they have been having a nightmare and that the events experienced did not really happen. The sufferer from sleep paralysis is much less certain that the events experienced did not really happen.

Many of the symptoms of panic attacks are the same as those experienced during sleep paralysis, including intense fear, a rapidly beating heart, sweating, and catastrophic thinking (e.g., "I am going to die"). Furthermore, susceptibility to sleep paralysis is associated with susceptibility to panic attacks, and panic attacks can occur when an individual is emerging from sleep. Thus, the two disorders would be easy to confuse. One big difference is, of course, that someone suffering a panic attack is not paralyzed and will not experience symptoms such as hallucinations and a sense of presence. Finally, the panic felt during a panic attack is unexpected and hits the sufferer in a great rush, whereas that experienced during sleep paralysis typically grows more slowly. Often it is felt to be in response to the other symptoms of sleep paralysis.

Another sleep-related phenomenon that must be distinguished from sleep paralysis is that of night terrors (*pavor nocturnis*). On closer consideration, night terrors can be seen to be very different from sleep paralysis. Indeed, the only thing they seem to have in common is terror. The sufferer from night terrors will typically scream, leap out of bed, and appear to flee with all their might away from some unnamed terror. The first and most obvious difference between this phenomenon and sleep paralysis is that the sufferer is very definitely not paralyzed. Second, the sufferer typically cannot recall much, if anything, of the episode, whereas the hallucinations experienced during sleep paralysis are all too memorable. Finally, it is extremely difficult for benevolent others to rouse the sufferer from night terrors back into normal wakefulness, whereas sleep paralysis sufferers are easily roused from their torment—and typically extremely grateful to be rescued.

The final sleep-related anomalous experience to be described here is that of the colorfully named phenomenon of *exploding head syndrome* (EHS).[10] EHS refers to a brief hallucinatory experience that occurs just as the sleeper is drifting off into, or emerging from, sleep. Typically, a loud noise is heard, such as an explosion, fireworks, a gunshot, a scream, or a clash of cymbals, but visual sensations (e.g., a flash of light) are also reported in about 10 percent of cases. There are several characteristic differences between EHS and sleep paralysis. First, whereas EHS sufferers report that the jarring sensations wake them up, sleep paralysis sufferers are convinced that they are awake during their ordeal. Second, EHS episodes are short, sharp shocks lasting at most a few seconds; sleep paralysis episodes may last for minutes. Third, the sensations experienced during EHS are nondistinct, whereas the perceptions during sleep paralysis are elaborate and detailed.

DEFINING ISSUES

Sleep paralysis is one of four common symptoms of narcolepsy. Together, these four symptoms are often referred to as the *narcoleptic tetrad*. The other three symptoms are *cataplexy* (sudden loss of muscle tone often triggered by strong emotion), overwhelming feelings of drowsiness, and vivid *hypnagogic hallucinations* (that is, vivid hallucinations experienced at sleep onset).

Sleep paralysis can also be caused by acute intoxication or as a side effect of withdrawal from certain medications. However, it can also occur in the absence of any of these medical conditions or any other sleep disorder, in which case it should technically be referred to as *isolated sleep paralysis* (or *recurrent isolated sleep paralysis* if multiple episodes are being referred to). Brian Sharpless has argued that a useful distinction can be made in the clinical context between those who suffer significant levels of distress or fear as a result of their isolated sleep paralysis and those who do not. The former, he suggests, should be referred to as suffering from *fearful isolated sleep paralysis* (or, in the case of multiple episodes, *recurrent fearful isolated sleep paralysis*).[11] Most, but not all, of the cases described in this chapter would be examples of this type.

As stated, many members of the general public have never heard of sleep paralysis. If they have a frightening episode themselves, they often have no idea what to make of it. Similarly, if a friend or a family member tells them about an episode that they have experienced, they are at a loss to offer them any explanation or advice. Even worse, as we have seen, many medical professionals have themselves not heard of sleep paralysis, despite it being fairly common among the general population.

Specialists in sleep disorders will be familiar with the condition. The third edition of the *International Classification of Sleep Disorders* includes the following criteria for the diagnosis of recurrent isolated sleep paralysis:[12]

A. A recurrent inability to move the trunk and all of the limbs at sleep onset or upon awakening from sleep.
B. Each episode lasts seconds to a few minutes.
C. The episodes cause clinically significant distress including bedtime anxiety or fear of sleep.
D. The disturbance is not better explained by another sleep disorder (especially narcolepsy), mental disorder, mental condition, medication, or substance use.

In contrast, the fifth edition of the *Diagnostic and Statistical Manual of Mental Disorders* contains no mention at all of sleep paralysis.[13] As a consequence, the condition is sometimes misdiagnosed by psychiatrists and clinical psychologists, leading to the recommendation of entirely inappropriate courses of treatment (e.g., prescribing antipsychotic medication).

HOW PREVALENT IS SLEEP PARALYSIS?

As discussed elsewhere, the reported lifetime prevalence rates for sleep paralysis vary enormously across studies carried out in different countries.[14] For example, a study of prevalence rates in 359 American adults reported a rate of 5 percent.[15] In contrast, 62 percent of a sample of sixty-nine adults from Newfoundland, Canada, reported having experienced sleep paralysis at least once in their lives.[16] Brian Sharpless and Jacques Barber assessed lifetime prevalence by aggregating data across thirty-five studies, resulting in a total sample size of 36,533.[17] They estimated that 7.6 percent of the general population experience sleep paralysis at least once in their lives. However, two particular subgroups report markedly higher prevalence rates: for students, the prevalence rate is 28.3 percent, and for psychiatric patients, it is 31.9 percent. This probably reflects the fact that, in those with an underlying susceptibility to sleep paralysis, it is most likely to actually manifest if sleep patterns are irregular. Both students and psychiatric patients are likely to have irregular sleep patterns, albeit for different reasons.

It is not entirely clear why prevalence rates vary so much from one study to another. It may be that there are genuine differences in prevalence rates across different types of respondents in different countries, but the reported rates are also likely to be affected by the wording of items in surveys or interviews. This possibility was investigated directly by Kazuhiko Fukuda in a sample of 593 Japanese university students.[18] Those students who were given a questionnaire asking if they had ever experienced *transient paralysis* reported the lowest prevalence (26.4 percent) compared to a group who received a questionnaire referring to sleep paralysis as *kanashibari*, the traditional Japanese folkloric term for sleep paralysis (39.3 percent). Presumably, the students were more willing to respond positively when the question did not seem to imply some kind of medical condition. If the neutral word *condition* was used in the questionnaire, an intermediate 31 percent of students responded positively. It has been suggested that there may be a general trend for higher prevalence rates to be reported in cultures where the phenomenon is widely recognized to be nonmedical in nature, such as Japan and Newfoundland.

THE SCIENCE OF SLEEP PARALYSIS

Although more research is needed into this perplexing phenomenon, science can already provide a convincing explanation in general terms. To understand what is going on, we must first understand the neurophysiology of normal sleep. At the start of a normal night's sleep, the sleeper typically progresses through stages, predictably labeled as Stage 1, Stage 2, and Stage 3 sleep.[19] As the sleeper moves from one stage to the next, a number of clearly identifiable physiological changes occur, including changes in heart rate, breathing rate, brainwaves, and so on. The sleeper will then begin to reemerge from the deepest stages of sleep into the lighter stages and ultimately into what is referred to as *REM sleep*. REM stands for rapid eye movement. During this stage of sleep, an observer can actually see the sleeper's eyes moving rapidly beneath the closed lids. It is during this stage of sleep that the sleeper, if awoken, is most likely to describe being in the middle of a vivid dream. It is also the case that, during REM sleep, the muscles of the body are actually paralyzed—presumably to stop the sleeper actually carrying out the actions of the dream. After spending some time in REM sleep, the cycle begins again.

The cycle typically lasts around ninety minutes and is repeated several times throughout the night. The relative amount of REM sleep to non-REM sleep increases throughout the night. Sleep paralysis episodes can be thought of as glitches in the normal sleep cycle during which, to put it simply, the mind wakes up but the body does not. Thus, the sleeper may be able to open their eyes and clearly see that they are in their bedroom but be completely unable to move.[20] Furthermore, they may experience a unique altered state of consciousness in which dream imagery is experienced in combination with the contents of normal waking consciousness.

As described above, normally the sleeper spends some time in non-REM sleep prior to going into the REM stage, so why are sleep paralysis episodes often experienced just as someone is drifting off to sleep? It appears that such episodes are more likely to occur if the sleeper atypically enters straight into the REM stage immediately upon falling asleep. Sleep researchers refer to this as a sleep-onset REM period (SOREMP). Narcoleptics who suffer from sleep paralysis are known to be prone to SOREMPs.[21] When narcoleptics prone to sleep paralysis were woken up during a SOREMP, they regularly reported an episode of sleep paralysis, but they did not do so if they were awoken from non-REM sleep or from REM sleep that followed a non-REM period. The claim that the sleep paralysis state is a unique altered state of consciousness that combines both REM consciousness and normal wakeful consciousness is strongly supported by the results of a study by Michele Terzaghi and colleagues.[22] They analyzed electroencephalographic (brainwave) data from a fifty-nine-year-old narcoleptic during wakefulness and during different phases of sleep. The pattern recorded during a sleep paralysis episode corresponded very closely to the pattern produced by combining the patterns recorded during eyes-closed wakefulness and that recorded during REM-stage sleep.

J. Allan Cheyne, professor emeritus at the University of Waterloo in Canada, is one of the world's leading authorities on the topic of sleep paralysis. In 1999, he and colleagues published an influential model of the neuropsychology of sleep paralysis, which, although somewhat speculative, provides a plausible account of the phenomenology of such episodes.[23] The model was based on survey responses collected from a student sample and two online samples. These data were analyzed using a statistical technique known as factor analysis, which allows responses that tend to co-occur to be grouped together. This analysis suggested that the various sensations experienced during sleep paralysis episodes could be grouped together into three main factors, which Cheyne and colleagues labeled Intruder, Incubus, and Unusual Bodily Experiences. In addition to identifying these three factors, Cheyne et al. speculated about their neurophysiological bases.

Our brains have evolved to alert us to potential threats in the environment. Even though we may not be consciously aware of it, during wakeful consciousness we are constantly scanning the world around us for potential threats; should such a potential threat be detected, attention will be directed away from ongoing activities and toward the possible threat to allow a proper assessment and the initiation of appropriate approach or avoidance behavior if required. The amygdalae play a major role in dealing with such emergency situations. They are two small, almond-shaped clusters of nuclei, one deep within each cerebral hemisphere, that play a major role in the initiation of appropriate fight-or-flight behavior. Normally, the whole sequence of events, from detecting a possible threat, through identification and assessment, to initiation of appropriate action, takes only a fraction of a second. Cheyne and colleagues argue that heightened activation of the amygdalae may explain the first of the factors that they identified in their study. This possibility is supported by the fact that neuroimaging studies indicate that the amygdalae show significant levels of activation during REM-stage sleep.[24]

The Intruder factor consists of the sensed presence, extreme fear, and visual and auditory hallucinations. Cheyne and colleagues argue that heightened activation of the amygdalae as one is drifting into or emerging from sleep causes a state of alert apprehension that lasts much longer than it would during wakefulness because there is actually no potential threat to be identified by scanning the surroundings. Because no potential threat can be identified and then properly assessed, the amorphous sense that there is *something* dangerous nearby can last for seconds or even minutes—hence the overwhelming sense of presence that is such a common feature of sleep paralysis episodes. It is argued that efforts to identify the threat will continue based on both external inputs (such as shadows, ambient sounds, and so on) and internally generated imagery (dreamlike intrusions). The end result is often terrifying visual and auditory hallucinations involving threatening monsters and demons.[25]

Cheyne and colleagues identified a second factor, consisting of pressure on the chest, difficulty breathing, and pain, which often co-occurred with the Intruder factor. They labeled this factor Incubus. In traditional folklore and mythology, an incubus was a male demon who would typically lie atop a sleeping woman to have sexual intercourse with her. The explanation offered for this group of symptoms focuses on the fact that under normal circumstances breathing can be controlled voluntarily, even though such voluntary control of breathing is not actually necessary. Most of the time we continue to breathe without conscious awareness of our breathing, thanks to involuntary muscles that do the job quite satisfactorily. However, during an episode of sleep paralysis, the sufferer may become aware of the fact that they cannot take control of their breathing, leading them to think that they may be suffocating or being strangled. In fact, there is no real danger of this, as the involuntary system will continue to operate, but the ensuing panic is quite understandable. Strenuous attempts to breathe deeply may cause painful spasms.

Both the Intruder- and Incubus-type experiences imply the presence of another being, one that is perceived as a threat to one's very survival. It is not unreasonable to suggest that sleep paralysis episodes may be a factor underlying beliefs in demons and other spirit beings that attack individuals while they sleep, as discussed later in this chapter.

The third factor identified, Unusual Bodily Experiences, consists of flying and floating sensations, out-of-body experiences, and sometimes feelings of bliss. In the words of Cheyne et al.:

> Respondents indicated that these experiences were rarely the rather passive sensations suggestions [sic] by the term "floating," but often consisted of more vigorous sensations of flying, acceleration, and even wrenching of the "person" from his or her body. In response to questions about floating sensations respondents spontaneously reported a variety of inertial forces acting on them, which they described as rising, lifting, falling, flying, spinning, and swirling sensations or similar to going up or down in an elevator or an escalator, being hurled through a tunnel, or simply accelerating and decelerating rapidly. . . . Several people reported feeling forcibly pulled or sucked from their bodies, sometimes through the forehead and sometimes through the feet, and one person described a sensation of "falling out of" his body.[26]

The psychology of out-of-body experiences is explored in more detail in chapter 7, but for now it will suffice to note that the explanation offered by Cheyne and colleagues fits well with recent thinking regarding the most likely explanation of such experiences, whether experienced during sleep paralysis or in other contexts. During wakeful consciousness, the vestibular system coordinates head and eye movements along with other proprioceptive feedback to produce our sense of balance and spatial orientation. Brain areas controlling the sleep-wake cycle are known to be closely associated with such vestibular nuclei in the brainstem. Cheyne and colleagues proposed that activation of these nuclei, in the absence of correlated visual input and head movements, is interpreted as floating or flying. This conflicting interpretation of simultaneously floating or flying above one's bed while at the same time lying motionless on it is resolved by splitting the phenomenal self and the physical body: in other words, an out-of-body experience. Some people report seeing themselves lying motionless, a phenomenon known as *autoscopy*. Interestingly, although sleep paralysis episodes are typically reported as being extremely stressful and frightening, those that involve unusual bodily experiences are more likely to be reported as being pleasant and, sometimes, even blissful.

A team led by postdoctoral researcher Dan Denis, including myself and the aforementioned Alice Gregory, has conducted systematic reviews of variables associated with sleep paralysis.[27] Among other findings, it was noted that stress and trauma are strongly correlated with incidence of sleep paralysis, as was post-traumatic stress disorder and, to a lesser extent, panic disorder. It is not clear whether such factors have a direct effect on frequency of sleep paralysis episodes, or if they have an adverse effect on sleep quality and thereby have an indirect effect on sleep paralysis frequency. Poor subjective sleep quality has been shown to be associated with sleep paralysis in several studies. In examining this relationship more closely, it was found that sleep latency (that is, how long it takes to get to sleep) and daytime dysfunction (for example, excessive daytime sleepiness) were predictive of sleep paralysis.[28] A study by the same group established, as had long been suspected, that there is a moderate genetic influence with respect to susceptibility to sleep paralysis.[29]

CROSS-CULTURAL INTERPRETATIONS OF SLEEP PARALYSIS

Although there is still much to learn about the science of sleep paralysis, the general outline presented above is well supported by empirical evidence. However, the science of sleep itself is a relatively recent development in human history. Prior to that, it would be natural for sufferers to explain these bizarre episodes in terms of the actions of spirits. This still applies today in many parts of the world and even among those with strong religious beliefs in modern Westernized countries.

Historically, episodes of sleep paralysis were typically described as being nightmares, but, as previously stated, it is clear that the term had a much more specific meaning in past centuries than it does today. Whereas today the term is used to refer to any type of scary dream, in the past it was reserved for experiences involving inability to move, intense fear, and so on—in other words, for episodes of sleep paralysis.

It would be going too far to claim that episodes of sleep paralysis fully explain human belief in spirits, but it is reasonable to suggest that they may contribute to the development and maintenance of such beliefs. The strong sense of presence that is so characteristic of episodes is typically felt to be a sentient being of some sort with definite intentions, usually malign, toward the sufferer. In many cases, strange beings are not just sensed but are actually seen, heard, felt, and even smelled. Furthermore, these beings appear to be able to appear and disappear without leaving any physical traces. What stronger proof is needed of the otherworldly nature of these beings?

In addition to the fact that some episodes of sleep paralysis are readily interpretable as encounters with supernatural beings, they can sometimes, as described earlier, turn into full-blown out-of-body experiences. In such cases, experiencers would feel that they themselves are no longer inhabiting physical bodies but instead have become free-floating spirits, often able to pass through walls and travel huge distances in an instant. An alternative, nonspiritual explanation for the characteristics of the out-of-body experience is presented in chapter 7.

Although such spiritual interpretations of sleep paralysis can be found throughout history, it is also true that there have always been commentators who argued for a more naturalistic explanation. For example, Samad Golzari and colleagues discuss a manuscript written by Persian scholar Akhawayni Bokhari in the tenth century in which Akhawayni argues that sleep paralysis "is caused by rising of vapors from the stomach to the brain."[30]

When considered from a cross-cultural perspective, we can see that the same core characteristics of sleep paralysis are reported throughout the world and throughout recorded history. However, the favored interpretation of the experiences and the label applied to them can vary considerably. Brian Sharpless and Karl Doghramji present no fewer than 118 different terms from around the world that refer to sleep paralysis.[31] While it might be seen as inevitable that these bizarre experiences would often naturally be interpreted within the framework of the sufferer's dominant belief system, it also appears that the belief system can sometimes influence the actual content of the hallucinatory images themselves. This is illustrated beautifully by the articles included in a special issue of the journal *Transcultural Psychiatry* published in 2005. As the editors noted, "As a paradigmatic case for the study of the interaction of culture and biology in psychopathology, sleep paralysis illustrates how cultural elaboration of experience can occur before, during, and after a biologically patterned event."[32]

Owen Davies argues convincingly that the records of witch trials and other writings on witchcraft from the early modern period indicate that accusations of witchcraft were sometimes supported by testimony that appears to modern eyes to be describing sleep paralysis episodes.[33] For example, consider these testimonies presented during the Salem witch trials:

> Robert Downer's experience occurred after the accused witch, Susan Martin, had said "some She-Devil would shortly fetch him away." That night, "as he lay in his bed, there came in at the window, the likeness of a cat, which flew upon him, took fast hold of his throat, lay on him a considerable while, and almost killed him." Bernard Peach also testified that, one night, "he heard a scrabbling at the window, whereat he then saw Susanna Martin come in, and jump down upon the floor. She took hold of this deponent's feet, and drawing his body up into an heap, she lay upon him near two hours; in all which time he could neither speak nor stir." When the paralysis finally began to wear off he bit Martin's fingers and she "went from the chamber, down the stairs, out at the door." Bridget Bishop was similarly accused. Richard Coman stated that eight years before, while he was in bed, she had "oppressed him so, that he could neither stir himself, nor wake anyone else, and that he was the night after, molested again in the like manner." John Louder also testified that, one night, after having argued with Bishop, "he did awake in the night by moonlight, and did see clearly the likeness of this woman grievously oppressing him; in which miserable condition she held him, unable to help himself, till near day."

It was also commonly believed during the Middle Ages that sex-crazed demons would sometimes come and have their wicked way with their sleeping victims. As stated previously, the male version of such demons was known as an *incubus* (from the Latin *incubare*, "to lie upon") and the female version was known as a *succubus* (Latin for "prostitute"). It was further believed that the demons could transform themselves from the female form to the male form and back again. The idea was that the succubus would collect sperm from a male sleeper, and then transform into an incubus and use the sperm to diabolically impregnate a female victim. As already noted, there is often a strong sexual element to episodes of sleep paralysis, whether the sufferer is male or female. This can sometimes be so intense that women report feeling as though they have been raped during the night.

In many modern societies, spiritual interpretations are still common. David Hufford's ground-breaking book *The Terror That Comes in the Night* describes the widespread belief in Newfoundland that sleep paralysis episodes are best explained as an attack by the Old Hag, who sits astride the sleeper's chest. The unfortunate victims report that they cannot move and feel as though they are being strangled. They are said to have been "ridden by the Hag" or "Hag-rid" (which may be of particular interest to fans of Harry Potter).

In Japan, sleep paralysis episodes are referred to as *kanashibari* (which literally translates as "bound in metal").[34] In contrast to many modern Western societies, the phenomenon is widely discussed and featured in the media in the form of films, TV programs, books, and even computer games. Anna Schegoleva interviewed a group of Japanese children aged ten to twelve regarding their knowledge and experience of kanashibari.[35] She found that almost all of them were familiar with the term and were generally very excited to have the opportunity to talk about it. Perhaps this is not surprising, given the love that many Japanese people have for horror films and ghost stories. A third of the children reported that they themselves had experienced kanashibari. As would be expected, these episodes always involve paralysis, difficulty breathing, and a sense of pressure on the chest, but the details of the hallucinations experienced varied somewhat. In Schegoleva's words:

Most common visions include: Sadako [a film character], ghosts (*obake*), unfamiliar people. Some other examples: spiritual photographs, suicides, zombies, burglars, frightening things seen on TV. One of the girls (aged 10) saw sleeping pills falling down from the ceiling; there were so many she could hardly stand the weight. The most impressive is Sadako, the one from the *Ringu* movie. A female figure in white clothes, limping up to the sleeper, with her face enclosed by long hair hanging down from her head . . .

The appearance of Sadako in many accounts is noteworthy and is reminiscent of the appearance of one of the Gentlemen as described at the beginning of this chapter. It is clear that terrifying creations from the imaginations of fiction writers are capable of taking on a life of their own. The popularity of horror fiction is a clear indication that many people enjoy the thrill of being scared out of their wits. Indeed, when Schegoleva asked the children if they used any strategies to try to prevent such episodes, she was quite surprised by their responses—far from trying to prevent kanashibari, many had tried different techniques to induce it.

These are just a few examples of the various culture-specific interpretations of the same core experience across time and space. There are dozens more, as listed by Sharpless and Doghramji. For example, in China, sleep paralysis is referred to as *ghost oppression*.[36] In Germany, it has been referred to as *alpdrück* ("elf pressure") and *hexendrücken* ("witch pressing").[37] Mexicans speak of *se me subio el muerto* ("a dead body climbed on top of me").[38] Norwegians are sometimes plagued by *Svartalfar* (black elves who paralyze their victims with arrows, sit on their chests, and whisper horrible things),[39] while the Catalan Spanish are familiar with *pesanta* (an enormous cat or dog that enters homes and sits on sleepers' chests).[40] One of my personal favorites, because it is so creepy, is the interpretation of sleep paralysis found in parts of the West Indies, where the phenomenon is referred to as *kokma*. Kokma is said to be caused by the spirits of unbaptized children crawling onto the sleepers' chests and throttling them.

As will be discussed in more detail in chapter 5, sleep paralysis episodes are also implicated in many cases of alleged alien abduction. In most cases, the aliens are not actually seen during the sleep paralysis episode itself, but, for reasons to be discussed, the episode is taken as evidence that the individual has indeed been abducted by extraterrestrials who then wiped the victim's memory of any further details. Those details, including vivid images of the aliens themselves, may only be recovered via the application of hypnotic regression and are almost certainly based on false memories rather than events that actually occurred.

The most amusing interpretation I ever came across of the strong sense of malign presence that often occurs during a sleep paralysis episode occurred some years ago. I had been taking part in a radio program about sleep paralysis and was being driven home afterward. My driver asked me what I talked about on the show, so I gave him a brief description of the main features of the phenomenon. "I've had that," he said. At the risk of being accused of being influenced by politically incorrect, somewhat misogynistic stereotypes beloved by British comedians in the 1970s, I could not help but smile when he said, in all seriousness, that he had interpreted that evil presence as being his mother-in-law.

On closer consideration, though, this turned out to be a lovely example of an interpretation based on the interaction between culture and physiology. As the driver clarified, he had been convinced that the strong malign presence, staring at him with evil intent, was not actually his mother-in-law at the time the incident occurred—she was his mother-in-law-to-be. When the episode had taken place, he was lying in bed next to his future wife—in her parents' bed, having taken advantage of the fact that his future in-laws were away on holiday. Furthermore, he was a Cypriot and, within his extended family, premarital sex was definitely frowned upon. So when he woke up in the early hours, unable to move and terrified, even though he could not turn over to look in the direction of the presence, he was convinced it was his very angry future mother-in-law.

IS SLEEP PARALYSIS ALWAYS HARMLESS?

I am keen to spread the message far and wide that sleep paralysis, although it can be absolutely terrifying, is generally harmless. That is not to minimize the very real stress that can be caused when sufferers are unaware of the true nature of the phenomenon and either believe that they are losing their minds or are on the receiving end of bizarre supernatural attacks. In my experience, simply discovering that there is a third possible explanation, one that is scientifically and medically recognized, provides huge relief to sufferers. Sometimes that knowledge alone can be enough to reduce sleep-related anxiety, resulting in more regular sleep patterns, which in turn may reduce the frequency of sleep paralysis episodes. I am therefore somewhat hesitant to point out that it has been argued that, in certain very rare and specific circumstances, sleep paralysis may result in death.

This possibility was put forward by Shelley Adler as an explanation of the phenomenon of *sudden unexpected nocturnal death syndrome* (SUNDS).[41] Between the late 1970s and early 1990s in the United States, over 100 Southeast Asians died in their sleep without any obvious cause. The deaths occurred predominantly among male Hmong refugees, usually within two years of arriving in the country. Several possible causes were investigated, including toxicology, genetics, metabolism, and nutrition, but none of these investigations provided a solution to the mystery. It was found that the victims appeared to have some abnormalities of the cardiac conduction system, but this did not explain why the deaths only occurred following arrival in the United States.

Adler argued that the explanation may involve the conjunction of a number of very specific factors, one of which was sleep paralysis. The Hmong believe that one is at risk of spiritual attack from *dab tsog* (pronounced "da cho") during sleep. Dab tsog is a traditional nocturnal pressing spirit, and hence such attacks take the form of sleep paralysis episodes. Whereas one might survive one or two such attacks, repeated attacks were thought to weaken and ultimately kill the victim. Back in their countries of origin, the Hmong would have traditional remedies to deal with such attacks. They would consult the local shaman and engage in various rituals, including animal sacrifice, to ward off the evil spirits. They also believed that ancestor spirits would protect them from evil spirits so long as religious practices had been properly followed. Their belief in the effectiveness of the rituals would be likely to reduce their anxiety, improving the quality of their sleep, and ultimately having the desired effect.

As refugees in the United States, however, these traditional remedies were no longer available. Also, it appeared that whereas the protective ancestral spirits had been unable to relocate, the dab tsog had somehow managed to follow the refugees to their new homes. Other factors also came into play to explain why it was particularly the male heads of households who were prone to SUNDS. Back in their home countries, society was organized in a patriarchal manner, and spiritual protection of the family was one of the responsibilities of the male head of the household. In the context of a new unfamiliar society, the older Hmong men found themselves unable to fulfill their traditional roles. Typically, they were unemployed and struggled to learn English, whereas their children learned to adapt much more quickly. The stress that this situation caused would disturb sleep, making sleep paralysis episodes more likely. Ultimately, according to Adler, it was this deadly combination of a culture-specific belief system, a stressful new context, an underlying cardiac abnormality, and susceptibility to sleep paralysis that resulted in SUNDS.

I cannot emphasize too strongly that, even if Adler is correct in her explanation of SUNDS, in the vast majority of cases, sleep paralysis episodes are essentially harmless, albeit potentially terrifying. It was the tragic and unlikely combination of factors described above that may have had deadly consequences for specific Hmong refugees, factors that, I sincerely hope, will not apply to any readers of this book.

SLEEP PARALYSIS AS ARTISTIC INSPIRATION

It is not surprising that artists of all kinds have been inspired by episodes of sleep paralysis, given the vivid and nightmarish imagery that is so often a feature of the phenomenon. Without doubt, the most famous painting representing sleep paralysis is *The Nightmare* by Henry Fuseli (figure 2.2). Painted in 1781, it was enormously popular with the public, inspiring Fuseli to produce at least three other versions of the same scene. The painting presents a sleeping woman, reclining in a supine position with her head and arms dangling over the edge of the bed. A grotesque demon is perched on

Figure 2.2

Henry Fuseli's painting *The Nightmare*, the first version of which was painted in 1781, memorably captures many aspects of sleep paralysis, including the nightmarish imagery, the sense of being watched, and pressure on the chest.

her chest, his full weight pressing down on her. In the background, a weird horse stares at the scene through eyes that lack pupils. The scene successfully conjures up the frightening and oppressive atmosphere of a nightmare. Interestingly, those who are susceptible to sleep paralysis are more likely to actually experience an episode if they sleep on their backs (although it might be felt that sleeping in the rather extreme position depicted on this canvas is simply asking for trouble).

Fuseli is far from the only artist to produce visual representations of sleep paralysis. Type "sleep paralysis images" into your favorite search engine and an almost endless array of terrifying depictions will appear. The work of photographer Nicolas Bruno is particularly striking.[42] His surreal compositions, reminiscent of the works of René Magritte, are beautiful, disturbing, and fascinating in equal measure and are clearly directly inspired by his own experiences of sleep paralysis. They illustrate many of the main features of the phenomenon, including the sense of unreality, impending threat, immobility, and struggling for breath.

Inevitably, filmmakers have also been drawn toward this topic. The documentary films of Carla MacKinnon and Rodney Ascher have already been mentioned, but sleep paralysis has also inspired dozens of fictional horror movies and TV shows, including *The Conjuring, Dead Awake, Between the Darkness, The Haunting of Mia Moss, Shadow People, Mara,* and *Slumber,* not to mention, of course, *The X-Files.* Wes Craven's hugely successful 1984 horror film, *A Nightmare on Elm Street,* is said to be directly inspired by press reports at that time reporting on SUNDS. Ironically, as Corinne Purtill wrote in an article for *Quartz News,* sleep paralysis episodes featuring a threatening "man in a hat" are now common around the world, possibly yet another example of a fictional creation—in this case, the notorious killer, Freddy Krueger—making his way from a writer's imagination into our own nightmares![43]

Arguably one of the most bizarre manifestations of sleep paralysis—in a very strong field—was that which occurred to a woman in Moscow in 2016, as described by Matthew Tompkins.[44] The woman had been playing Pokémon Go on her smartphone prior to drifting off to sleep. Later, she woke to find herself being sexually assaulted by an actual Pokémon character. Struggling against the pressure holding her down but unable to scream to rouse her boyfriend, who was sleeping beside her, she was eventually able to get up. The Pokémon had gone and was nowhere to be found. She called the Moscow police to report the attack.

Many famous literary works have included descriptions of sleep paralysis, including Herbert Melville's 1851 classic *Moby-Dick*, Thomas Hardy's (1888) story *The Withered Arm*, F. Scott Fitzgerald's (1922) novel *The Beautiful and Damned*, and Ernest Hemingway's (1938) short story "The Snows of Kilimanjaro." Here is one literary example, a short extract from Guy de Maupassant's (1887) story *Le Horla*:

> I sleep—a long time—2 or 3 hours perhaps—then a dream—no—a nightmare lays hold of me. I feel that I am in bed and asleep—I feel it and I know

it—and I feel also that somebody is coming close to me, is looking at me, touching me, is getting on to my bed, is kneeling on my chest, is taking my neck between his hands and squeezing it—squeezing it with all his might in order to strangle me.

And then suddenly I wake up, shaken and bathed in perspiration; I light a candle and find that I am alone, and after that crisis, which occurs every night, I at length fall asleep and slumber tranquilly until morning.[45]

There can be little doubt that Maupassant was all too familiar with the terror that comes in the night.

PREVENTION AND COPING STRATEGIES

Most people will never experience an episode of sleep paralysis. Of those who do, most will only experience the mild form, that is, a period of paralysis lasting a few seconds that spontaneously lifts. Such episodes may cause some consternation but are easily shrugged off and forgotten. A smaller percentage may experience the more florid form of sleep paralysis involving such symptoms as a strong sense of malign presence, hallucinations, difficulty breathing, and intense fear. Not surprisingly, these episodes may cause considerable stress and anxiety in the short term, but most people who experience them will probably only experience them once or twice in their lifetime. If so, they may always be remembered but are unlikely to seriously affect quality of life in the long term. Finally, there is a very small percentage of the population who suffer from what Brian Sharpless refers to as *recurrent fearful isolated sleep paralysis*. These poor unfortunates may experience the florid form of sleep paralysis on a regular, possibly nightly, basis. There is a strong chance that their quality of life will be seriously impaired by their nocturnal terrors. What, if anything, can be done to help them?

The sad truth is that there has been little systematic research into the best strategies to reduce or eliminate the frequency of sleep paralysis episodes or to cope with them when they do occur. Some general advice can be offered on the basis of previous research findings. Furthermore, numerous websites provide anecdotal accounts of different strategies that people have found to be effective. One of the very few studies to attempt any analysis of the self-reported effectiveness of these strategies was carried out by Brian Sharpless and Jessica Grom.[46] They collected data using clinical interviews from 156 undergraduate students who had experienced isolated sleep paralysis. Around three-quarters of the sample reported that they experienced fear during the episodes, with around 15 percent reporting clinically significant distress. About 20 percent reported that they had made attempts to prevent the attacks, and about 80 percent of these claimed some success. This relatively low percentage of respondents who actively sought to prevent the episodes probably reflects the fact that for most respondents the episodes were not frequent. A much larger proportion, around 70 percent, actively attempted to disrupt the episodes when they did occur, but their self-reported success rate was only 54 percent.

It is well established that an irregular sleep pattern makes sleep paralysis episodes more likely in those with the underlying susceptibility, so practicing good sleep hygiene is recommended. In addition to adopting a regular sleep pattern that is appropriate with respect to age, lifestyle, and health, here are some more general tips about sleep hygiene from the National Sleep Foundation:[47]

- Limit daytime naps to thirty minutes.
- Avoid stimulants such as caffeine and nicotine close to bedtime.
- Exercise to promote good-quality sleep.
- Steer clear of food that can be disruptive right before sleep.
- Ensuring adequate exposure to natural light.
- Establish a regular relaxing bedtime routine.
- Make sure the sleep environment is pleasant.

As noted, stress, anxiety, and depression are all known to have an adverse effect on sleep quality, and so sufferers from such conditions may well find that any treatment that is effective in improving their mental health also has the indirect effect of reducing the frequency of sleep paralysis episodes. For example, in one study it was found that five of eleven patients suffering from panic disorder and recurrent isolated sleep paralysis showed improvements in the latter even though only the former had been directly treated with cognitive-behavioral therapy.[48] For those with strong beliefs that the episodes really are some sort of spiritual attack, traditional remedies may well be successful. This is not to endorse such interpretations but simply to recognize that *belief* in such remedies may well be strong enough to reduce anxiety, thus improving sleep quality and thereby reducing the frequency of attacks. To date, there are no psychopharmacological treatments for isolated sleep paralysis with proven track records, although anecdotal evidence and case histories suggest that some antidepressants may have potential as treatments.

In addition, those prone to sleep paralysis should avoid sleeping on their backs, as it is known that more episodes occur in this sleep position than any other. In extreme cases, some sufferers try sleeping sitting up or sew walnuts or tennis balls into the back of their pajamas so that the supine position is too uncomfortable to adopt.

Sufferers are generally more likely to adopt strategies to cope with sleep paralysis episodes when they do occur rather than trying to prevent them. The most commonly reported technique is to attempt, by a huge effort of will, to move a finger or a toe. If successful, this will break the spell. It is also commonly reported that the sufferer may attempt to scream at the top of their lungs. However, the only observable result is typically a feeble croak or groan. Sometimes, however, the sufferer's partner is able to recognize these telltale signs and rouse the sleeper, thus rescuing them from their torment.

Some sufferers learn to recognize that they are having a sleep paralysis attack while the attack itself is occurring and attempt mental relaxation techniques, with varying levels of success. Others go a step further and, on recognizing that they are experiencing sleep paralysis, make no effort at all to disrupt the experience but instead simply let it happen and enjoy the experience in the same way that some people enjoy a good horror movie. This does not work for everyone. Some who try it report that, despite the fact that they are fully aware that what they are experiencing is not real, there is nothing they can do to control the overwhelming terror that grips them.

Brian Sharpless and Karl Doghramji have produced *A Cognitive Behavioral Treatment Manual for Recurrent Isolated Sleep Paralysis: CBT-ISP*, which is included as an appendix in their book *Sleep Paralysis: Historical, Psychological, and Medical Perspectives*. This is the most comprehensive, systematic treatment program for the condition currently available, covering such topics as self-monitoring, psychoeducation, sleep hygiene, disruption techniques, and coping with catastrophic thoughts and hallucinations. Although, at the time of writing, no outcome studies had yet assessed the success rate of this program, it does appear to provide a promising first step in the right direction.[49] If you suffer from frightening sleep paralysis episodes, I wish you well in finding strategies to reduce their frequency or even eliminate them—or, failing that, at least learning to live with them.

Although research shows that most people do not opt for a paranormal interpretation of sleep paralysis episodes, because the experience is relatively common, there is little doubt that many claims of encounters with ghosts, demons, and other spirits are plausibly explained in such terms. Having said that, there are many other reasons why someone may come to believe that they have had a ghostly encounter, as we will see in the next chapter.

3 HIGH SPIRITS, PART 1: GHOSTLY ENCOUNTERS

Belief in ghosts is common in societies throughout history, and yet conventional scientists continue to reject any notion that some essence of a person—the spirit, soul, or consciousness, call it what you will—survives bodily death. If ghosts do not exist, what alternative explanations might there be for such widespread belief and for the fact that, as we saw in chapter 1, a sizable minority of even modern Western societies claim to have had personal encounters with ghosts? Sleep paralysis provides a plausible explanation for many such cases, as described in the previous chapter. In this chapter, we will consider a range of other factors that might lead someone to believe that they have had an encounter with a spirit from beyond the grave.

If ghosts really are the spirits of the dead, this would have profound implications for our understanding of the nature of consciousness. Indeed, any evidence that supported the reality of some form of life after death would necessarily imply that *dualism* was the correct philosophical view of consciousness. Dualists follow the great French philosopher and mathematician René Descartes in maintaining that the universe consists of two fundamentally different kinds of stuff: the physical and the mental. Physical objects, such as chairs, cakes, and brains, have properties like mass, location, and size, whereas the mental realm does not. Thoughts, dreams, and desires cannot be located in space and measured in the same way that physical objects can.

Dualism is an appealing notion for many people, as it seems to be consistent with human experience. Our mental life does indeed intuitively feel altogether different from the processes that affect the atoms and molecules that make up the universe "out there." The problem is that no one has managed to come up with a plausible explanation of how an immaterial soul could interact with a material brain. In attempting to solve this problem, some philosophers of mind have considered the possibility that there are not, in fact, two fundamentally different kinds of stuff in the universe: there is just one. Such monist views take one of two forms: either everything is mind or everything is matter. Without going into details, suffice it to say that, to date, the nature of consciousness has not been adequately explained by either philosophers or scientists, although real progress has been made in recent decades.[1]

According to the vast majority of modern neuroscientists, however, consciousness is entirely dependent on the activity of neurons in a living brain and cannot be separated from that underlying neural substrate. This view is supported by a huge amount of evidence, including consideration of the effects of drugs, brain damage, and direct electrical and magnetic stimulation of the brain, as well as data from recordings of brain activity. Although it would be premature to claim that modern neuroscience can provide a comprehensive explanation of the nature of consciousness, very little evidence appears to support a dualist view—but convincing evidence of post mortem survival clearly would do so (as would some interpretations of out-of-body experiences, as discussed in chapter 7).

There are many reasons to doubt the traditional interpretation of ghosts and hauntings, as this chapter will demonstrate. First and foremost, almost all of the evidence put forward in support of the existence of ghosts is anecdotal in nature. Such hearsay evidence should be treated with caution for reasons outlined by Scottish philosopher David Hume in a famous essay titled "Of Miracles" published in 1748.[2] Hume addressed the question of whether one would ever be rationally justified, in the absence of any other evidence, to accept another person's claim that they had witnessed a miracle. He defined a miracle as an event that violates a law of nature, thus including paranormal events. He then argued that "no testimony is sufficient to establish a miracle unless that testimony be of such a kind that its falsehood would be more miraculous than the fact which it endeavours to establish."[3] He further argued essentially that, whereas evidence for the violation of laws of nature is thin on the ground and possibly even nonexistent, evidence that people sometimes lie and make mistakes is plentiful. Therefore, in practice it is always more reasonable to doubt reports of miracles than to accept them.

It is also worth noting that the reported appearance and behavior of ghosts vary considerably from one society to another as discussed by historian R. C. Finucane.[4] Consider his description of ghosts in ancient Greece:

> In Homer's day, ghosts deserting the mangled remains of warriors on the plains of Troy flitted and squeaked like crazed bats all the way down to Hades, where they spent eternity meekly standing around murmuring to each other in hollow tones. Their rather boring conversation ran to gossip concerning recent arrivals, arguments over family pedigrees, and long-winded recitations about famous battles.

Clearly, ghosts in ancient Greece were rather different from those reported in the modern era. As Finucane shows, ghosts from various other eras, such as the early Christian era, the Reformation, and the Victorian era, all had their own characteristic appearance and behavior. Such variation across cultures is best explained by the idea that ghosts are the products of particular belief systems that dominated at particular times and places as opposed to having any objective reality.

For modern Western readers, I suspect that the word *ghost* conjures up a mental picture of a transparent apparition, perhaps having just walked through a wall (figure 3.1).[5] Such depictions are common in works of fiction. In fact, however, reports of ghostly encounters of this kind are relatively rare. It is far more common for people to believe that a ghost is around on the basis of far less dramatic evidence, including such things as a sense of presence, shivers down the spine, sudden drops in temperature, dizziness, and so on, as well as inexplicable smells, noises, or the movement of objects.

HOAXES

Might it be the case that all reports of ghostly encounters are simply hoaxes, the accounts coming from either the victims or the perpetrators of such hoaxes? This is extremely unlikely, but it should always be borne in mind

Figure 3.1

For most Westerners the word *ghost* conjures up an image of a full-form apparition as in this illustration—but most claims of ghostly encounters are based on much less dramatic sensations.

that deliberate deception is indeed the correct explanation for some cases, albeit probably only a minority.

It is appropriate in this context to outline three possible explanations for so-called *poltergeist* activity (figure 3.2). The word *poltergeist* can be literally translated as *noisy spirit*, corresponding to the traditional interpretation of such reports as being due to the activity of a destructive ghost. Poltergeists are said to engage in a wide range of disturbing behavior, including making loud noises, levitating (and often smashing) objects, starting fires, causing flooding, and interfering with electrical equipment. In some cases, it is claimed that poltergeists may directly attack people.

Although some parapsychologists accept the traditional notion that poltergeist activity is caused by the spirit of a person who has died, others reject this account in favor of an alternative paranormal interpretation. It has often been noted that the disruptive activity typically only occurs around one specific individual, whom parapsychologists refer to as the *focus* of the activity. This individual is often someone who was experiencing severe mental health problems prior to the outbreak of poltergeist activity. The idea is that this inner psychological turmoil somehow becomes externalized in the form of psychokinetic energy, leading to the effects described. So, according to this view, although the disruption is a genuine paranormal effect, it is caused by the troubled mind of a living person rather than the spirit of a dead person.

Boring old skeptics like me agree that in such cases the cause of the disruption is a living person. They disagree, however, with the idea that anything paranormal is involved. Instead, they view such disruption as deliberate attention-seeking behavior on the part of the focus person—in other words, a hoax. It would be going too far to claim that all alleged poltergeist cases have been proven to be hoaxes. All too often, skeptics are not actually welcomed to investigate with a view to discovering possible nonparanormal explanations as a "genuine" ghost story is considered to be much more interesting than the exposure of a hoax. However, dozens of such cases can satisfactorily be accounted for as hoaxes. We will limit ourselves here to describing a couple of famous cases.[6]

Figure 3.2
An 1851 illustration of alleged poltergeist activity in the presbytery in Cideville, France.

The first is the case of the Columbus Poltergeist, which received international attention in 1984. The family of Tina Resch reported that objects had started flying around in their home shortly after fourteen-year-old Tina, an emotionally disturbed adoptee, had seen the movie *Poltergeist*. The *Columbus Dispatch* sent reporter Mike Harden to interview Tina, and he took photographer Fred Shannon along in the hope that they might catch some paranormal activity on camera. The resulting article included several dramatic photographs allegedly showing precisely that. One famous image shows Tina sitting in an armchair as a telephone flies through the air in front of her. She is apparently screaming in terror as this happens. Parapsychologist William Roll stayed in the Resch house to investigate the case and concluded that genuine recurrent spontaneous psychokinesis was occurring. However, he himself never actually saw any objects fly through the air.

Not everyone was convinced. James Randi went to Columbus to investigate for himself but was denied access to the house. By examining the photographs taken by Fred Shannon, Randi was able to present a strong case that the phenomena were faked by Tina herself.[7] Shannon had reported that no objects ever took off if he was observing them directly. So he had obtained his shots by looking away, holding his camera pointing in what he hoped would be the right direction, and pressing the shutter when he detected movement in the corner of his eye. The resulting photographs were, according to Randi, all consistent with the idea that Tina had simply waited until no one was watching her and then thrown the objects, pretending to be shocked and terrified as they flew through the air.

As Randi goes on to describe, even stronger evidence of fakery was to emerge, albeit by accident. The case had become something of a media circus, with reporters and TV film crews coming from far and wide to get in on the action. On one such occasion, a crew from WTVN-TV in Cincinnati had been packing up their kit after filming, but the cameraman had inadvertently left his camera running. The recording clearly shows Tina looking around. Thinking she was unobserved, she pulls a table lamp toward herself, screaming with feigned terror. When confronted with this undeniable evidence, Tina claimed that she had only been fooling around because, on this occasion, she wanted the cameraman to leave.

There is a tragic postscript to this story. Clearly, Tina did not have a happy life. It is reported that John and Joan Resch were physically abusive toward their adopted daughter, Tina. Following her time in the media spotlight, Tina subsequently married and divorced twice. She changed her name to Christina Boyer. In 1992, her three-year-old daughter, Amber, was beaten to death. Christina and her boyfriend were tried for her murder. She was sentenced to life plus twenty years.

An even more famous case is that of the Amityville Horror, which became the subject of an internationally bestselling book written by Jay Anson and published in 1977, *The Amityville Horror: A True Story*. The book allegedly describes a series of terrifying paranormal events that beset the Lutz family after they moved into 112 Ocean Avenue in Amityville in December 1975. Within a month, George and Kathy Lutz and their three children had fled the property in terror. The book was the basis for a series of films, the first of which was released in 1979.

The house certainly had a horrific past prior to being bought by the Lutzes. It was there that, in November 1974, Ronald DeFeo Jr. had shot dead six members of his own family. He subsequently claimed that he had heard voices telling him what to do. A year later, he was sentenced to six consecutive life terms, despite his plea of insanity.[8] Shortly after that, the Lutz family took up residence at 112 Ocean Avenue. Knowing the house's tragic history, the Lutzes took the precaution of calling in a family friend, referred to as Father Frank Mancuso in the book (real name, Father Ralph J. Pecararo) to bless the house. The priest claimed he heard a masculine voice demanding that he "Get out!" as he performed the ritual.

The series of bizarre events that allegedly took place in the short time that the Lutzes were in the house is summarized by Robert Morris, in his review of the book, as follows:

> Some were physical: a heavy door was ripped open, dangling on one hinge; hundreds of flies infested a room in the middle of winter; the telephone mysteriously malfunctioned, especially during calls between the Lutzes and Mancuso; a four-foot lion statue moved about the house; windows and doors were thrown open, panes broken, window locks bent out of shape; Mrs. Lutz lev-

itated while sleeping and acquired marks and sores on her body; mysterious green slime oozed from the ceiling in a hallway; and so on. Some phenomena were experiential: Mrs. Lutz felt the embrace and fondling of unseen entities; Mr. Lutz felt a constant chill despite high thermostat temperatures; the Lutzes' daughter acquired a piglike playmate; the Lutzes saw apparitions of a pig and a demonic figure; the children misbehaved excessively and the family dog slept a lot and avoided certain rooms; marching music was heard; et cetera.[9]

Despite Anson's assurance that "all facts and events, as far as we have been able to verify them, are strictly accurate," the evidence strongly suggests that this case is nothing more than a deliberate hoax. According to Morris, Anson made very little attempt to verify any of the Lutzes' claims. He appears to have never once visited the property himself nor to have directly interviewed the main witnesses, instead basing his account mainly on tape-recorded accounts from the Lutzes.

Morris discusses a number of factors that should lead us to be dubious regarding the accuracy of Anson's account, all of which apply to assessing accounts of anecdotal evidence for paranormal claims in general, as discussed in the following sections on the psychology of hauntings. In the present context, the important point is that there is also clear evidence that much of the account is pure fabrication on the part of the Lutzes. There are several examples of the Lutzes giving descriptions of the weather on specific dates that clearly do not match the actual weather on those dates. For example, they claim that a torrential rainstorm on January 13 prevented them from leaving the house, forcing them to spend another night there. Records show that no such storm took place. Furthermore, William Weber, the lawyer who defended Ronald DeFeo Jr. at his trial, is on record as saying, "I know this book is a hoax. We created this horror story over many bottles of wine."[10] In this case, the motive was not attention-seeking behavior, it was simple greed. It is worth noting that no subsequent owners of the property reported any paranormal activity.

Perhaps it is time to introduce French's First Law: "The more spectacular the claim of any kind of ghostly encounter, the more likely it is to be based on a deliberate hoax."

MISINTERPRETING NATURAL PHENOMENA

Although the possibility of deliberate hoaxing should always be borne in mind when assessing claims of ghostly encounters, especially the more sensational ones, it is almost certainly the case that most such claims are not attempts at deliberate deception. A number of potentially relevant psychological factors may underlie such claims.[11]

An individual may become convinced that their house is haunted is because they have had one or more experiences that they simply cannot explain any other way. Of course, just because one cannot explain something does not mean that there is no naturalistic, nonparanormal explanation—but, equally, for someone who already believes in ghosts, it is not completely irrational to speculate about the possibility of a paranormal explanation.

The allegedly inexplicable experience in question may be purely psychological in nature, such as a frightening episode of sleep paralysis as described in the previous chapter. However, sometimes the puzzling experiences may be related to events in the external physical world. Vic Tandy and Tony Lawrence list a number of obscure yet mundane phenomena that might produce unexpected physical effects that could lead someone to suspect that their home was haunted: "water hammer in pipes and radiators (noises), electrical faults (fires, phone calls, video problems), structural faults (draughts, cold spots, damp spots, noises), seismic activity (object movement/destruction, noises), . . . and exotic organic phenomena (rats scratching, beetles ticking)."[12]

A rather cute example of the mysterious movement of objects was reported in the British media in March 2019.[13] Stephen Mckears, a seventy-two-year-old man from Severn Beach, South Gloucestershire, was puzzled when he noticed that his garden shed was being tidied up overnight by unknown means and began to consider the possibility that a helpful ghost was responsible—a kind of anti-poltergeist who did the opposite of more traditional poltergeists by creating order out of chaos. After some months, and with the help of a friendly neighbor, Stephen decided to get to the bottom of the mystery by setting up a video camera to record what was happening when he left his shed. The answer came as something of a surprise—no ghost, but a tiny house-proud mouse spending several hours tidying away the metal objects that Stephen had deliberately left scattered around. These included not only nuts, bolts, and screws but even small metal tools that the determined mouse would strenuously tidy away. This is a wonderful example of an explanation for the mysterious movement of objects that no one would have guessed if we did not have the evidence of the video recording, reminiscent of that scene from Disney's *Cinderella*. Needless to say, the video went viral, enthralling viewers around the world.

Of course, if one is able to figure out the cause of the initially puzzling events, one may be able to stop worrying about them (although this often does not apply in cases of sleep paralysis). But if one is unable to come up with a satisfactory nonparanormal explanation, the idea that one's house is haunted may take hold, and thenceforth even relatively mundane events are interpreted within that context. You can't find your keys in the morning? Aha, the ghost must have moved them! Your TV malfunctions? Damn it, that ghost is causing problems again! I know from experience that if a light begins to flicker when I am giving a talk on ghosts, I am guaranteed to elicit a nervous giggle from my audience if I suddenly look puzzled and mutter, "Oooh, spooky!" Doing the same thing during a statistics lecture would not elicit the same response. Context is all-important.

CONTEXT AND PRIOR BELIEF

Indeed, I would go so far as to argue that the two most important psychological factors associated with reports of ghostly encounters are context and prior belief in ghosts. At the anecdotal level, many readers will be familiar with the effects of being told, prior to entering an old building such as a stately home, castle, or pub, "This place is said to be haunted." Almost at once, one is more alert to stimuli that one would probably not notice unless so primed. These stimuli may be external, such as creaking floorboards or cold drafts, or internal, such as a chill down the spine or the feeling that you are being watched.

Rense Lange and James Houran demonstrated this effect empirically.[14] They asked two groups of participants to walk around a disused movie theater and to note whether they experienced any cognitive, physiological, emotional, psychic, and/or spiritual responses in reaction to their surroundings. Half of the participants were simply told that the property was currently being renovated, and the other half were told that paranormal activity had been reported there. As predicted, the latter participants reported significantly more physical, emotional, psychic, and spiritual experiences than those in the former group.

In a somewhat similar study, Richard Wiseman and colleagues demonstrated the role played by prior belief in ghosts.[15] They collected data from 678 participants walking around Hampton Court Palace, reported to be one of the most haunted locations in England. It is alleged that the ghost of Catherine Howard still lingers in this historic building. Catherine, the fifth wife of Henry VIII, was found guilty of adultery and sentenced to death fifteen months after her marriage to Henry in 1540. It is said that on hearing the news of her sentence, she tried to run to Henry to beg for her life, but guards blocked her way and dragged her, kicking and screaming, back along a part of the palace now known as "The Haunted Gallery." Since then, it is claimed that inexplicable screams are sometimes heard in this area and a mysterious woman in white sometimes appears. A range of other anomalous experiences are reported in this and other areas in the palace, including dizziness, a strong sense of presence, and sudden cold spots. In Richard's study, as predicted, those visitors who believed in ghosts reported more unusual experiences as they walked around than did nonbelievers, and they were also more likely to attribute these experiences to ghostly intervention.

ON SEEING THINGS THAT ARE NOT REALLY THERE

One of the responses from skeptics to claims of ghostly encounters that causes the most anger and annoyance to claimants is, "Maybe you were just seeing things." What is clearly implied but not explicitly stated here is that maybe you were seeing things *that were not really there*. The reason that this suggestion causes such a defensive reaction is that, for most people, suggesting that they may have been hallucinating is akin to suggesting that they are "crazy." In fact, hallucinations occur far more frequently among the nonclinical population than is generally realized.[16] The previous chapter dealt with one particular type of hallucinatory experience—sleep paralysis—but hallucinations can occur in other contexts without necessarily being an indication of serious psychopathology. Furthermore, hallucinations may occur in all sensory modalities, not just "seeing things."

In recent years, views of mental illness have moved away from seeing psychosis as an all-or-none phenomenon to instead accepting that psychotic symptoms, including the tendency to hallucinate, lie on a continuum. Some people never experience a single symptom in their lives. Others experience such symptoms frequently and, in some cases, may find them distressing enough to require professional help from a psychiatrist or clinical psychologist. But many others are between these two extremes and may experience occasional psychotic symptoms without ever meeting the criteria for clinical psychosis. Indeed, some people even come to view such symptoms as a positive aspect of their lives.[17]

Hallucinations are not the only way we may perceive things that are not really there. Robert Todd Carroll defines the phenomenon of *pareidolia* as "a type of illusion or misperception involving a vague or obscure stimulus being perceived as something clear and distinct."[18] Thus, for example, we often see faces and forms in random visual stimuli such as in clouds, the grain of wood, or stains on a floor. Leonardo da Vinci offered the following advice to his followers as a method to facilitate the development of their imaginations and artistic expression:

> If you look at walls covered with many stains or made of stones of different colours, with the idea of imagining some scene, you will see in it a similarity to landscapes adorned with mountains, rivers, rocks, trees, plains, broad valleys, and hills of all kinds. You may also see in it battles and figures with lively gestures and strange faces and costumes and an infinity of things.[19]

One of the most commonly perceived categories as a result of pareidolia is faces. This is perhaps not surprising when one considers just how important faces are as a source of information in our day-to-day interactions. From an evolutionary perspective, it was vitally important that an individual could quickly recognize if another human being was a stranger or someone they knew. Were they a threat or not? What was their current emotional state and likely intentions? What were they likely to do next? Our brains have evolved to allow us to quickly decide whether we recognize someone and, if we do, to identify them and what we know about them, as well as their current emotional state. Different parts of our brains are involved in these different components of face recognition. Research using the noninvasive neuroimaging techniques of functional magnetic resonance imaging (fMRI) and magnetoencephalography (MEG)[20] suggests that illusory faces are processed in the brain in the same way that real faces are for the first quarter of a second of processing but are then processed as ordinary objects rather than faces.[21] This is consistent with the idea that our cognitive systems are wired to rapidly detect anything that looks like it may be a face even if it is then quickly determined that it is not. It is not the case, therefore, that pareidolia is the product of slower cognitive reinterpretation of stimuli.

For most of us, spotting faces in inanimate objects in the world around us is simply a source of mild amusement, but sometimes such perceptions are deemed to have greater significance.[22] It is notable that images of religious figures, such as Jesus and the Virgin Mary, crop up rather a lot. For example, as reported by Wikipedia, images of Jesus "have been reported in such varied media as cloud photos, Marmite, chapatis, shadows, Cheetos, tortillas, trees, dental x-rays, cooking utensils, windows, rocks and stones, painted and plastered walls, and dogs' hindquarters."[23] True believers are especially likely to be conclude that such images are divinely produced (with the possible exception of the last example listed).

It seems reasonable to suggest that pareidolia may well be responsible for some alleged ghost sightings, given the general human tendency to perceive meaningful forms and figures in ambiguous and degraded visual input. There is evidence supporting the idea that believers in the paranormal may be more susceptible to pareidolia than nonbelievers. Peter Brugger and colleagues presented brief displays of random dot patterns to volunteers.[24] The volunteers were asked to indicate any trial where they saw "something meaningful," having been informed that the experiment was an investigation of subliminal perception and that "meaningful information" was embedded in the images on some trials. In fact, all of the displays consisted entirely of random dot patterns. Believers in the paranormal reported seeing something meaningful on significantly more trials than did skeptics.

Considering the specific tendency to see faces that are not really there, as opposed to a general tendency to see "something meaningful," it has also been found that believers in the paranormal are more prone to illusory face perception when presented with images of landscapes and scenery, some of which did contain face-like areas and some of which did not.[25] Believers were better at spotting the face-like areas when they were there but also more likely to report seeing them when they were not there; in other words, they demonstrated a response bias toward positive responses. Interestingly, the same pattern was found when participants were divided into groups on the basis of their religious beliefs, but it should be noted that paranormal and religious belief were strongly correlated in this study.

ON NOT SEEING THINGS THAT ARE THERE

In the context of providing potential explanations for ghostly encounters, the possibility of claimants seeing things that are not really there has, perhaps understandably, received rather more attention than the opposite: that is, the failure to see things that really are there. However, this is a topic that has been the subject of a considerable amount of psychological research more generally in recent years. Psychologists refer to this failure to notice stimuli that are right before our very eyes when we are engaged in some other task as *inattentional blindness*.

The most famous study of inattentional blindness was reported by Daniel Simons and Christopher Chabris in 1999.[26] Participants were asked to watch a short video clip in which two groups of people, one wearing white shirts, the other wearing black shirts, throw a ball to each other. They were instructed to count the number of times that the people in the white shirts threw the ball to each other, ignoring the people in the black shirts. At the end of this simple task, the volunteers gave their answer to the question posed and were then asked if they had noticed anything unusual as they watched the video. In fact, halfway through the video, someone dressed in a gorilla suit had walked into the center of the action and stood there beating their chest for several seconds before walking out of shot. Surprisingly, around half of the participants in this study completely failed to spot the gorilla. This is a highly counterintuitive result, as most people are convinced, prior to hearing about this experiment, that something so odd and unexpected would be sure to grab their attention.

This reminds me of an incident from my personal life that occurred many years ago when my wife, Anne Richards, and I were looking to move house. We had called in at an estate agent in Blackheath, London, and Anne was busily concentrating on learning more about the available properties in the area. I must confess that I was paying rather less attention to this important task than she was. When we emerged back on the street, I said to her, "That was a bit weird. I wondered if maybe we were taking part in a psychology experiment." Looking puzzled, she replied, "What do you mean?" "Didn't you notice anything a bit odd in there?" I asked. She hadn't. I instructed her to pretend to be looking at the details of properties on display in the window but to actually look through into the office that we had just spent half an hour in. She did so and was surprised to see—I kid you not—a full-size stuffed bison in the middle of the room. Concentrating on the important task of considering possible new properties, she had completely failed to spot this large and totally incongruent object.

The importance of inattentional blindness was recognized immediately by psychologists. After all, many serious accidents occur because people fail to notice stimuli that are right in front of them. Literally hundreds of experiments have now been carried out to learn more about the phenomenon, and, indeed, it became a focus of Anne's research in the Department of Psychology at Birkbeck College, University of London. She assures me that her interest in this topic is totally unrelated to her inability to spot stuffed bisons in estate agents' offices!

Some years later, Anne called to ask me if I had a copy of a scale to measure the personality variable known as *absorption*. If you are the kind of person who would get a high score on such a measure, then you are the kind of person who, when reading a novel, watching a film, or doing a crossword puzzle, would be very difficult to distract from that activity; that is to say, you would become absorbed in whatever you were focusing your attention on. Although the general view among inattentional blindness researchers at the time was that susceptibility to this phenomenon did not correlate with any particular personality variables, we both felt it was likely to correlate with absorption. Furthermore, I was intrigued because I knew from my wider reading that absorption reliably correlated with paranormal belief and the tendency to report paranormal experiences.

The obvious next step was to carry out a study to investigate the relationship, if any, between all three of these variables.[27] We did not use the gorilla video in our study but instead used a conceptually similar task. Volunteers were instructed to watch a computer screen displaying black and white letters moving around the screen and sometimes "bouncing" off the edge of the screen. They were asked to count the number of times the white letters bounced off the edge of the screen while ignoring the black letters. Halfway through, a red cross moved slowly across the screen from one side to the other. At the end of the task, participants reported how many times the white letters had bounced off the edge of the screen and were then also asked if they had noticed anything unusual during the task. As expected, around half of the participants did not see the red cross. These inattentionally blind participants had significantly higher scores on both a measure of absorption and on a measure of paranormal belief and experience, results that we replicated in a second experiment.

How might inattentional blindness relate to reports of ostensibly paranormal events? One possibility is that some reports of allegedly ghostly encounters (but by no means all) may have a mundane nonparanormal explanation that had been missed by the claimant because they failed to take in some relevant information at the time. As a hypothetical example, suppose someone were to claim that a book moved from one location to another when no one else was in the vicinity. If one were to suggest that maybe there was someone else around who may have moved the book and that they just did not notice the other person, they would probably indignantly insist that they were absolutely sure that they would notice if anyone else was nearby.

But would they? After all, lots of people fail to see a gorilla beating its chest when it is right in front of them.

Although the term *inattentional blindness* did not appear in print until the late 1990s, the phenomenon had in fact been inadvertently demonstrated almost four decades earlier in a couple of quirky studies by parapsychologist Tony Cornell.[28] Cornell appeared to accept the idea that spirits of the dead really do sometimes appear to the living. He wanted to study how people react to such appearances. To do so, he covered himself from head to toe with a muslin cloth and presented himself to unsuspecting individuals in a variety of settings. In his first study, he emerged from behind a small mound in a cow pasture, strolled to the next mound, and then abruptly vanished behind it. To his chagrin, none of the eighty or so passersby gave any sign of having seen him (although apparently the local cows found him fascinating). He hoped that the setting of his second attempt would produce better results: the graveyard of a church. Sadly, he was again disappointed. Of over 140 potential witnesses to his "Experimental Apparition," only four appeared to notice anything unusual. Matthew Tompkins describes their reactions as follows:

> Under questioning, it emerged that none of them believed they had witnessed anything remotely paranormal. The first person described the apparition as "a man dressed as a woman, who surely must be mad" another assumed that it was "an art student walking about in a blanket." Two witnesses, when questioned together, did realise that the Experimental Apparition was probably *intended* to simulate a paranormal event, but went on to note that the effect was spoiled because "we could see his legs and feet and knew it was a man dressed in some white garment."[29]

Undeterred, Cornell had one last attempt to collect the data he wanted. This time, his Experimental Apparition took fifty seconds to walk from one side of a cinema screen to the other and back again in front of the audience for an X-rated (that is to say, adult) movie. He chose this setting partly because he felt that this would ensure that his witnesses were focusing their attention in the right direction, and partly to ensure that no children would be traumatized by the experience. He need not have worried. In Tompkins's words:

None of the audience reported anything remotely paranormal. Many saw nothing unusual at all: 46% of the respondents had failed to notice the Experimental Apparition when Cornell first passed in front of the screen, and 32% remained completely unaware of it. Even the projectionist, whose job was to watch for anything unusual, reported that he had completely failed to notice the apparition. Those that did see 'something' were not particularly accurate in their descriptions. One person reported seeing a woman in a coat, another thought they had seen a polar bear, and another believed that they had observed a fault in the projector. Only one person accurately described a man dressed in a sheet pretending to be a ghost.

Cornell was disappointed by his results, having no inkling that the phenomenon he had so strikingly demonstrated would become a major focus of psychological research several decades later. Of course, in his case, inattentional blindness had caused people to miss the ostensibly paranormal phenomenon itself, as opposed to any potential explanations for such phenomena. In fact, on the basis of his results, Cornell suggested that there may actually be many more ghosts around than is generally appreciated but that people simply fail to notice them!

THE FALLIBILITY OF MEMORY

As Cornell's experiments, along with literally thousands of others, clearly demonstrate, eyewitness testimony can be extremely unreliable. This is a fundamental lesson to bear in mind whenever one is presented with anecdotal evidence in support of a paranormal claim. As David Hume asked, is it more likely that an event occurred that violated the laws of nature as we currently understand them or that the claimant is mistaken (or lying)?

One of the main reasons that the unreliability of eyewitness testimony has been the subject of so much research is because of its obvious implications for the legal system. The reliance on eyewitness testimony in many criminal trials when making life-changing decisions regarding guilt or innocence continues despite the findings from countless psychological experiments clearly demonstrating the frailty of human memory.[30] Of course, the findings from such studies generalize beyond the forensic context to autobiographical memory in general, including memory for anomalous experiences.[31]

Such research has confirmed a number of commonly held beliefs that coincide with our intuitive expectations of how memory works. For example, it comes as no surprise that memory for peripheral detail is poorer than memory for that which is the focus of our attention, that memory is worse for things seen only briefly or under imperfect viewing conditions, and that memory is less accurate when we are underaroused (e.g., sleepy) or overaroused (e.g., terrified). It is noteworthy that many of the factors just listed often apply to reports of ghostly encounters. Lest the reader conclude that memory science has achieved nothing more than confirming what everyone already knew, it should be noted that many other commonly held beliefs about memory have been shown to be completely false, a fact that has worrying implications for our legal system.[32]

One of the most widely held misconceptions about memory is the notion that memory works like a video camera, accurately recalling every detail of everything we ever experience.[33] In fact, both memory and perception are constructive processes. With respect to perception, despite our intuition that we take in and process all of the available sensory information around us at every instant, we actually process only a small fraction of it. How then are we able to build up and maintain that mental model of the world around us and our place within it that is the basis of our sense of being a single unified consciousness?

It is generally accepted that we perform this near-miraculous feat on the basis of two sources of information. On the one hand, we do of course process information coming in via our all of our senses. This is sometimes referred to as *bottom-up processing*. But, as already stated, we actually only fully process a fraction of the sensory information available to us from moment to moment. How then does our brain, without our conscious awareness, fill in the gaps to give us the impression that we have a complete mental model of the world around us? This is achieved via *top-down processing*. We are able to fill in the gaps in our mental model on the basis of what we already know about the world through our prior experience. Thus, our mental model is based on the constant interaction between bottom-up and top-down processing. The model is constantly updated on the basis of new incoming sensory information.

In general, this system works well and provides us with a mental model that is usually reliable enough for us to safely interact with the world. If we had to process every single detail of every single sensory channel before we had a reliable mental model, the system would simply overload. However, this process of filling in the gaps in our mental model on the basis of expectations built up from prior experience means that there is the potential for illusory perceptions, especially if the current input is degraded or inherently ambiguous. Thus, we may end up perceiving things that are not there.

It may be more accurate to describe memory as a *re*constructive process rather than a constructive process. Every time you recall something, you do so on the basis of more or less accurate memory traces laid down at the time the event occurred, but once again you will unconsciously fill in any gaps on the basis of top-down processing. We may thus end up with memories based on what we think must have happened rather than what actually did happen.

As a striking illustration of this, see if you can answer the following question: Without actually looking at any clocks or watches, can you remember how the number four is represented on clocks and watches with Roman numerals on them? As I know from asking this question in countless public talks, most people confidently reply that it is represented as "IV"—and are quite surprised to learn that on the vast majority of such timepieces it is, in fact, represented as "IIII." Everywhere else it is indeed represented as "IV"— but not on clocks and watches.[34]

That particular example of a memory error caused by top-down processing has a special place in my heart. My wife, Anne Richards, and I carried out an experimental investigation of this effect that was published in the *British Journal of Psychology*.[35] That paper was, without a doubt, the one that required the least effort of all the papers that we ever published. We collected the data on a single afternoon from volunteers attending an open day at Goldsmiths. Data were collected from three groups. The first group was shown an ordinary clock (our kitchen clock, in fact) with Roman numerals on its face for one minute. The clock was then removed from view and participants were instructed to draw it from memory. The procedure for the second group was identical, with the exception that this group was informed in advance that they would be required to draw the clock from memory. The third group was simply asked to draw the clock with the clock remaining in full view throughout. The results were striking. In both of the memory groups, most participants incorrectly represented the four as "IV." In the copy group, no one did.

In our paper reporting these results, we described how this curious phenomenon first came to our attention. We were visiting my wife's parents. Our eldest daughter, Lucy, was about eight years old. Her attention was caught by a clock on the mantlepiece. Here, extracted from our paper, is the conversation that then took place:

> **Lucy** On the clock, why does "V" come after "IIII"?
> **CCF** (*without looking at the clock*) It doesn't say "IIII." It says "IV" for four.
> **Lucy** It doesn't. Look.
> **CCF** (*looking at clock*) Incredible! You'd think clock-makers of all people would know Roman numerals! But this is how it should be. (*Shows his wrist-watch.*) Would you believe it, they've got it wrong on here as well!

These days, Anne insists that she had the above conversation with Lucy. She is wrong, of course, thus providing us with another nice example of memory distortion (hers, not mine).[36] Regardless of that, we wrote up the paper in a couple of hours, submitted it, and were requested to revise only a single sentence before it was suitable for publication. If only it was so straightforward to get all scientific papers published! Our one regret is that we did not christen this particular example of memory distortion as the *Lucy effect* (sorry, Lucy!).

Once one appreciates the reconstructive nature of memory, one can understand why one's memory of an event depends not only on what was happening when the experience itself was happening but also on events that preceded and followed it. Your prior experiences will lead you to have a particular set of beliefs and expectations that will then influence how you perceive what is going on around you, especially if those events are ambiguous and unclear (as in pareidolia, for example). You will subsequently recall your interpretation of the event, not the event itself. As the event itself is taking place, you will only recall those aspects that were encoded at the time as demonstrated by studies of inattentional bias. You will also be affected by suggestions made at the time. Once the event has taken place, it can still be distorted further by events that take place subsequently. All of these effects have been demonstrated in thousands of psychological studies, including many in the area of anomalistic psychology.

MEMORY FOR FAKED SÉANCES

In order to systematically study the reliability of memory, we need to consider alternatives to reports of spontaneous ghostly encounters. The latter occur at unpredictable times and places, and objective records of what actually happened are typically unavailable. If we need to assess how accurate someone's recall is, we need to know exactly what was taking place at the time so that we can compare this with the individual's report. The experiments reported by Cornell provide one approach to this issue, but there are alternative approaches. One of these is to assess how well people recall the events that take place during a faked séance.

Séances were all the rage in the Victorian era, a period during which Spiritualism grew enormously in popularity.[37] We can date the origins of the Spiritualist movement very precisely. In March 1848, in a house in Hydesville, New York, two young sisters by the name of Kate and Margaretta Fox (figure 3.3) reported hearing strange rapping noises and eventually discovered that they could communicate with "the other side" using a simple code. They claimed that the noises were made by the spirit of a peddler who had been murdered and buried in the cellar. Soon others also claimed to be able to communicate with the dead, and séances spread like wildfire across America and Europe.

Séances eventually came to involve other phenomena too, including movement of tables and objects; the playing of musical instruments by unseen hands and lips; strange lights in the dark; levitation of objects, the table, or even the medium; the disappearance or materialization of objects; the materialization of hands, faces, or even complete spirit forms (ostensibly composed of "ectoplasm"); disembodied voices; spirit paintings and

Figure 3.3
The Fox sisters (from left to right): Margaretta, Kate, and Leah. Elder sister Leah acted as manager for her younger sisters for a time.

Figure 3.4
During typical séances in the Victorian era, sitters were instructed to hold hands and remain in their seats throughout the proceedings.

photographs; and written communications from the spirit world (figure 3.4). All of these phenomena were said to be produced by spirits summoned by the medium.

Unfortunately, it was a very rare medium indeed who was not caught at some point engaging in trickery to produce these effects. In 1888, Margaretta Fox admitted that the original rapping noises produced forty years earlier, when she was just a girl, were not in any way psychic. They were produced in a variety of ways but mainly by cracking her toe and ankle joints, a skill she demonstrated in public. What had begun as a prank got out of hand, and the sisters felt unable to own up. Spiritualists simply refused to believe the confession, and the movement continues to this day.

Interestingly, the first systematic study of the unreliability of eyewitness testimony took place in the context of a faked séance. It was reported by S. John Davey in the *Proceedings of the Society for Psychical Research* in 1887.[38] Ray Hyman provides the following account:

Davey had been converted to a belief in spiritualistic phenomena by the slate-writing demonstrations of the medium Henry Slade. Subsequently Davey accidentally discovered that Slade had employed trickery to produce some of the phenomena. Davey practised until he felt he could accomplish all of Slade's feats by trickery and misdirection. He then conducted his well-rehearsed seance for several groups of sitters, including many who had witnessed and testified to the reality of spiritualistic phenomena. Immediately after each seance, Davey had the sitters write out in detail all that they could remember having happened during his seance. The findings were striking and very disturbing to believers. No one realized that Davey was employing tricks. Sitters consistently omitted crucial details, added others, changed the order of events, and otherwise supplied reports that would make it impossible for any reader to account for what was described by normal means.[39]

In 1932, Theodore Besterman reported similar levels of memory distortion in the recall of his faked séance.[40] Sitters reported objects as having moved during the séance when they had not as well as failing to report important events such as the experimenter actually leaving the room. This line of research was continued more recently by Richard Wiseman and colleagues with a particular emphasis on investigating the power of suggestion.[41] In one experiment, the actor playing the part of the medium suggested the table was moving when in fact it was not. Afterward, one-third of the participants incorrectly reported that the table had moved. Believers in the paranormal were more susceptible to this suggestion than nonbelievers. In a second experiment, it was found that once again believers in the paranormal were more susceptible to the effects of suggestion, but only if the suggestion was consistent with their belief. Although all of the effects demonstrated in the séances were produced by trickery, many participants reported that they believed that they had witnessed genuinely paranormal phenomena.

As an aside, it is also worth noting that the anomalous sensations reported by participants in this study were not limited to the mysterious movement of objects. As Wiseman and colleagues report:

> Many people reported the type of quite dramatic phenomena often associated with 'genuine' seances, including being in an unusual psychological state (e.g. 'Feeling of depersonification and elation when the objects moved'); changes

in temperature (e.g. 'Cold shivers running through my body when I concentrated hard on moving the objects'); an energetic presence (e.g. 'A strong sense of energy flowing through the circle which increased'); and unusual smells (e.g. 'A smell of hot plastic, combination of sweet and acrid smell'). Thus, the fake seances caused participants to report many of the experiences described by those attending 'genuine' seances, suggesting that such effects are the result of psychological processes (e.g. psychosomatic experiences brought about by participants' heightened expectations or strong beliefs), rather than being caused by paranormal, psychic or mediumistic mechanisms.[42]

MEMORY FOR OTHER OSTENSIBLY PARANORMAL PHENOMENA

The effects of prior belief in the paranormal were also demonstrated in an investigation by Richard Wiseman and Robert Morris of the recall of what they called "pseudo-psychic demonstrations"—or what you and I might call "conjuring tricks."[43] Participants were misled into believing that an individual had presented himself to the Koestler Parapsychology Unit at the University of Edinburgh claiming to possess genuine paranormal abilities. They were then shown videos of this individual performing conjuring tricks that appeared to demonstrate effects that are often presented as being paranormal. Finally, they were asked to rate the degree to which they thought the phenomena were genuinely paranormal and also to answer a number of recall questions.

Perhaps not surprisingly, the believers in the paranormal rated the demonstrations as more "paranormal" than did the skeptics. Of greater interest were the results of the memory tests. Wiseman and Morris had asked about recall for both "important" and "unimportant" details of the demonstration. By this they meant that some aspects of the demonstrations were completely unrelated to the methods used to achieve the apparently paranormal effects. These were the unimportant details. But some aspects of the demonstration did indeed provide potential clues to the trickery that was being employed. For example, a couple of the video clips appeared to demonstrate psychokinetic metal-bending of the kind made famous by Uri

Geller, in one case the apparent bending of a key.[44] This was achieved by using sleight of hand to surreptitiously switch the originally straight key for a pre-bent copy partway through the performance. This can only be achieved if the key passes out of sight for a split second at some point. This was precisely the kind of detail that the skeptics were more likely to recall than the paranormal believers. Skeptics and believers did not differ in the accuracy of their recall for unimportant details.

The most obvious explanation for this effect is that the two groups of participants approached the task with different intentions. The skeptics were more likely to be in problem-solving mode, assuming the effects would be achieved by trickery and actively trying to spot the techniques being used. They would, therefore, be more likely to actually spot the clues and recall those details accurately. The believers, on the other hand, would be more likely to assume that they were about to witness genuine paranormal effects and therefore would just sit back and enjoy the show. As we know from research into inattentional blindness, if information is not encoded at the time an event occurs, it will simply not be available for recall later.

Although most magicians are convinced that apparent psychokinetic metal-bending is always the result of trickery, many people who witnessed Uri Geller performing this feat early in his career were not convinced by this scientific explanation. One of their main reasons for rejecting this explanation was, they claimed, that the metal object in question *continued to bend even after Geller had put it down!* How could that possibly be achieved by sleight of hand?[45]

I suspect that by now readers may already have guessed the explanation for these claims. Was it possible that the reports of spoons, forks, and keys that continued to bend after they had left Geller's hands were based on nothing more than the power of suggestion, as demonstrated in the fake séance studies? Suspecting that this may well be the case, Richard Wiseman and Emma Greening put this explanation to the test under properly controlled conditions.[46] Participants were once again presented with a video clip showing an alleged psychic claimant apparently performing a psychic key-bend. In fact, of course, the effect was achieved using sleight of hand. In one condition, after the key had been bent, it was placed on the table and, with the key still in full view, the "psychic" can be heard suggesting, "It's still bending."[47] Participants in the other condition were shown an identical video with the single exception that no suggestion was made that the key continued to bend. Around 40 percent of the participants who received the suggestion reported that they thought the key carried on bending compared to only one (out of twenty-three) in the no-suggestion condition.

This effect was replicated in a second experiment. In this second experiment, it was found that those who reported that the key continued to bend were more confident in the accuracy of their own memory than those that did not and also were less likely to recall that the "psychic" had explicitly suggested that the key continued to bend. Surprisingly, given previous similar studies, believers in the paranormal were not found to be more susceptible to suggestion.

Not long after this report was published, I was involved in demonstrating the effect for a TV company. In this case, we wondered how impactful it might be to witness such an effect live as opposed to watching a prerecorded video. In this case, fellow skeptic and amateur magician Tony Youens played the role of our psychic, using a nail instead of a key. By the end of the day, we had enough shots of our volunteers expressing their amazement at the nail that, they claimed, they had seen continuing to bend as it lay in Tony's palm following his psychokinetic exertions.

One of the most memorable aspects of that day's filming was the reaction of our cameraman. Before we started filming, he told me that he had seen Geller performing his metal-bending feat back in the 1970s, and he had seen with his own eyes that the metal had indeed continued to bend even after Geller put it down. So, by the end of the day, was he now convinced that this was due to nothing more than the power of suggestion? Not on your life!

Given the overwhelming evidence that eyewitness testimony can be so unreliable, there are reasonable grounds to be cautious in taking the uncorroborated testimony of a single eyewitness at face value. Does it make sense to give more evidential weight to those situations when there is more than one eyewitness and they all give similar accounts? The short answer is yes—but even here, there is the potential for serious memory distortion. One factor that may lead to such distortion is referred to as *memory conformity*. This refers to the phenomenon whereby one eyewitness's account may influence the memory of a second eyewitness. If the first person's account is inaccurate in some ways, those inaccuracies may be unintentionally incorporated into the second person's memory.

Memory conformity is one example of the more general phenomenon of memory distortion due to post-event misinformation. Psychologists have for many decades successfully distorted the memories of witnesses by subtly presenting them with misinformation following the viewing of, say, a simulated crime scene. Elizabeth Loftus and colleagues carried out research using this technique.[48] For example, in one study participants were shown a sequence of slides depicting events before, during, and after a car accident in which a pedestrian was injured. Half of the participants saw a slide showing a red Datsun at an intersection with a stop sign in the background. The other half saw the same scene but with a yield (give way) sign in place of the stop sign.

After seeing the slides, participants were interviewed using a standard questionnaire. One of the questions was deliberately misleading for some participants. For example, half of the participants who had viewed a stop sign were asked, "Did another car pass the Datsun while it was stopped at the stop sign?," but the other half were misled by being asked, "Did another car pass the Datsun while it was stopped at the *yield* sign?" (italics added). Similarly, half of the participants who had actually originally viewed a yield sign were asked a question implying that a stop sign had been viewed, with the other half being asked the question in a nonmisleading form. Presentation of misleading information in the post-event question increased the probability that participants would subsequently report seeing a sign that was different than the one that they had actually seen. Since that early work, there have been innumerable variations of the original technique leading to one indisputable conclusion: exposure to misleading information about an event after the event has taken place often distorts a witness's memory of it.

Some more recent post-event misinformation studies are characterized by the presentation of misleading information via a social channel, typically discussion with a co-witness. Witnesses to an unusual and unexpected event, such as a crime, a possible ghostly encounter, or a UFO sighting, are likely to spontaneously discuss what they have seen prior to any formal interview.

Studies of memory conformity often involve getting volunteers to watch a video of a staged crime and then discuss it in pairs prior to giving an independent report of what happened. In some studies, one of the members of the pair is in fact a confederate of the experimenters who has been instructed to introduce a few specific items of misinformation (e.g., what the perpetrator was wearing) into the discussion.[49] This misinformation is often incorporated into the account subsequently given by the genuine participant.

My friend Fiona Gabbert, now a professor of psychology at Goldsmiths, and colleagues used a novel technique whereby pairs of volunteers watched a short film of the same staged crime but viewed from different angles.[50] The filming was carried out in such a way that version A included some details that could not be seen in version B and vice versa. The pairs of volunteers were under the impression that they had both watched the same video. After watching the video, the pairs discussed what they had seen and then independently answered questions regarding their recall of the events. Regardless of the technique used, studies of memory conformity routinely demonstrate that information (or misinformation) only available via a co-witness's report is incorporated into accounts provided by a significant proportion of witnesses.

Krissy Wilson, at that time one of my postgraduate students, and I decided to investigate memory conformity effects in an anomalistic context.[51] To do so, we used the same video clip that Richard Wiseman and Emma Greening had used in their "It's still bending" study described above. We incorporated a memory conformity element into our study by having participants watch the video in pairs and then discuss it together prior to giving independent reports of what they had seen. As in previous studies, one member of each pair was in fact a confederate who had been instructed to say either that the key continued to bend after it had been placed on the table or to say the key did not continue to bend. In a third condition, the confederate remained neutral, not commenting on it at all. Our results showed that both the suggestion from the fake psychic and the comments from the stooge co-witness had an effect on the reports of the genuine participants. Indeed, around 60 percent of the participants who were in the condition in which they received both the suggestion from the psychic and the reinforcement of this suggestion from the stooge reported that they thought the key carried on bending. In our study, believers in the paranormal were indeed found to be more susceptible to suggestion, possibly as a result of us using a different measure of paranormal belief than that used by Wiseman and Greening.

As demonstrated in the experiments described above, as well as literally thousands of other published studies, eyewitness testimony can be notoriously unreliable insofar as memory for an event can be distorted by experiences that take place before, during, and after the event in question. In fact, we can even have apparent memories in our minds that are not just distorted memories of events we have witnessed but are apparent memories for events that never actually happened. We shall defer further discussion of these *false memories* until a later chapter where they are arguably more relevant. For now, it should be borne in mind that any individual paranormal anecdote may not just be a distorted version of an event that really happened—it may be a complete fabrication from start to finish, albeit one that was unintentionally generated and is now sincerely believed.

ENVIRONMENTAL FACTORS ASSOCIATED WITH HAUNTINGS

There is, as already discussed, evidence that when people are told that a location is haunted, they tend to report more anomalous experiences there than people who are not primed in this way. This is to be expected given what we know about the power of suggestion. But is it also possible that some locations are just inherently spookier than others? This possibility was investigated in two studies, one at Hampton Court Palace, the other at the South Bridge Vaults in Edinburgh, carried out by a team led by—yes, you've guessed it—Richard Wiseman.[52] Both locations have considerable reputations for being haunted, but, interestingly, within each some areas are reported to be more haunted than others; that is to say, that whereas visitors report a high number of anomalous experiences in some areas, very few are reported in others.

Visitors were asked to walk around the locations and to note any anomalous experiences they had and where those experiences had occurred. One might expect that visitors who already knew something about each location in terms of which areas were reputed to be the most haunted would, by a process of what we might call *self-priming*, duly report more anomalous experiences in the "haunted" than the "nonhaunted" areas, whereas this pattern would not be observed for visitors who did not possess such knowledge. In fact, this is not what was found. Instead, there was a general tendency for more anomalous experiences to be reported in the "haunted" areas than the "nonhaunted" areas by all visitors, regardless of their prior knowledge. This strongly supports the notion that some locations are just inherently spookier than others, possibly as a consequence of environmental factors.

Fortunately, Wiseman and his team collected data relevant to investigating this notion. Analysis of the data from the Edinburgh study showed that people reported more anomalous experiences when going from a relatively well-lit exterior to the dark interior of a vault than when entering a vault with less discrepancy between exterior and interior lighting levels. This is perhaps not a surprising result, as it is a natural human response to feel more vulnerable in the dark. From an evolutionary perspective, our ancestors were at much greater risk from potential threats in conditions of dim illumination. The same argument might apply to another finding from this study; that is, more such experiences were reported in vaults with higher ceilings. Again, from an evolutionary point of view, one might argue that it makes sense for humans to be more alert on entering caves with high ceilings given the possibility that an attacker might suddenly drop on them from above.

Another suggestive finding, however, cannot be explained in such terms. In the Hampton Court Palace investigation, it was found that, although there was no significant difference in the strength of the local magnetic fields between the "haunted" and the "nonhaunted" areas, the variance in the fields was significantly greater for the former.[53] Furthermore, the variance correlated with the number of anomalous experiences reported in each area. No such pattern of results was found in the Edinburgh investigation, however, suggesting that this may be a spurious finding.

Some readers may wonder why these investigators were measuring magnetic fields in the first place. What on earth could they have to do with reports of ghostly encounters? They were gathering data relating to an interesting, albeit controversial, theory put forward by the late Canadian professor of psychology, Michael Persinger.[54] For several decades, Persinger had been arguing that a range of ostensibly paranormal and religious experiences might actually be the result of unusual activity in the temporal lobes of the brain, causing hallucinatory experiences of the type often associated with reputedly haunted locations. Furthermore, such neuronal activity, he argued, could be induced in susceptible individuals if they were exposed to certain types of weak, complex electromagnetic fields.

Before we go any further, it should be pointed out that there is, in fact, absolutely no doubt that exposing brains to strong magnetic fields can have immediate and unambiguous effects. A technique known as *transcranial magnetic stimulation* (TMS) has become increasingly popular among neuropsychologists as another method to investigate brain function. TMS involves the application of simple, high-intensity magnetic pulses near the cranium that are known to be strong enough to induce currents within areas of the brain that are close to the source of the field. The biophysics of the technique is well understood. The technique provides a way of studying brain function by essentially temporarily "knocking out" specific parts of the brain so that the effects on cognition, behavior, and subjective awareness can be assessed. The technique is also used in diagnosis and treatment.

What is happening when TMS is applied is very different from the kind of effects claimed by Persinger, as pointed out by Jason Braithwaite.[55] Whereas TMS involves the use of simple, high-intensity magnetic pulses, Persinger claims that anomalous experiences may be induced in the presence of weak yet complex electromagnetic fields. Indeed, it has been pointed out that Persinger's fields are about 5,000 times weaker than that produced by a typical fridge magnet![56] There is currently no plausible biophysical mechanism whereby such weak magnetic fields could have an effect on brain function. Whereas the effects of TMS on brain function are instantaneous, definite, and specific, Persinger claims that the effects of the weak fields that he is referring to take time to build up (some twenty to forty minutes) and are rather nebulous in nature. Furthermore, TMS works on everyone, whereas according to Persinger you will only have anomalous experiences as a result of exposure to weak, complex electromagnetic fields if you already have sufficiently labile temporal lobes. Persinger gauges the lability of an individual's temporal lobes by administering his Temporal Lobe Signs Inventory, which is a subscale on his Personal Philosophy Inventory.[57]

Braithwaite has comprehensively reviewed the evidence for and against Persinger's claims of a link between electromagnetic fields and anomalous experiences.[58] The evidence put forward in support of these claims is primarily of three types. First, Persinger claimed the number of anomalous experiences reported is significantly correlated with changes in the Earth's background electromagnetic field. This would appear to be highly unlikely, given that one would experience much greater fluctuations in electromagnetic fields walking through the average home as a result of the use of electronic devices. It is likely that the significant correlations reported by Persinger are simply the result of data-trawling through massive data sets until spuriously significant correlations were found.

The second line of evidence is the claim that magnetic anomalies are more frequently found in reputedly haunted locations than in appropriately matched control locations. It turns out, however, that only two examples of the type of temporally complex fields that Persinger claims are required to induce hallucinatory experiences have been reported from around fifty locations investigated. Whether the fields in these two cases played any direct role in the anomalous experiences reported in these cases is a moot point, but these studies do not provide support for the idea that weak, complex electromagnetic fields are a major factor underlying reports of ghostly encounters.

The third and final strand of evidence comes from laboratory studies. Persinger claimed that he could induce the required pattern of unusual activity in the temporal lobes of susceptible volunteers by getting them to wear a specially constructed helmet that could generate the appropriate weak-intensity, temporally complex electromagnetic fields near their craniums. Although Persinger referred to this device as a "Koren helmet," others tend to call it the "God helmet," as Persinger claimed that the experiences induced were often interpreted as religious experiences.

Persinger published many papers with results from experiments using the God helmet, reporting a range of anomalous experiences.[59] It was even claimed that application of the technique had led to the subjective experience of a full-blown apparition in a middle-aged man with a prior history of such experiences.[60] As one might expect, this line of research has received much media attention. It has been featured in two *Horizon* documentaries produced by the BBC. In the first, in 1994, renowned skeptic Susan Blackmore volunteered to don the helmet as she lay back in a dentist's chair in a soundproof room in Persinger's laboratory. For the first ten minutes, nothing happened. Here is her description of what happened next:

> Then it felt for all the world as though two hands had grabbed my shoulders and were bodily yanking me upright. I knew I was still lying in the reclining chair, but someone, or something, was pulling me up.
>
> Something seemed to get hold of my leg and pull it, distort it, and drag it up the wall. It felt as though I had been stretched half way up to the ceiling. Then came the emotions. Totally out of the blue, but intensely and vividly, I suddenly felt angry—not just mildly cross but that clear-minded anger out of which you act—but there was nothing and no one to act on. After perhaps ten seconds, it was gone. Later, it was replaced by an equally sudden attack of fear. I was terrified—of nothing in particular.[61]

A decade or so later, *Horizon* featured Persinger's God helmet again, this time in a program titled *God on the Brain*. On this occasion, the presenter was Richard Dawkins. After trying the God helmet, Dawkins reported, "It pretty much felt as though I was in total darkness, with a helmet on my head and pleasantly relaxed." In the words of Craig Aaen-Stockdale, "Not exactly a road to Damascus experience, but Dawkins is, of course, a damned sceptic and Persinger simply argued that he wasn't temporal-lobey enough."[62]

The main problem with this body of evidence is that, by and large, it has all been produced by a single laboratory—that of Michael Persinger himself. Furthermore, concern has been expressed over the degree to which Persinger employed proper double-blind methodologies in his published research. In studies using double-blind methodologies, neither the participant nor the experimenter is aware of whether the participant is in the experimental or the control condition until data analysis has been completed. That way, there is no possibility of bias, either intentional or unintentional, affecting the results.

There is an urgent need for independent replication of Persinger's results by other laboratories using appropriate double-blind methodologies. One attempt to do this was reported by Pehr Granqvist and colleagues.[63] They used equipment borrowed directly from Persinger's laboratory but were unable to replicate his finding of a link between the application of weak, complex electromagnetic fields and reports of a sensed presence. Instead, they found that scores on various measures associated with suggestibility, including absorption, temporal lobe signs, and New Age lifestyle orientation, correlated with the tendency to report such experiences. Their report was highly critical of Persinger's research, arguing that his findings may reflect little more than the effects of priming, suggestibility, and poor methodology. Predictably, Persinger rejected these criticisms, but it remains the case that, without more supporting evidence from independent researchers, the wider scientific community is unlikely to accept Persinger's claims.[64] Braithwaite concludes his examination of the evidence by advocating a position that "acknowledges the possibility of an effect, but an effect which is rare and requires considerable further investigation."[65]

A second invisible environmental factor has also been suggested as possibly producing similar anomalous experiences: infrasound. Infrasound is sound energy below the audible frequency range (i.e., lower than 20 Hz). The late Vic Tandy, inspired by his own ghostly encounter, put forward the idea that infrasound is capable in some circumstances of inducing hallucinatory experiences, including a sense of presence, feelings of depression, cold shivers, and even apparitions.[66] On more than one occasion, I had the pleasure of listening to Vic recount the story of how he first came up with his theory in front of rapt audiences. It is a good story, well worth summarizing here.

Vic used to work for a company producing medical equipment. The laboratory had a reputation for being haunted, but Vic never took the reports of strange experiences seriously until he had one himself. Working late one night, alone in the lab, he started to feel uncomfortable. He felt depressed. He was sweating even though it was cold. He also had a strong sense of presence. As he wrote at his desk, the feeling that someone—or something—was watching him grew stronger. He slowly became aware of a blurry gray figure on his left, in the periphery of his vision. Terrified, he managed to pluck up the courage to turn and face the apparition—at which point it simply faded away. Convinced that he was cracking up, he went home.

The next day he went into work early to use the lab equipment to work on a fencing foil ahead of a competition. He fixed the blade into a vise and noticed that it was vibrating wildly up and down. Trying to suppress thoughts of "Oh god, not again!" the rational part of his brain stepped in. His engineering background meant that he quickly deduced that there must be a source of energy that was varying in intensity at the resonant frequency of the blade. To cut a long story short, his investigations revealed that a newly installed fan system was producing a standing wave in the lab with a frequency of around 19 Hz. It is suggested that this standing wave may be the resonant frequency of the human eyeball, and thus the standing wave may have caused Vic's eyeball to vibrate, resulting in visual "smearing" that was perceived as the apparition. In a follow-up study, Vic claimed he found evidence of infrasound of a similar frequency in a reputedly haunted cellar in Coventry.[67]

Once again, Jason Braithwaite has assessed this theory and the evidence put forward to support it.[68] It is fair to say that he is not convinced on either front. Regarding the idea that the hallucination may have been caused by vibrating eyeballs, Braithwaite argues that this would be more likely to cause distortion across the entire visual field rather than just in the periphery of the left visual field. He also opines that such vibration would be unlikely to result in the perception of a hallucination resembling a human figure or form. Moreover, he criticizes the studies for making no attempt to measure infrasound sound levels appropriately and for not taking measurements in baseline control areas for comparison purposes.

To the best of my knowledge, only one study has attempted to investigate the effects of both weak, complex electromagnetic fields and infrasound within the same study—and that was the one that I carried out with Usman Haque, a designer of interactive architectural systems, and a team from the Anomalistic Psychology Research Unit.[69] I cannot remember exactly when I was first approached by Usman with what may have sounded like a pretty crazy idea. I do know that it was before I had read Braithwaite's critiques of the theories and evidence linking electromagnetic fields and infrasound to anomalous experiences. So I was very receptive to Usman's suggestion that it would be pretty neat to build an artificial "haunted" room by manipulating these two environmental factors.

Of course, it is one thing to come up with an exciting idea for a study like this, and quite another to actually make it a reality. There were many decisions to make at the planning stage. Not least among them was whether we should make our haunted room look like, say, an ordinary bedroom or take a minimalist approach and go for a simple featureless enclosed space. After much discussion, we decided to go for the latter approach and set up a circular featureless chamber using white canvas, approximately ten feet wide and thirteen feet high with constant temperature and illumination levels. The next question was, where would we actually set up our chamber? It would have to be in place for quite a long time in order for us to collect enough data. I am not sure what type of bribery, if any, was involved, but Usman's mum kindly agreed to let us take over her living room in North London.

We made sure that the levels of infrasound and the patterns of electromagnetic activity to which we would be exposing our volunteers (and Usman's mum) were at safe levels. Of course, we had to get ethical clearance for our study, which necessitated having to inform our volunteers in advance that they may be exposed to infrasound, to complex patterns of electromagnetic activity, to both, or to neither, which may result in them having some unusual sensations. We ensured, via pilot testing, that neither the infrasound nor the electromagnetic activity would be consciously detectable.

Seventy-nine volunteers each spent fifty minutes alone inside the chamber. During this time, they recorded on a floor plan of the area where they were when any anomalous experience occurred, along with a brief description of the experience and the time at which it occurred. When they emerged, they completed three questionnaires. The first, our EXIT scale, was a list of specific anomalous sensations (such as "Felt a presence," "Tingling sensations," and so on). The second was the Australian Sheep-Goat Scale (ASGS), a standard measure of paranormal belief and experience. The third was Persinger's Temporal Lobe Signs (TLS) scale, which Persinger used in his own research to identify individuals with particularly labile temporal lobes.

Many of our volunteers did indeed report anomalous sensations of various kinds while in our "haunted" room. As reported in our paper, the results from our EXIT scale were as follows:

> 63 (79.7%) of the participants felt dizzy or odd, 39 (49.4%) felt like they were spinning around, 33 (41.8%) experienced recurrent ideas, 29 (36.7%) felt tingling sensations, 26 (32.9%) felt that they were somewhere else, 25 (31.6%) felt pleasant vibrations through their bodies, 20 (25.3%) heard a ticking sound, 18 (22.8%) felt detached from their bodies, 18 (22.8%) felt a presence, 9 (11.4%) experienced sadness, 8 (10.1%) remembered images from recent dreams, 8 (10.1%) experienced odd smells, 7 (8.9%) experienced terror, and 4 (5.1%) experienced sexual arousal.[70]

On the basis of those results alone, we felt we had some justification for claiming that we had indeed succeeded in building an artificial haunted room (we were a bit surprised about the sexual arousal!). But now came the really exciting bit. Analysis would reveal the degree to which the anomalous sensations reported were the result of our manipulation of electromagnetic fields and infrasound. The answer was . . . drumroll, please . . . not at all. It did not matter whether the electromagnetic field was on or off. It did not matter whether the infrasound was on or off. The frequency of anomalous sensations reported was pretty consistent across all of the conditions in our experiment. So what was going on here?

The answer is hinted at by one of the significant results that our analysis did reveal: the number of anomalous sensations reported was significantly correlated with scores on both the TLS scale and, to a lesser extent, the ASGS. Scores on the TLS are known to correlate with suggestibility.[71] So the most parsimonious explanation of our results is simply that if you say to people before they go into an enclosed space for fifty minutes that they might have some anomalous experiences, some of the more suggestible ones will indeed do so. That is quite an interesting result from a psychological perspective, but it would have been much more exciting to have found a correlation between the presence of our manipulated environmental variables and anomalous sensations. *C'est la vie.*

It is only fair to point out that the results of our single study do not rule out the possibility that these environmental variables really might play some role in inducing the types of sensations that are typically reported at reputedly haunted locations. It remains possible, for example, that such factors only play a significant role in doing so in appropriate contexts—in other words, in locations that already look and feel "spooky" for other reasons. Maybe our decision to opt for a featureless chamber rather than an ordinary bedroom was a mistake? Indeed, maybe some of the sensations reported (such as "feeling dizzy") were unintentionally produced by those unusual surroundings? My personal opinion is, however, that the available evidence would suggest that these environmental factors are nowhere near as important as the psychological factors previously discussed in explaining ghosts and hauntings.

This chapter has primarily concentrated on describing psychological factors that may be relevant in explaining spontaneous ghostly encounters. In the next chapter, we shall turn our attention to a consideration of the psychological factors involved in deliberate attempts to communicate with the spirit realm.

4 HIGH SPIRITS, PART 2: COMMUNICATING WITH THE DEAD

MENTAL MEDIUMSHIP

Reports of ghostly encounters are not the only evidence put forward to support the existence of spirits. Throughout history, there have been many reports of communication with the dead. Of course, the important thing here is that the communication is two-way. William Shakespeare makes this point nicely in his play, *King Henry IV, Part 1*. Glendower boasts to Hotspur, "I can call the spirits from the vasty deep." Hotspur's reply: "Why, so can I, or so can any man; But will they come, when you do call for them?"

We have already considered one such claim. The Fox sisters claimed that the rapping sounds heard in their house in Hydesville, New York, were produced by a spirit, and they claimed to be able to communicate with that spirit using a simple code based on the number of knocks. As the Spiritualist movement grew, séances became more elaborate, often involving a wide range of physical effects. Mediums who specialized in this type of séance are referred to, unsurprisingly, as *physical mediums*, in contrast to *mental mediums*, who claim that they are able to pass messages between the living and the dead. These days, the latter vastly outnumber the former.

Mental mediumship can sometimes involve one-to-one readings, sometimes be done for a small group, and sometimes be performed in front of large audiences in packed theaters. Professional mediums typically charge for their services, and some become celebrities in their own right, with their own TV shows, bestselling books, and online services. Famous modern examples in the United States include Sylvia Browne, John Edward, and James Van

Praagh, while in the United Kingdom the list includes Derek Acorah, Colin Fry, Sally Morgan, and Doris Stokes.

Mental mediums typically appear to go into a trance during their séances. It is claimed that they are then able to "channel" messages from the spirit world. They may do this directly or by means of a spirit guide who acts as an intermediary between the medium and the spirit in question. Sometimes they will even appear to be temporarily possessed by a spirit, resulting in dramatic changes in their voice, mannerisms, and occasionally, it is claimed, their very appearance.

Séances reached the height of their popularity in the Victorian era. Indeed, their popularity was one of the factors that led to the founding of the Society for Psychical Research in the United Kingdom in 1882, followed by the founding of the American Society for Psychical Research a few years later. The other main factor was as a reaction to an obvious implication of Charles Darwin's *On the Origin of Species*, published in 1859. Darwin's work clearly cast doubt on the idea that humans were God's special creation—and therefore the idea that humans had an immortal soul. The hope was that by applying the scientific method to the question of postmortem survival, proof would be forthcoming that the Christian belief in life after death was valid. The obvious place to start was by testing the claims of the numerous mediums of the day.

Initially there was cooperation between the scientists and the mediums, but this waned as the former were keen to apply greater controls in their investigations. After all, the conditions that typically applied during the séances, of dim illumination and a prohibition on sitters moving around, were ideal conditions for the employment of trickery by the medium and any accomplices. As previously stated, it was indeed a rare medium who was not caught red-handed in executing such deception. Many scientists were dismissive from the start and refused to even look at the evidence. Many of their more open-minded colleagues eventually decided that this was not a fruitful line of inquiry and concluded that no genuinely supernatural phenomena were taking place in the séance room. However, a few notable scientists of the day were convinced of the genuineness of the phenomena that they witnessed, including Alfred Russell Wallace, who had independently come up with the idea of evolution by natural selection at about the same time that Darwin did, and Sir William Crookes, the discoverer of thallium.

Early attempts to evaluate the accuracy of mediums were typically not methodologically sound. For example, an investigator who is keen to find evidence in support of life after death may well be inclined to count many more of the mediums' often-ambiguous pronouncements as accurate compared to someone without that motivation. Furthermore, a simple tally of accurate versus inaccurate statements is of very limited value. After all, if I simply came out with a lot of statements that had a very high probability of being true for everyone ("Your grandpa had two eyes, one on either side of his nose"), this would not constitute impressive evidence that I was in direct contact with your dearly departed ancestor, would it?

A better approach is to get the medium to make a number of readings for different individuals without actually meeting them. The sitters can then be presented with the reading that was actually produced specifically for them along with one or more readings produced for other sitters. If the medium's claim to paranormal ability is true, the reading that really was done for that specific sitter should contain more accurate, personal, and specific details than the other readings. This can be evaluated either by getting each sitter to pick the reading that is most applicable to them or by giving an accuracy rating for each reading. The results can then be subjected to statistical testing. Such techniques were first developed and applied in the 1930s.

In 1994, Sybo Schouten published a review of such studies of both mediums and psychics up to that date.[1] It should be noted here that the main difference between mediums and psychics is simply their technique. Whereas mediums claim to obtain information via the spirits of the dead, psychics may claim to use ESP. The readings produced are very similar. However, it should also be noted that in general it would prove impossible to decide whether information produced by a medium had actually been obtained via communication with a spirit as opposed to ESP.

Here is what Schouten concluded from his thorough review:

> The main question asked in most of these studies was whether a significant number of correct statements deviated significantly from chance expectation. Another question, less often addressed, was whether psi ability was necessary

to explain the correct statements. The present study indicates that the number of studies with significant positive results is rather small. Moreover, in most of these, one or more potential sources of error were present that might have influenced the outcome. It seems, therefore, that there is little reason to expect psychics to make correct statements about matters unknown at the time more often than would be expected by chance.[2]

A more recent review by Marco Aurélio Vinhosa Bastos Jr. and colleagues assessed studies of mediums carried out this century.[3] Assessing accuracy, only five studies were considered to have adequately controlled for potential information leakage and to have employed a suitable triple-blind methodology. This methodology ensures that (a) the mediums are blind to the identities of both the sitters and the deceased persons known to the sitters; (b) the researchers are also blind to these aspects; and (c) the sitters do not know which readings were intended for them as opposed to being decoy control readings. Of these five studies, two reported significantly higher accuracy ratings for target ratings compared to decoy ratings and three did not.[4] The two studies reporting significant positive results across three experiments, involving 28 mediums and 102 readings, were all carried out by Julie Beischel and colleagues at the Windbridge Institute in Tucson, Arizona. In contrast to many previous studies, there are no obvious major flaws in the methodology used or the statistical analysis employed. Once again, however, independent replication by other investigators is required before such controversial results are more widely accepted by the scientific community.

COLD READING

The jury may still be out regarding the possibility that mediums and psychics, not to mention other diviners, really are able to tap into information sources in a way that cannot be explained by conventional science. However, there is no doubt at all that many members of the general public are convinced that they can. What possible alternative explanations are there for this?

One possibility is that all of these practitioners are deliberate frauds using deceptive techniques to persuade people to part with their money.

Such techniques do indeed exist, and foremost among them is the practice of *cold reading*. Cold reading can be used to give complete strangers the impression that you know all about them even though you have never met them before. I used to joke with the students on my final-year course on anomalistic psychology that I would introduce them to the art of cold reading so that if they could not get a job once they graduated, they would have something to fall back on.

I will only be able to present a brief summary of cold reading here, but for any readers who would like to learn more (whatever their reasons!), I strongly recommend two publications. The first is a classic article, published in the first volume of *The Zetetic* in 1977 (before it changed its name to the *Skeptical Inquirer*), by Ray Hyman with the straightforward title "'Cold Reading': How to Convince Strangers That You Know All about Them."[5] It provides a great introduction to the topic as well as presenting Hyman's "Rules of the Game." The other is Ian Rowland's book *The Full Facts Book of Cold Reading*, full of useful practical tips.[6]

Strictly speaking, it would be more accurate to describe cold reading as a set of techniques that work together to produce the desired effect. One such technique relies on a phenomenon known to psychologists as the *Barnum effect*. It is also sometimes known as the *Forer effect*, after a classic demonstration of this effect by psychologist Bertram R. Forer in 1949.[7] Forer told his class of thirty-nine psychology students that he would provide each of them with a brief personality assessment based on their responses on a psychology test. In fact, unbeknownst to the students, each of them received identical feedback that was not based in any way on their responses on the test. Here are the statements that the students were given:

1. You have a great need for other people to like and admire you.
2. You have a tendency to be critical of yourself.
3. You have a great deal of unused capacity which you have not turned to your advantage.
4. While you have some personality weaknesses, you are generally able to compensate for them.
5. Your sexual adjustment has presented problems for you.

6. Disciplined and self-controlled outside, you tend to be worrisome and insecure inside.
7. At times you have serious doubts as to whether you have made the right decision or done the right thing.
8. You prefer a certain amount of change and variety and become dissatisfied when hemmed in by restrictions and limitations.
9. You pride yourself as an independent thinker and do not accept others' statements without satisfactory proof.
10. You have found it unwise to be too frank in revealing yourself to others.
11. At times you are extroverted, affable, sociable, while at other times you are introverted, wary, reserved.
12. Some of your aspirations tend to be pretty unrealistic.
13. Security is one of your major goals in life.

The students were impressed with the accuracy of the statements as descriptions of themselves, giving an average accuracy rating of 4.30 on a scale where 0 = very poor and 5 = excellent. The point is, of course, that the statements had deliberately been chosen to be vague and general enough to apply to pretty much everyone while also sounding perceptive about the students' innermost personalities. The Barnum effect has been the subject of a great deal of psychological investigation, not least because it proves that it would be unwise to judge the validity of any kind of psychological test purely on how accurate those tested reported it to be.[8]

Although Forer told his students that the profile was based on their responses on a psychology test, the effect is just as powerful if people are misled into believing that it is based on some form of divination. Indeed, it is worth noting here that Forer mainly used statements taken from an astrology book in the original experiment.

I have often combined those statements into a single paragraph and used them in demonstrations of the Barnum effect in the classroom and occasionally on TV. I have found that, particularly for public demonstrations where someone is asked to read the profile aloud before rating it for accuracy, it is wise to exclude statement no. 5 from the profile. This might indicate that people do not struggle with their sexual adjustment as much nowadays as they did in the 1940s, but I suspect it is more likely that most people would prefer not to endorse this statement in public even if it is true!

It is not that difficult to come up with your own Barnum-type statements. One that I am fond of is "You have a good sense of humor." Of course you do, don't you? Because if other people don't laugh at things that make you laugh, they are the ones with the poor sense of humor, aren't they? Furthermore, the statements need not only apply to your personality. I think it was Richard Wiseman who suggested throwing "I'm getting something about an unfinished book" into a reading. This is a clever one, because if that person happens to actually be writing a book, they will be incredibly impressed. How could you possibly know that? But if they're not, they will think of a book that they are halfway through reading or one that they gave up on. Either way, you can't lose.

The type of Barnum statements described above can be sprinkled liberally throughout any reading to pad it out, but a reading can be made somewhat more specific by taking into account demographic factors such as age and gender. After all, one would be unlikely to give similar readings for, say, a fourteen-year-old girl and an eighty-year-old man. Whereas the former is likely to want information about relationships and exams, the latter is more likely to be concerned about health and finances, for example. Furthermore, you may be able to deduce much from someone's appearance and their accent regarding such factors as socioeconomic class.

There is, of course, much more to cold reading than simply trotting out generalities. Sitters often report being told specific details about themselves that could only, to their minds, be explained in terms of the ability to tap into some paranormal information source. Mediums and psychics are helped by the fact that sitters do not expect them to be 100 percent accurate. Indeed, it would probably arouse suspicion if the reader was too accurate. Given this, it is always worth throwing out a few statements that, although sounding very specific, apply to a surprisingly large percentage of the population.

In 1994, Susan Blackmore published ten statements in a national newspaper, asking readers to indicate which ones were true for them.[9] From a sample of 6,238 respondents, over a quarter endorsed each of the following statements: "I have a scar on my left knee" (33.5 percent), "I have a cat" (28.7

percent), "I own a CD or tape of Handel's *Water Music*" (28.3 percent), "I have been to France in the past year" (27.1 percent), "My back is giving me pain at the moment" (26.9 percent), and "I am one of three children" (26.4 percent). If such statements had been included in a reading back in the 1990s, there is a very good chance that one or more of them would have been endorsed by the sitter. It is likely that believers in the paranormal would obligingly remember the hits and forget the misses. They may even count near misses as hits. I recall that once, when I was appearing on a live radio phone-in with a medium, the caller agreed with the medium's assertion that his father had died from lung cancer. Further probing revealed that the cause of death was actually stomach cancer, but clearly, for this caller, lung cancer was close enough. The idea that believers will tend to recall the hits and forget the misses is certainly plausible, but it would be nice to see it tested empirically. I suspect that skeptics may well show the opposite bias.

Selective recall can also result in readings being recalled as being more specific than they actually were. I once took part in a TV program in which volunteers were sent out to have readings done by different types of diviner. The readings were recorded as they were being delivered, and afterward the clients were asked to assess their reading. One young woman reported being particularly impressed that the psychic who had done her reading had correctly stated that her mother's name was Sheila. No doubt she would subsequently relate this amazing psychic feat to friends and family, with no one able to provide an alternative, nonparanormal explanation. The only problem was that in fact the psychic never said that her mother's name was Sheila. Instead, he came out with many names during her reading, most of which meant nothing to her. At one point, he said, "I'm getting the name Sheila . . ." To the young woman, it was obvious he was referring to her mother, who was indeed named Sheila. But he never actually said, "Your mother's name is Sheila." If her mother had not been called Sheila, she may well have thought of someone else she knew of that name, be it a friend, a neighbor, or whoever. The point is that the cooperative sitter was doing her best to make sense of what the psychic was saying and had put a very specific interpretation on one of the many vague utterances with which she was presented. Subsequently, she recalls that part of the reading as being much more specific than it actually was.

This observation directly inspired a study of the post-event misinformation effect described previously.[10] Krissy Wilson and I presented volunteers with a video of an alleged psychic doing a reading for a client, followed by a postreading interview with the client. Or at least that is what we told our volunteers. In fact, both the reading and the postreading interview were scripted. Furthermore, there were two versions of the postreading interview, with half of the volunteers viewing one version, half viewing the other. During the reading, the "psychic" had said at one point, "I'm getting a name beginning with an 'S' or an 'F' . . . erm, is it? It's Sharon . . . or Shelley . . . or Sandra . . . or Sheila?" The "client" replies, "Yes, yes"; that is, she does not explicitly state during the reading that Sheila is her mother's name.

The two postreading interviews are identical with the exception of a single sentence. In one, the "client" correctly says, "He mentioned the name Sheila, which is my mother's name." In the other, she incorrectly says, "He said my mother's name was Sheila, which it is." Following the video, our volunteers were asked to indicate their level of agreement, on a 7-point scale (where 1 = strongly agree and 7 = strongly disagree), with a number of statements. The crucial one for our purposes was: "The psychic said that the client's mother was called Sheila." Given that this statement is in fact false, higher scores would indicate greater accuracy of recall.

Our results were interesting, albeit not in accordance with our original hypotheses. We had expected to find that the misinformation would distort the memories of all volunteers but that a stronger distorting effect of the misinformation would be found for believers in the paranormal compared to disbelievers, given that the misinformation made the reading seem more impressive than it was, in line with the belief that psychics have genuine paranormal ability. Instead we found that the paranormal believers inaccurately remembered what the "psychic" had said, thus recalling it as being more impressive than it really was, with or without the post-event misinformation. The nonbelievers, in contrast, recalled what the "psychic" had said pretty accurately when no misinformation was presented after the reading. When misinformation was presented, however, their recall became as inaccurate as that of the believers.

There are a couple of common misconceptions about how cold reading works that we should mention here. The first is that it largely depends on body language, also referred to as *nonverbal communication*. The idea is that people unconsciously give away lots of information about themselves through gestures, body position, eye movements, and so on. Such ideas have often been oversold, particularly in books about how to attract potential sexual partners or how to win in negotiations and so on. In fact, although such cues may occasionally provide some clues to enhance a reading, they play a minor role at best.

The second is what we might call the *Sherlock Holmes fallacy*. This is the idea that by applying flawless logic and keen observation one can deduce amazing amounts of specific background information about a person simply by looking at them. In fact, on some occasions, Holmes could allegedly perform such amazing feats before ever seeing the object of his deductions. In *The Hound of the Baskervilles*, he accurately deduces all of the following about Dr. James Mortimer simply by looking at his walking stick: "There emerges a young fellow under thirty, amiable, unambitious, absent-minded, and the possessor of a favourite dog, which I should describe roughly as being larger than a terrier and smaller than a mastiff."[11] Although such feats are hugely entertaining in the context of Holmes's wonderful adventures, they are regrettably purely the stuff of fiction.

On one occasion, I was talked into passing myself off as a psychic on a popular daytime TV show in the United Kingdom (for British readers, I think it was *Richard & Judy*). When the researcher contacted me to take part in a discussion about psychics for the next show, I naturally gave her my spiel about cold reading and how it worked. "Great!" she replied. "Can you come on the show and demonstrate it for us?" I was very reluctant to do so. Just as not every client is impressed with the reading they get from a psychic, not all readings based on cold reading impress the person they are done for. But the researcher was persistent, and eventually I agreed to give it a go.

When the day arrived for me to go on the program, I was genuinely nervous. The idea was that I would be introduced to the volunteer sitter as a genuine medium, I would do a reading for her, and she would then comment on how impressive (or not) she had found the reading to be. This would all be prerecorded, and an edited version would be included as part of the live show. Even so, if my reading was a total flop, there would be no way to pretend it wasn't.

One of Ray Hyman's "Rules of the Game" is "Get the client's cooperation in advance." I decided to use my genuine nervousness to my advantage, explaining to the volunteer sitter that I was very nervous because this was the first time I had ever done a reading for TV. I also explained that I usually spend at least an hour with each client, but, as we only had ten minutes, it would really speed things up if she could tell me in advance if there was anyone in particular that she would like me to try to contact. So, I knew before I even started that granddad was "in spirit," as we mediums say.

Fortunately for me, the reading went well. I knew from the many interactions between psychics and sitters that I had analyzed over the years that psychics tend to ask an awful lot of questions during a reading. When you think about it, this is a little bit odd given that they are supposed to be telling you stuff. But the thing is, they often do it in a rather clever way that makes it look like they are only asking for confirmation of something they already know. I decided to use this ploy during my reading. "Was your granddad a tidy man?" I asked. "Oh yes," came the reply. "I thought so," I said. "Everything had its place for your granddad, didn't it?" Of course, if she had replied, "Oh no," my response would have been, "I thought not. He never used to put things away after himself, did he?" My sitter was duly impressed.

I followed up with "And what about yourself? Are you tidy?" Once the sitter is convinced that you can read them like a book, they will often open up completely, as in this case. "Oh no," she replied, "I'm terrible!" "I thought so," I said again. "Your granddad's saying that you're always so messy. He's saying it with love and a smile." I went further. Following the advice in my recently reread copy of Ian Rowland's book, I told the sitter that I had a mental image of her room and that I could see stacks of old photos that had not yet been put into albums. The truth is, of course, that before digital cameras became the norm, virtually every home would have such stacks of photographs—but this trivial detail really blew the socks off my sitter!

Now, as you might imagine, I felt a bit mean deceiving my volunteer in this way. The point was certainly not to make her look stupid or gullible but purely to demonstrate how effective the technique can be. Once the reading was over and her glowing review of my paranormal abilities had been recorded, she was informed that I did not really claim to have psychic powers and that I was using a technique called cold reading. I apologized for the deception and explained that we were not out to embarrass her in any way. She was given the option of vetoing the reading before it was broadcast if she wanted to. Fortunately, she was happy for it to be broadcast.

Let me make it very clear at this point that I do not believe for one second that all people claiming to be mediums or psychics—or indeed any other kind of diviner—are deliberate con artists using cold reading and other deceptive techniques to fleece their gullible victims of their hard-earned cash. My personal opinion, for what it is worth, is that most people who claim to have paranormal abilities genuinely believe that they do possess such abilities. They are fooling themselves as much as they are fooling others. There is also good evidence that many Spiritualist mediums really are hearing voices that they attribute to spirits of the dead. For example, Spiritualist mediums have higher levels of both absorption and proneness to auditory hallucinations compared to the general population.[12] Research suggests that for many such individuals the belief system of the Spiritualists provides an explanation for anomalous experiences that might otherwise be distressing. Interpreting the hallucinated voices as a gift rather than as a potential sign of mental illness allows the mediums to live happy and productive lives.

Having said that, it is equally clear that some of those who claim paranormal abilities are deliberate fraudsters. At this point, I would like (with tongue firmly in cheek) to propose French's Second Law: "The higher the profile of the psychic claimant, the more likely it is that they knowingly use fraudulent techniques." Sincere but deluded low-profile claimants are typically unable to consistently deliver impressive performances that would convince the majority of their clients that they genuinely possess paranormal powers. They are, however, able to deliver a sufficient number of performances that allow them to just about earn a living (or at least a bit of extra pocket money) that way. The high-profile psychic or medium, in contrast, has to consistently deliver impressive performances on stage or in front of TV cameras. The pressure to resort to trickery is high.

The deliberately fraudulent psychic is unlikely to rely solely on cold reading to practice their dark art. While cold reading can sometimes produce amazingly impressive results, it can also sometimes fail to do so. To guarantee impressive readings, it is best to also use *hot reading* techniques. Whereas cold reading is the only option for someone who is genuinely a complete stranger, hot reading requires gathering information about your client in advance.

There are numerous ways to achieve this, but there is no doubt that the widespread sharing of personal information via social media has made the fake psychic's task a lot easier. Another technique widely employed during stage performances is for the fake psychic to have confederates queue up with the genuine audience members as they are waiting to go in, engage them in friendly conversation, and surreptitiously feed any information gleaned back to the psychic before the show begins.

This tendency of fraudulent psychics use information fed to them in advance can sometimes be used against them by skeptics determined to reveal their trickery. For example, psychologist Ciarán O'Keeffe exposed the trickery of the late Derek Acorah by deliberately feeding him false information on numerous occasions.[13] O'Keeffe worked as the resident skeptic on the popular ghost-hunting British TV show, *Most Haunted*. Acorah was the medium who would frequently appear to become possessed by the spirits at the various haunted locations featured. O'Keeffe suspected that the information apparently being psychically channeled by Acorah had in fact been gathered in advance. He proved his point by feeding the medium information about fictional characters to see if Acorah would become "possessed" by them. He did. These characters included a South African jailer named Kreed Kafer (an anagram of "Faker Derek") and highwayman Rik Eedles (an anagram of "Derek lies"). On a shoot at Craigievar Castle, near Aberdeen, O'Keeffe "made up stories about Richard the Lionheart, a witch, and Richard's apparition appearing to walk through a wardrobe—the lion, the witch and the wardrobe!"[14] Sure enough, Acorah repeated all of these stories. In fact, Richard I had reigned 500 years before the castle was even built.

OUIJA BOARDS AND TABLE-TILTING

Of course, you don't necessarily need a medium to communicate with the dead. Several decades ago, when I was a final-year undergraduate at the University of Manchester, I lived in a house with five other male students. One of our favorite things to do on returning from the pub on a Friday night was to set up a Ouija board and converse with the spirits of the dead. Well, maybe that's not strictly accurate. For one thing, only one of us really believed that these sessions might have anything to do with discarnate spirits, and he steadfastly refused to take part. He would scuttle off to his room and we would not see him again until the next morning.

The second inaccuracy is that we did not have a proper Ouija board.[15] Instead, we used the approach, beloved of curious teenagers around the globe, of using an upturned wineglass resting on a smooth table. Before the session, we would write all of the letters of the alphabet plus the numbers from zero to nine on scraps of paper and arrange them in a circle on the table. We would then place the upturned wineglass in the center of the circle, and each of us would place one finger on the base of the glass (which now faced upward). Then one of us would solemnly ask, "Is there anybody there?"

Typically, it would take a while for a session to really get going. Initially, the glass, with our fingers still touching it, would appear to move hesitantly, spelling out short answers to our questions, but after a while the movements appeared to become stronger, almost as if "the spirits" were becoming more familiar with the positions of each letter on the board. During a really good session, the glass would whizz from letter to letter, spelling out messages from the Great Beyond.

Now, the obvious nonparanormal explanation for what was going on here is simply that one of us was knowingly pushing the glass around in response to the questions asked of the "spirits." However, it was noticeable that, during a long session, each of us would on occasion take our fingers off the glass and just sit back and watch, as if to say, "Well, that proves that I am not pushing it." I cannot really remember now, but I do not think I had a good explanation for what was going on, but I do know that, most of the time at least, I did not think it was communication from beyond the grave.[16] We engaged in this activity for entertainment purposes only, often with weird and hilarious results. For example, on one memorable occasion, it looked like contact had been made with "King Michael of Denmark." We asked him how he died. "Love of toast," came the reply. Confused, we asked for clarification. The glass slowly spelled out, "Poison in the marmalade."

In fact, the movement of the glass around the table is best explained in terms of the *ideomotor effect*. First used in 1852 by William Benjamin Carpenter, the term is defined in the *Skeptic's Dictionary* as "the influence of suggestion or expectation on involuntary and unconscious motor behavior."[17] In other words, we really were pushing the glass around the table, we just were not aware that we doing so. It follows, if the explanation in terms of the ideomotor effect is correct, that no meaningful messages would be produced if all sitters were blindfolded—and that is precisely what is found.[18]

This is the explanation favored by most scientists and virtually all skeptics, who would view the use of Ouija boards as nothing more than harmless fun. But what about claims from fundamentalist Christians (and some self-styled paranormal investigators) that the use of Ouija boards is dangerous and can open the door to powerful and evil supernatural forces? It is easy to scoff at such notions, and, in my personal opinion, fears of that sort are generally unfounded—but not always. If someone is psychologically vulnerable to begin with, messages spelled out by the Ouija board could potentially feed in to other delusional beliefs with disastrous results.

The ideomotor effect provides an explanation for another allegedly paranormal phenomenon, that of table-tilting (also known as table-turning, table-tapping, and table-tipping; figure 4.1). Table-tilting was a craze that caught on in America and Europe at the height of the popularity of séances, back in the Victorian era. It involved having sitters place their hands on a small round wooden table as an alleged means to communicate with spirits. Questions would be put to the spirits, who would respond by making the table move. The movements might consist of nothing more than simple shuddering, but a really good session might result in the table moving quickly around the room.

Table-tilting has a special place in the history of anomalistic psychology because it caught the attention of the great English physicist Michael Faraday

Figure 4.1
Communicating with the spirits through table-tilting could occasionally get out of hand.

PROFESSOR FARADAY'S LECTURE AT THE ROYAL INSTITUTION.

Figure 4.2
The great English scientist Michael Faraday, pictured here giving a lecture at the Royal Institution in London, carried out investigations into the phenomenon of table-tilting.

(figure 4.2). In 1852, Faraday, with admirable open-mindedness, was so intrigued by the phenomenon that he designed a number of ingenious tests to investigate whether the table really was being moved by some mysterious external force or whether, in fact, those taking part were pushing the table without being aware of it. You can guess what he found.

ELECTRONIC VOICE PHENOMENON

One other means of spirit communication that is particularly popular with many self-styled paranormal investigators is known as the *electronic voice phenomenon* (EVP). This technique was discovered, if that is the right word, by Swedish artist Friedrich Jürgenson in 1957. He claimed that other human voices were to be found in a recording he had made of his own voice and then, a couple of years later, in a recording he had made of birds singing. He believed these voices to be those of alien life-forms as well as of his deceased mother. However, it was the Latvian psychologist Konstantin Raudive who brought the phenomenon to the attention of a wider public with the publication of his book *Breakthrough* in 1971.[19]

The basic technique, as currently employed by numerous amateur paranormal investigation groups, is to leave a recording device in record mode in a reputedly haunted location. Typically, this device will be left running for several hours in, for example, an empty room. It is claimed that when the recording is listened to carefully, voices can be heard. These voices are usually interpreted as being produced by spirits of the dead, but some EVP enthusiasts also entertain the idea that the voices may be those of aliens or even beings from other dimensions.

There is also much variation in the details of how the technique is employed. The recording device is not always left in an empty room. Instead, some investigators prefer to direct questions toward the spirits and then leave a gap of silence in the hope that the response will be recorded. Although no reply is typically heard by those present at the time, it is claimed that clear and coherent replies can be heard when the recording is played back. In the days prior to digital technology, it was claimed that EVP could be recorded from radios not tuned to a particular channel.

Paranormal investigators typically claim that the messages received are very clear and meaningful. There are dozens of websites where you can listen to these messages for yourself, as well as many YouTube videos featuring EVP from both popular paranormal TV series and amateur groups. On the websites, the message is typically displayed on the screen as the audio recording is played. When investigators report their findings, they typically tell the viewer what they think the message is before they play the EVP and often, just for good measure, display it in writing on the screen as the audio clip is repeatedly played.

How can these voice-like sounds be explained if, in fact, they are not being produced by discarnate entities? I think a clue lies in the fact that the recordings fall into two fairly distinct groups. The first set consists of pretty clear recordings of easily discernible messages from what sound like human voices. So the most obvious explanation is that these sounds are, in fact, the voices of living human beings that have been recorded inadvertently. Despite the best efforts of the investigators, it is inevitable that the recording device will occasionally pick up snippets of speech from people who just happen to be within auditory range. Other possibilities, depending on the specific EVP method used (and in which era), include interference from broadcasts and artifacts arising from the technology used (e.g., incomplete erasure of previous recordings).

The second group of EVP clips is much less clear. They are typically very short, poor-quality recordings, often with a lot of background hiss. In fact, it is usually not possible to make out the message unless you read it for yourself or someone tells you. The interpretations given to these vague and ambiguous sounds vary considerably from one person to another. When I discuss EVP in public talks, I like to play examples to the audience without telling them what the message is and ask them to guess. Audience members typically have no idea what the message is. When I tell them what it is supposed to be and play it again, many of them agree that they can now "kind of hear it." Sometimes I will throw in a trial where I get them to hear exactly the same audio clip in different ways simply by priming them with different messages having a generally similar vowel structure.

Michael Nees and Charlotte Phillips played samples of EVP, actual speech, acoustic noise, and degraded speech to participants who were told either that the experiment was about paranormal EVP or that it was about speech intelligibility without any mention of the paranormal.[20] Interestingly, despite the participants generally having quite low levels of belief in the paranormal, they indicated that they could detect more voices in both the EVP and degraded speech trials in the paranormal context. However, when a voice was reported, there was very little agreement between the participants on trials regarding what the voice actually said. As noted by many commentators,

EVP samples appear to be nothing more than auditory pareidolia arising from speech-like background noises.[21]

Some paranormal investigators interpret some of the sounds they hear as nonspeech sounds, such as coughing or even animal noises. One paranormal investigator I know of claimed to have recorded a ghost horse neighing. It had been recorded in a part of a building that used to be a stable. The noise did not sound like a horse to me, but I am no expert so I asked my wife, Anne, and one of my daughters, Alice, to listen to the short clip and tell me what they thought it was. Both were keen horse-riders at the time. Neither of them said it was a horse neighing. In fact, when I suggested that that is what someone thought it was, they dismissed the idea outright.

On another memorable occasion, I was taking part in a TV series called *Haunted Homes* when apparently a ghostly sneeze was recorded on an EVP recorder belonging to our paranormal investigator, Mark Webb. This took place in a reputedly haunted radio station, Radio Beacon, in the United Kingdom. The recording caused great excitement, as a ghostly sneeze had been reported on many previous occasions in the building. As this was the only time in two series that anything objective had been recorded during one of our investigations, the program makers were keen to make the most of this amazing piece of hard evidence.

I did an interview to camera giving my reaction: "In the case of the sneezing, which Mark actually thought he heard himself last night, he's played that recording back to me, and I have to say, to me it sounds like something that might be a sneeze but it might be a hundred and one other things as well. And this is a general problem with the EVP, the electronic voice phenomenon, that it's very, very easy for people to read into those very ambiguous sounds whatever it is that they think they are supposed to be hearing."

On our second night at that location, I decided to pop to the toilet before filming began on the first floor, near to where the ghostly sneeze had been recorded. As I emerged from the cubicle, Mark was there with a somewhat disgruntled look on his face, pointing at the tiled wall. As I looked toward where he was pointing, the mystery was solved—there on the wall was an automatic air-freshener. Within a couple of minutes, sure enough, it operated, making a noise that sounded just like a sneeze. Mark and I both did interviews to camera the next morning, explaining that we now knew what the noise was, and it was definitely not a ghost. When the program was actually broadcast, a lot of coverage was devoted to the ghostly sneeze. Sadly, however, the editors did not appear to have enough time to include our explanation in the program.

BEFORE WE MOVE ON . . .

Many years ago, I was lying on the bottom bunk bed with my daughter Kat, about six years old at the time, lying above me in the top bunk. (I can't remember the details, but I assume her younger sister, Alice, must have been poorly and sleeping with her mum.) All of a sudden, Kat let out a heart-wrenching sob. Panicked, I shot out of bed and asked her what was wrong. "I don't want to die," she wailed.

Trying my best to be reassuring, I said, "You don't need to worry about that. You're young, you've got your whole life ahead of you."

"Yes," she replied. "But I will die, won't I?"

Desperate, I found myself saying, "Well, some people think that when you die you go to heaven . . ."

Interrupting me, she tearfully said, "Yes, but you don't, do you?"

Sometimes it's hard to be an atheist.

I have mainly concentrated in the last two chapters on anomalous experiences such as sleep paralysis and various cognitive biases that may be associated with belief in and experience of ghosts. But we should not neglect emotional and motivational factors. Very few people can honestly claim not to be frightened by the idea of their own mortality—or, perhaps even more so, the mortality of their loved ones. Most of us want to believe in some form of life after death, an idea that is an integral part of the world's major religions. Most of us would like the death of our physical bodies to not mean the end of our subjective consciousness. If there is an afterlife, then there must be something more to us than brains encased in meat machines—each of us must have a soul or a spirit, call it what you will.

However, if that it is indeed the case, it opens up the possibility that sometimes those souls may on occasion fail to move on to heaven (or wherever else they are supposed to go) and instead hang around on the earthly plane. Thus, even though many of us might find the idea of ghosts scary, there is no denying that evidence of ghosts supports the idea of an afterlife.

Confirmation bias is a ubiquitous cognitive bias. It affects everyone. It refers to the fact that we are all more impressed by evidence that appears to support what we would like to be true or that we already believe to be true. We tend to notice more of such evidence compared to evidence that contradicts our belief. We find it more compelling. We think of reasons why the contradictory evidence can be ignored.

Given the strength of our primordial existential fear regarding our own mortality, it is no surprise that so many people find even the weakest evidence for life after death to be so convincing.

5 FANTASTIC MEMORIES OF ALIEN ENCOUNTERS

Surprisingly, I do not recall ever thinking as a child, "I want to be an anomalistic psychologist when I grow up." Neither did I want to be a train driver, a footballer, a cowboy, or a pirate. I wanted to be an astronomer. Just thinking about the vast scale of the cosmos blew my tiny mind. I was particularly enthralled by those "if the Earth was the size of a pea" comparisons.[1]

If the Earth was the size of a pea (0.5 cm), the Sun would be the size of a large inflatable beach ball almost 200 feet away. Pluto (which back then was definitely a planet!) would be less than a 0.1 centimeters in diameter, and its orbit would vary between 1 and 1.8 miles from the Sun. And that is just our own puny solar system!

Excluding the Sun, the nearest star to the Earth is Proxima Centauri, which is about 4.25 light-years away. A light-year is the distance that light travels in one year at a mind-boggling speed of 186,000 miles per second (or, if you prefer, 299,792,458 meters per second). As this makes clear, a light-year is a very, very, very long way. If the Earth was the size of a pea located in central London, the red dwarf star Proxima Centauri would be three inches in diameter and 9,800 miles away somewhere in Southwest Australia.

In a probably forlorn attempt to keep things even remotely comprehensible when it comes to galactic and intergalactic distances, let us rescale our model. Imagine now that it is the Sun that is the size of a pea (with the Earth—including almost everything and everyone you have ever known—a correspondingly tiny twentieth of a millimeter in diameter). On this scale, Proxima is a mere ninety miles away. Even on this scale, our galaxy, the Milky Way, is 2.1 million miles in diameter (as opposed to the 100,000 light-years that it really is), and the nearest major galaxy to our own is some fifty-three million miles away. I think we can all agree with Douglas Adams's immortal words: "Space is big. You just won't believe how vastly, hugely, mind-bogglingly big it is. I mean, you may think it's a long way down the road to the chemist's, but that's just peanuts to space."[2]

Our Sun is one of over a hundred billion stars in the Milky Way, and our galaxy is but one of billions and billions of galaxies in the universe as a whole. Inevitably, the question arises: Are we alone in this vast universe or has life—possibly intelligent life—arisen elsewhere? This question has been considered by curious minds for centuries.[3] One notable attempt to address this issue was the development of an equation by Dr. Frank Drake in 1961. The Drake equation, as it is known, attempts to estimate the number of advanced civilizations in our own galaxy by considering such factors as the rate of star creation, the fraction of those stars that have planets, the average number of planets per star that might support life, and so on.

The Drake equation was originally put forward simply to provide a framework for thinking about the possibility of intelligent life arising elsewhere in the Milky Way. The range of possible answers generated varied enormously, depending on the estimates used, but it is fair to say that some quite reasonable guesstimates suggested that our galaxy could be teeming with intelligent life. If that is the case, the question posed by physicist Enrico Fermi naturally arises: Where is everybody? To date, conclusive evidence that life has evolved anywhere other than on our planet is lacking according to most scientists.

Many attempts have been made to resolve what has become known as Fermi's paradox. One obvious possibility is that intelligent life is, for reasons that we do not yet understand, extremely rare. It is even possible that we really are alone in the universe. Another possibility is that advanced civilizations arise but do not survive long enough to perfect interstellar travel. But the answer preferred by many millions of people is that conventional scientists have simply got it wrong. Not only, they would argue, is there plenty of evidence for the existence of extraterrestrial life, there is evidence that such life-forms are regular visitors to Earth. In this chapter, we will take a closer look at such claims and the psychological factors that may underlie them.

LEVELS OF BELIEF IN EXTRATERRESTRIALS

Opinion polls regularly reveal relatively high levels of belief in the existence of aliens and alien visitation to our planet. For example, a 2019 Gallup telephone survey of a random sample of 1,522 American adults indicated that a third of respondents believe that some UFO sightings are alien spacecraft visiting the Earth, and 16 percent said that they had themselves witnessed a UFO.[4]

Similarly, around the same time a poll by Ipsos of over a thousand American adults showed that over half (52 percent) believed that aliens exist and over a quarter (29 percent) believed they visit the Earth.[5] Most respondents (88 percent) had at least heard of Area 51, a top-secret US Air Force base in Nevada, with over half claiming to be "somewhat" or "very familiar" with it. Of those who had at least heard of Area 51, 26 percent claimed that crashed UFOs are held there, and 21 percent believe that aliens (alive or dead) and alien technology are held there. A YouGov poll revealed similar levels of belief in extraterrestrial life in the United States (54 percent), Germany (56 percent), and the United Kingdom (52 percent).[6]

Many factors contribute to the belief that aliens have visited the Earth in the past and continue to do so. One such factor is the widespread, and often uncritical, media coverage of certain classic accounts of UFO encounters that have survived many attempts at debunking by skeptics, including two of the most famous UFO cases ever: Roswell and Rendlesham Forest.

THE ROSWELL INCIDENT

The term *flying saucer* entered the English language as a consequence of press reports of Kenneth Arnold's sighting of what he believed were technologically advanced aircraft flying near Mount Rainier in Washington state on June 24, 1947. A couple of weeks later, on July 9, 1947, the *Roswell Daily Record* carried a memorable headline: "RAAF Captures Flying Saucer on Ranch in Roswell Region."

Some days before, rancher William "Mac" Brazel had found some strange debris in a field near Roswell, New Mexico. The debris consisted of tinfoil, rubber strips, sticks, and paper. The following day, on learning of Arnold's report of strange craft, Brazel wondered if that was what he had found in the field. He reported it to Sheriff Wilcox, who in turn reported it to Roswell Army Air Field's Major Jesse Marcel. Following the recovery of more debris, the RAAF issued a press release stating that a "flying disc" had been recovered. Needless to say, the story caused great excitement. A few days later, however, after the debris had been examined at Fort Worth Army Airfield by General Roger Ramey, the military announced to a disappointed world that the debris was, in fact, nothing more than the remains of a crashed weather balloon. Photographs of the foil-like material were enough to convince most people that this was indeed the correct explanation, and that appeared to be the end of the story (figure 5.1).

Figure 5.1
The debris from an alleged flying saucer that crashed in Roswell, New Mexico, in 1947, consisting of tinfoil, rubber strips, sticks, and paper.

Some three decades later, however, interest in the story was rekindled as various ufologists, most notably physicist Stanton Friedman, reinvestigated the case. This resulted in numerous bestselling books with a seemingly endless series of new witnesses coming forward for each one.[7] Rather than clarifying the situation, the resulting contradictory accounts only muddied the water. The central claim, that the debris was the remains of a crashed alien craft, was arguably one of the few things that commentators agreed on. But there was disagreement over how many UFOs had crashed, where they had crashed, the number of aliens, whether any of the aliens had survived the crash, and so on. To any readers interested in critically assessing the conflicting claims behind this modern myth, I recommend two books published in 1997, the fiftieth anniversary of Roswell, one by Philip J. Klass and the other by Kal K. Korff.[8]

So, three-quarters of a century later, are we finally able to provide a definitive explanation for that strange debris found in a remote field thirty miles northwest of Roswell? Strict application of Occam's razor would suggest that the original description and photographs of the debris are more consistent with the idea that it came from an ordinary weather balloon as opposed to an interstellar alien spaceship.[9] In this particular instance, however, Occam's razor may have led us slightly astray, as the balloon in question was almost certainly far from "ordinary." According to a report released by the US Air Force in 1994, the debris was the remains of a high-altitude balloon being used in a top-secret project known as Project Mogul. This involved sending microphones high up into the atmosphere with the aim of detecting sound waves produced by Soviet atom bomb tests. Needless to say, ufologists have rejected the report as being yet another cover-up.

THE RENDLESHAM FOREST INCIDENT

As with the Roswell incident, it is probably wise to place more trust in the original reports of what happened in Rendlesham Forest in the United Kingdom in late December 1980 than to determine which of the mutually contradictory "eyewitness" reports that appeared later have the most credibility. At first glance, this appears to be a very strong contender as potential proof for a visit by extraterrestrials, including written statements from multiple military witnesses, one of which is an official memo from a senior officer in the US Air Force, and actual physical evidence of a UFO landing in the forest. It took a while for this case to become widely known, but on October 2, 1983, the story was reported by a British newspaper, the *News of the World*, under the striking headline, "UFO lands in Suffolk. And that's official!" Since then, it has been the subject of countless books, documentaries, and articles.

There are too many twists and turns in this tale to go into all of the details, but here are the basic facts. The events took place in or close to Royal Air Force (RAF) Woodbridge and RAF Bentwater, bases that at the time were being used by the US Air Force. In the early hours of December 26, 1980, personnel at RAF Woodbridge spotted strange lights apparently descending into Rendlesham Forest. Suspecting that an aircraft may have crashed in the forest, Sergeant Jim Penniston led a security patrol to investigate. According to the official memo to the UK Ministry of Defence written by deputy base commander Lieutenant Colonel Charles Halt (dated January 13, 1981, and wrongly dating the events themselves to December 27, 1980):

> The individuals reported seeing a strange glowing object in the forest. The object was described as being metallic in appearance and triangular in shape, approximately two to three meters across the base and approximately two meters high. It illuminated the entire forest with a white light. The object itself had a pulsing red light on top and a bank(s) of blue lights underneath. The object was hovering or on legs. As the patrolmen approached the object, it maneuvered through the trees and disappeared.[10]

Subsequently, Penniston claimed to have witnessed a "craft of unknown origin" and even to have touched it. This is not corroborated by the other witnesses who were present at the time. Later that day, the servicemen returned to the area and located three small impressions in a triangular pattern on the ground that they assumed to be landing marks, as well as broken branches and burn marks on the trees nearby.

In the early hours of December 28, 1980, Halt and several other servicemen took radiation readings from these depressions. It was subsequently claimed that these readings were higher than the expected readings for background radiation, thus proving that something extraordinary had taken place. Halt made a real-time recording of this investigation on a microcassette recorder. During the investigation, the UFO apparently returned in the form of a moving, pulsing red light. Subsequently, three "star-like objects" were seen in the sky.

Thanks to the work of a number of investigators, most notably writer and broadcaster Ian Ridpath, plausible explanations exist for each and every element of this complex case.[11] The strange lights that appeared to be descending into Rendlesham Forest in the early hours of December 26 were almost certainly a bright meteor that was spotted by several other independent witnesses that night over south England. As I was writing this chapter, a bright meteor was seen by several observers in England and caught on several doorbell cameras.[12] Looking at the footage obtained, it is easy to see how such an event could be interpreted as a crashing aircraft. The flashing light reported on both nights was almost certainly Orford Ness lighthouse, as first suggested by local forester Vince Thurkettle. He also had an explanation for the "landing marks" reported by the servicemen—they were in all likelihood spots where rabbits had been digging. But what about the excessive levels of radiation found at those spots? There weren't any. The manufacturer of the instrument used to take the readings confirmed that they were "of little or no significance." The burn marks on the trees? They were in fact marks made with an axe to mark some of the trees for felling. Given the appearance and location of the "star-like objects" reported by Halt, they were probably simply actual stars.

CLOSE ENCOUNTERS OF VARIOUS KINDS

Looking back, it strikes me as odd that it never occurred to me to wonder what the title of Steven Spielberg's classic movie *Close Encounters of the Third Kind* actually meant when I first saw it back in the late 1970s. It was not until many years later that I discovered that the title was based on the classification of UFO encounters put forward by J. Allen Hynek (figure 5.2). Hynek was an astronomer who had been an advisor to the US Air Force on

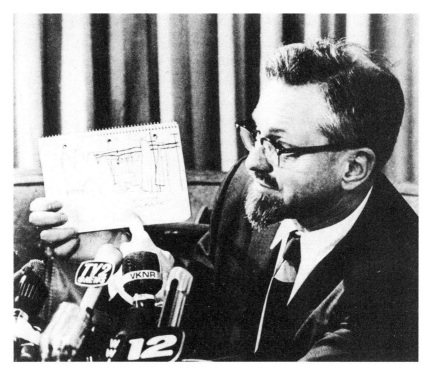

Figure 5.2
The original classification scheme for UFOs in terms of "close encounters" was proposed by J. Allen Hynek.

two projects undertaken to investigate the nature of UFO sightings: Project Sign (1947–1949) and Project Blue Book (1952–1969). Famously, Hynek began as a skeptic of the *ET hypothesis* (that is, the idea that UFOs were extraterrestrial in origin) but ended up defending both that hypothesis and the even more controversial idea that UFOs may be evidence of beings from other dimensions. He was hired as a consultant on Spielberg's film and even puts in a cameo appearance at the end of the film as the aliens disembark from the mother ship.

Records dating back literally thousands of years show that throughout history people have sometimes seen things in the sky that they could not identify.[13] Hynek devised a sixfold classification system to categorize different types of sighting and close encounters. The first category is *nocturnal lights*, the relatively common phenomenon of seeing lights in the night sky and being unable to identify them. Of course, if people only ever used the abbreviation "UFO" to refer simply to an *unidentified* flying object, there would be no problem. But these days many people automatically equate "UFO" with "ET." It has not always been the case that people made this huge inferential leap so readily. Indeed, in the very first survey of the public's opinion on what UFOs might be, released by George Gallup in August 1947, the option of "extraterrestrial craft" was not even included (and the most popular response, at 33 percent, was a refreshingly honest "No answer, don't know"). A second Gallup survey in 1950 only included the possibility of extraterrestrial crafts within a more general heading of "Comets, shooting stars, something from another planet"—and a mere 5 percent of respondents opted for this as their favored explanation.[14]

Ufologists agree that at least 95 percent of such sightings can be explained in mundane terms. Common causes for reports of UFOs include aircraft seen from unusual angles, bright celestial objects (especially the planet Venus), meteors, and laser displays. Such causes can often be identified as a result of proper investigation if the exact time, location, and direction of the sighting are available. Is there any good reason to assume that the few cases that cannot be readily explained are of extraterrestrial origin? No, there is not. Sometimes there will simply not be sufficient evidence available after the event upon which to base an explanation.

The constructive nature of perception and memory has already been discussed. Perception of a stimulus is greatly influenced by what an individual believes they are looking at, especially under the less than perfect viewing conditions that are typical of most UFO reports. For example, Allan Hendry, who at the time was the managing editor of the *International UFO Reporter*, analyzed reports that were known to have arisen as a result of the misidentification of aircraft towing advertising displays.[15] He concluded:

> In the three hundred calls that IUR has dealt with that were based on confirmed ad planes at night, 90 percent of the witnesses described *not* what was perceptually available, but rather that they could see a disc-shaped form *rotating* with 'fixed' lights; many of these people imagine that they see a dome on top and, when pressed, will swear that they can make out the outline with confidence.

Hynek's second category is *daylight discs*. Although most of the reports of unusual flying objects in this category are indeed disc-shaped, it is also used to include reports of daytime sightings of crafts of other shapes, such as cigar-shaped, triangular, or spherical. As already mentioned, the term *flying saucer* first entered the English language in 1947 following Kenneth Arnold's report of seeing strange craft flying at amazing speeds near Mount Rainier. However, it should be noted that Arnold did *not* describe the craft he saw as being saucer-shaped. He described them as being shaped more like boomerangs (figure 5.3). When he referred to saucers, he was describing the motion of the craft as being like a saucer skimming across water. But the media loved the expression "flying saucer" and so that became the common phrase. Lo and behold, in another wonderful example of the influence of top-down processing on perception, most subsequent sightings were indeed described as being saucer-shaped!

Hynek's third category was the relatively rare *radar-visual* type. This refers to a situation where a sighting is reported along with a corresponding radar reading. Interestingly, Hynek rejected radar-only reports in recognition of the fact that "clutter" could often appear on radar screens for a variety of reasons. Such sightings became rarer as radar technology improved.

Close encounters of the first kind are visual sightings of UFOs seemingly less than 500 feet away. There is an immediate and obvious problem with this category. The conditions under which UFOs are typically observed are far from perfect. There are few, if any, cues to size, speed, and distance of objects seen in the sky. The image on the retina of a small object near to the observer moving relatively slowly is identical to that of a much bigger version of the same object much farther away, moving at speed. This does not seem to prevent people from giving very confident estimates of the size, speed, and distance of UFOs that they spot. Ufologists seem to believe that certain professional groups, such as pilots, military personnel, police officers, and astronomers, are incapable of making mistakes in estimating such factors, despite evidence that conclusively proves this assumption to be wrong.[16]

Close encounters of the second kind are those that involve physical effects on animate and inanimate objects. The former might include people experiencing paralysis or animals being frightened. The latter include electronic

Figure 5.3
Although Kenneth Arnold's report in 1947 of seeing strange craft moving at incredible speeds led to the term *flying saucer*, he described craft shaped more like boomerangs.

equipment malfunctioning, cars stalling, and physical traces, including impressions in the ground or scorch marks where a craft has allegedly landed. These phenomena are, of course, open to alternative explanations as shown in the Rendlesham case.

Photographic and video evidence of UFOs would clearly fall into this category. The history of photography has from its outset been intertwined with the history of paranormal investigation based on the dubious assumption that "the camera never lies."[17] This was never true, of course, as proven by the number of fraudulent "spirit photographers" back in the Victorian era. Nowadays, readily available inexpensive software allows hoaxers to produce superficially convincing photographs and videos of a range of paranormal phenomena, including UFOs, with relative ease. It makes one nostalgic for the good old days when the hoaxers had to use other techniques, ranging from Frisbees to realistic-looking models suspended on string, to achieve their effects.[18]

Of course, not all UFO photographs and videos are deliberate hoaxes. In fact, I suspect that the vast majority are not. There are several means by which someone could sincerely believe that they have captured genuine evidence of extraterrestrial visitation. Several common causes of misinterpretation have already been described. In the absence of any serious attempt to investigate further and consider all possible alternative explanations, anyone capturing such phenomena on camera may remain convinced that they have recorded evidence of a close encounter with extraterrestrials.

Another possibility is that the "UFO" in a photograph was not noticed at the time the photograph was taken but spotted later when it was being examined. This should not come as a surprise, given that research into inattentional blindness shows conclusively that we often fail to notice other stimuli in the vicinity if we are concentrating on another task. Given that a static photograph captures the scene at a particular instant in time, this allows for many additional possible misinterpretations of natural stimuli over and above those already mentioned. For example, you may capture an image of a bird flying in the background, seen from an unusual angle. If your attention had been drawn to the bird at the time the photograph was taken, you would have easily recognized it, but that may be far from obvious in a still photograph.[19]

As I was in the final stages of writing this book, I inadvertently took one such photograph myself. Our house is opposite the Thames in Greenwich, England, and we are often lucky enough to witness fantastic sunsets over the river that result in the sky appearing to be filled with dramatic hues of orange, red, yellow, pink, blue, and purple. I often capture such moments by opening the window and taking a picture on my mobile phone. On one recent occasion, I decided not to open the window as it was too cold (see figure 5.4). Sadly, the monochrome version of the photograph reproduced in this book does not show the wonderful colors of the sky that evening—but it does appear to show what might be a large alien mothership hovering ominously over the Thames in the top right corner. I will leave it to the reader to figure out what it actually shows.

Given that these days virtually everyone has a good-quality camera and video recorder in their pocket, not to mention the CCTV cameras everywhere, it is odd that we do not have thousands of crystal-clear images of UFOs, ghosts, and cryptids if these phenomena genuinely exist. This was brought home to me about ten years as I crawled along in very slow-moving traffic on my way home from Goldsmiths one day. I did a double-take at the very unusual sight in front of me. There, on the back of the truck in front of me, was a large terra-cotta head of Queen Elizabeth II with flowers for hair (see plate 5). A similar representation of her husband, Prince Philip, was also being transported. Presumably they were something to do with the Queen's Diamond Jubilee celebrations, but they still made for a somewhat surreal sight. What struck me forcibly was the reaction of the numerous pedestrians as the truck passed slowly by. Almost without exception, they all had their phones out taking photographs and videos of the heads. I am sure the reaction would be identical if a flying saucer were seen hovering above any city or town. If extraterrestrial craft were really visiting Earth on a regular basis, we ought by now to have numerous examples of high-quality images and videos of those craft taken at the same time by numerous independent witnesses. That does not appear to be the case.

Figure 5.4
Is that an alien mothership hovering over the river Thames? Or is there another explanation? Photograph by the author.

CONTACT!

Things get even more interesting when we consider Hynek's final category: *Close encounters of the third kind*. This category consists of cases of alleged human-alien contact. The first and most famous of the so-called *contactees* in the modern UFO era was George Adamski. On November 20, 1952, Adamski claimed to have met a beautiful being from Venus named Orthon in the Mojave Desert, California (figure 5.5). Orthon is said to have communicated with Adamski via telepathy and hand signals. Adamski went on to write bestselling books about his ongoing interactions with the Venusians, including claims that they were Nordic in appearance. They, as well as aliens from other planets within our solar system, were said to be regular visitors to our planet, and they expressed concern that humans might destroy all life on Earth through the development of nuclear weapons. He also claimed to have been taken on a trip around the solar system, including to Venus, where the reincarnation of his late wife was located.

Many other contactees came forward throughout the 1950s, following in Adamski's footsteps with equally bizarre and unfounded claims. It is notable that the aliens of this era were generally said to be spiritually and technologically advanced compared humans. They wished only to help us to develop and to protect us from ourselves. Although many of the contactees did indeed make a successful living from their claims, they were generally not taken seriously even by the ufologists of that era who feared that their unbelievable claims would bring the field of ufology into disrepute.

ABDUCTED!

Although Hynek originally proposed just three categories of close encounter, some ufologists felt the need to add additional ones. *Close encounters of the fourth kind* are incidents in which a human is abducted by aliens. *Close encounters of the fifth kind* refer to situations where a witness communicates directly with aliens, via telepathy or other means. It is debatable whether extending this system beyond three types of close encounter has added much clarity.

Figure 5.5
George Adamski standing next to a painting by Gay Betts depicting a Venusian he allegedly met in the Mojave Desert, California, in November 1952.

The first reported case of alien abduction to receive wide attention was that of Antonio Villas Boas in Brazil. He claimed that on the night of October 15–16, 1957, when he was twenty-three years old, he was plowing his fields on his tractor when he noticed what looked like a large red star descending from the sky. He then realized that it was, in fact, an egg-shaped alien craft. As it landed, Villas Boas tried to drive away on his tractor, but the engine died. He was grabbed by three humanoid creatures, about five feet high, who dragged him into their spaceship. Once inside, he was stripped naked and covered in a strange gel. He had blood samples taken from his chin and became violently ill when some kind of gas was pumped into the room in which he was being held. He was then joined by a very attractive female humanoid. She was naked and the pair had intercourse (twice). Before leaving, the female pointed at her stomach and then upward, as if to indicate that the product of this unlikely coupling would be born on her home planet (plate 6). Although many people suspected that this was nothing more than a hoax, Villas Boas insisted throughout his life that he was telling the truth.

The first abduction case to really grab the public's attention was that of Betty and Barney Hill. On the night of September 19, 1961, the Hills were driving back from a short holiday in Canada to their home in Portsmouth, New Hampshire, when, at around 10:30 p.m., Betty spotted a bright light near the moon. When it appeared to get bigger and brighter, they stopped the car and observed the light through binoculars. Betty thought she saw a huge, oddly shaped craft that might be a UFO. Barney initially thought that it was an aircraft. However, he quickly changed his mind.

A little farther into their journey, they stopped again, and Barney got out of the car to observe the craft through the binoculars. He walked toward the craft until he was only fifty feet from it as it hovered at about tree height. He later reported that he saw a huge, pancake-shaped craft with a row of windows behind which were at least a dozen occupants in dark, Nazi-like uniforms. Barney panicked, got back into the car, and drove home. The Hills heard two sets of beeping sounds that appeared to cause their car to vibrate as they drove home, eventually arriving home in daylight at 5:15 a.m. The last 200 miles of their journey had taken seven hours, much longer than it should have done. They did recall leaving the highway onto a dirt road and coming across a roadblock of some kind, as well as seeing some sort of glowing orb.

When he arrived at home, Barney noticed that his shoes had scuff marks and the leather strap on his binoculars was torn. He felt an urge to examine his groin in the bathroom but noted nothing unusual on doing so. Betty noticed an odd pink powder on her torn dress. They found strange polished patches on their car that seemed to affect a compass needle. They also found that they could recall very little of what had happened between hearing the first set of beeping sounds and the second.

Ten days later, Betty began having a series of dreams, lasting over five nights, that seemed to fill in some of the gaps in memory. In her dreams, she and Barney were stopped by a group of humanoids at the roadblock and then taken aboard the alien vessel. The aliens were about five feet tall with large eyes, mouths like thin slits, and no protruding ears. They communicated via a mixture of telepathy and broken English. Similar aliens, now often simply referred to as *greys*, were described in many subsequent reports of alien abduction. On board the craft, the Hills were separated and each was medically examined.[20] This involved the collection of skin, nail, and hair samples and, in Betty's case, the insertion of a long needle into her navel, causing excruciating pain. Also, Betty was shown a star map allegedly showing the major trade routes used by the aliens.

A few months later, Barney was referred for psychiatric treatment. He was suffering from stress and exhaustion and had developed a ring of warts around his groin. After one year of treatment, he requested to undergo hypnotic regression in the hope that this would reveal what had really happened on that fateful journey from Canada to New Hampshire. He had his first hypnotic regression session with Dr. Benjamin Simon in December 1963, and not long after that Dr. Simon also hypnotically regressed Betty. The memories "recovered" during hypnotic regression were very similar to the narrative revealed in Betty's dreams. One additional detail was that Barney recalled a cuplike device being placed over his genitals, which he believed had been used to extract sperm. Also, as a result of a posthypnotic suggestion by Dr. Simon, Betty was able to draw a copy of the star map that she had been shown. School teacher and amateur astronomer Marjorie Fish subsequently

Figure 5.6
Betty and Barney Hill photographed in 1966 holding a copy of John Fuller's bestseller, *The Interrupted Journey*.

analyzed the map and claimed that it could only match stars seen from the vantage point of the Zeta Reticuli system, suggesting that this was where the aliens called home.

The Hills' story was told by writer John G. Fuller in his bestselling book, *The Interrupted Journey*, published in 1966 (figure 5.6) and then in the TV movie *The UFO Incident* in 1975.[21] Many of the elements that featured in this classic case were to be commonly reported in subsequent claims of alien abduction, including the sighting of a UFO, missing time, dreaming of aliens, and the use of hypnotic regression to "recover" memories.

As you might expect, the story was subjected to detailed critical analysis by skeptics.[22] Evidence suggests that the bright light that initially caught their attention and then appeared to follow them was probably the planet Jupiter. The "missing time" was not noticed until weeks after the incident, following questioning by ufologists. In fact, the Hills had taken an indirect route home, leaving the main highway for part of the journey. Many of the details were not recalled immediately afterward but instead first appeared in Betty's dreams. These details were then included in the accounts given during hypnotic regression sessions. It should be noted that although the Hills believed their accounts of being taken on board an alien spaceship were accurate, Dr. Simon believed them to be a fantasy based on Betty's dreams, the content of which would have been known to Barney. Fuller's account conveniently omits the psychiatrist's skepticism on this point.

Two bestselling books were published in 1987 that brought the phenomenon of alien abduction to an even wider audience. The first was by Whitley Strieber.[23] *Communion* was allegedly a true account of the author's own bizarre and horrifying interactions with alien beings, including having needles inserted into his head and anus. It topped the nonfiction bestseller lists of the *New York Times*, the *Washington Post*, and *Publishers Weekly*.

US journalist and UFO researcher Philip J. Klass critically assessed Strieber's story and was unconvinced of its authenticity.[24] Among numerous telling criticisms, Klass points out that prior to writing *Communion*, Strieber earned a living as a writer of horror stories. He claims to have had many bizarre experiences in his life but also admitted that he often just made claims up. For example, for years he claimed to be present at a real-life sniper attack, describing the horrific scene in gory detail—only to subsequently admit that he was never there. He has an obsession with intruders and appears to have difficulty separating fantasy from reality. His wife reports that he sees things that other people, including her, cannot see.

Here is an example of one of Strieber's encounters with aliens:

> Some time during the night I was awakened abruptly by a jab on my shoulder. I came to full consciousness instantly. There were three small people standing beside the bed, their outlines clearly visible in the glow of the burglar-alarm panel. They were wearing blue coveralls and standing absolutely still . . .

I thought to myself, My God, *I'm completely conscious and they're just standing there*. I thought that I could turn on the light, perhaps even get out of bed. Then I tried to move my hand, thinking to flip the switch on the bedside lamp and see the time.

I can only describe the sensation I felt when I tried to move as like pushing my arm through electrified tar. It took every ounce of attention I possessed to get any movement at all . . . Simply moving my arm did not work. I had to order the movement, to labor at it. All the while they stood there . . . I was overcome at this point by terror so fierce and physical that it seemed more biological than psychological . . . I tried to wake up Anne but my mouth wouldn't open . . . Again it took an absolute concentration of will . . . but I did manage to smile.

Instantly everything changed. They dashed away with a whoosh and I was plunged almost at once back into sleep.[25]

The other alien abduction bestseller published in 1987 was *Intruders* by Budd Hopkins (this was Hopkins's second book on the topic, the first being *Missing Time*, published in 1981).[26] Hopkins, who died in 2011 at the age of eighty, was convinced that the aliens were using abductees in a sinister crossbreeding project to produce human/alien hybrids. He viewed the alien visitors as callous in the extreme, showing no compassion toward their human victims. He was impressed with accounts from women who reportedly had become pregnant despite having no memory of having had normal sexual intercourse.[27] Some months later, these women would, equally mysteriously, find themselves no longer pregnant. According to Hopkins, these women had been abducted by aliens, who then artificially impregnated them on board their spaceship. Subsequently, before the pregnancy reached full term, they were abducted again, and the hybrid fetus was removed and given to its alien parents. Memory for both of these abductions, Hopkins believed, was wiped by the aliens, typically resulting in episodes of missing time. However, Hopkins believed, as we shall see, that the abductees could sometimes recall certain telltale signs that they had been abducted even if they could not recall the details of the abduction itself.

Despite having no relevant formal training, Hopkins routinely used hypnotic regression to "recover" the traumatic abduction memories. Some abductees claimed during these sessions that the aliens had implanted tiny devices into their unwilling victims (plate 7). The purpose of these implants was unclear, but it was speculated that aliens might use them for tracking their victims or for mind control. Such devices could potentially provide strong support for the ET hypothesis. If they could be obtained and subject to proper scientific analysis, they might reveal, for example, previously unknown alien technology. Unexplained scars and bruises were taken by Hopkins as evidence of alien abduction.

Budd Hopkins is the man who introduced Professor John E. Mack to the alien abduction phenomenon, resulting in a lengthy book on the topic by the latter, published in 1994.[28] This was a very significant development, given that Mack was a Pulitzer Prize winner and the head of the Department of Psychiatry at Harvard Medical School. The blurb of *Abduction* claimed that the book would "persuade every reader with an open mind that these accounts are not hallucinations, not dreams, but real experiences." Mack had a somewhat different take on the phenomenon. Whereas Hopkins viewed abductions in totally negative terms, Mack believed that the experience ultimately led to spiritual enlightenment and increased concern for the environment. He also felt that the experiences should be viewed as examples of visionary encounters similar to religious and mystical encounters through the ages.

There is absolutely no doubt that some claims of human-alien contact are nothing more than deliberate hoaxes.[29] However, most serious investigators of UFO-related claims, whether sympathetic toward or skeptical of the ET hypothesis, accept that most claimants are sincere. Supporters of the ET hypothesis claim that there is strong independent evidence to support the claims of contact. Skeptics are unimpressed, pointing out that all such evidence can be explained in more plausible prosaic terms. For example, very few "alien implants" have been subject to proper scientific analysis as they often seem to disappear mysteriously. Those that have been properly analyzed have not been revealed to be based on advanced alien technology. Some turn out to be everyday organic matter, such as cotton, that has somehow become encysted under the skin. A metallic "implant" that Susan Blackmore had analyzed turned out to be a dental filling that had fallen out.[30]

The unexplained scars and bruises that abductees find on their bodies have a number of potential mundane explanations. I suspect that if most people examined every square inch of their bodies, they would find some examples of such marks and be unable to account for how they got there. Alien abduction would not be top of the list of possible explanations for most people. Another possibility is injuries sustained during episodes of sleepwalking that are not recalled the following morning.

When it comes to unexplained pregnancies that subsequently just disappear, it should be noted that, despite repeated claims from ufologists, no properly documented cases have ever been presented.[31] Some such cases may well be examples of the well-documented phenomenon of false pregnancy (or *pseudocyesis*) in which many of the physical symptoms of pregnancy, such as abdominal growth, tender breasts, and delayed menstrual periods, occur in an individual who is not actually pregnant.

THE PSYCHOLOGY OF ALIEN CONTACT AND ABDUCTION CLAIMS

It seems highly probable that the majority of those who claim to have experienced alien contact are sincere, if mistaken, in their claims. The findings of a study by Harvard University psychologist Richard McNally and colleagues is relevant here.[32] Previous research has demonstrated that individuals suffering from post-traumatic stress disorder (PTSD) show heightened psychophysiological activity, indicating increased emotional arousal, when recalling the traumatic events that caused the PTSD.

In their experiment, McNally and his team recorded levels of psychophysiological arousal as abductees listened to recorded accounts of their own alien encounter as well as other stressful, neutral, and positive experiences. As expected, higher levels of psychophysiological reactivity were recorded when abductees listened to the abduction and stressful scripts compared to the neutral and positive scripts, and this effect was more pronounced for the abductees compared to a control group.

Such heightened arousal is typically seen when abductees recall details of their abduction during hypnotic regression sessions. They appear to be not just recalling the details of their abduction but actually reliving it, often leading observers to conclude that the events reported must have really happened. McNally and his team, in contrast, concluded that it was only necessary for the abductees to sincerely *believe* that their reported abduction had taken place for the strong emotional responses to be evoked.

If it is accepted that it is highly unlikely that claimants really have had contact with extraterrestrials but that they are sincere in their claims, is the explanation that they are suffering from some sort of serious psychopathology? This appears not to be the case. Several studies have compared claimants with control groups, and the results overall do not suggest that higher levels of serious mental illness are found in the former compared to the latter.[33]

On the other hand, those claiming to have had alien contact do tend to show significant psychological differences compared to the rest of the population. They often display symptoms of PTSD, reporting poor sleep patterns and high levels of unhappiness and loneliness.[34] In one study, over half of those reporting alien abduction had attempted suicide.[35] June O. Parnell and R. Leo Sprinkle concluded that those who claimed to have communicated with extraterrestrials "had a significantly greater tendency to endorse unusual feelings, thoughts and attitudes; to be suspicious or distrustful; and to be creative, imaginative, or possibly have schizoid tendencies."[36] A study by my own group found that nineteen individuals claiming to have experienced alien contact had higher levels of the tendency to hallucinate compared to a matched control group.[37]

One group of personality variables that are more pronounced in those with conscious memories of alien contact is of particular interest insofar as all of the variables concerned have been shown to correlate with susceptibility to false memories. Absorption, dissociativity, and fantasy-proneness have all been shown to intercorrelate with each other as well as with paranormal belief and the tendency to report a range of ostensibly paranormal experiences.[38] This raises the possibility that at least some reports of paranormal experiences might be based on false memories of events that never actually happened. Is it possible, then, that most, if not all, reports of alien contact are based on false memories?

THE RELEVANCE OF FALSE MEMORY RESEARCH

Before considering this possibility in greater detail, it is worth briefly discussing false memories in general. Of course, there is no clear dividing line between an apparent memory of an event that never took place and a grossly distorted memory of an event that did actually take place. Consider a hypothetical example. Suppose you have a clear subjective "memory" of going to the movies to watch *Close Encounters of the Third Kind* with friend A. It then becomes apparent that this never actually happened. In fact, you went to the cinema to watch that film with friend B. Is that best conceived of as a false memory of an event that never happened or as a distorted memory of an event that did happen? Although research into the unreliability of eyewitness testimony is sometimes treated as if it were totally distinct from more recent research into false memories, this example makes clear that the two phenomena actually exist on a continuum.

Research into false memories took off back in the 1980s and 1990s in response to the growing number of allegations of childhood sexual abuse being made solely on the basis of memories allegedly recovered during therapy.[39] The typical scenario involves an individual going into therapy suffering from a fairly common psychological problem such as depression, low self-esteem, insomnia, and so on. At this stage, the sufferer does not believe themselves to have been the victim of childhood sexual abuse (evidence suggests that genuine victims of childhood sexual abuse do not repress their memories of the experience[40]).

After a few months of therapy involving certain types of "memory recovery" techniques, such as hypnotic regression and guided imagery, some patients become convinced that they were indeed so abused and often have vivid and horrifically detailed memories to support that belief. In some cases, the memories involve ritualized satanic abuse of the most extreme kind, including memories of bizarre sexual perversion, human sacrifice, and cannibalism. The evidence strongly suggests that such memories are false memories. However, they have often been accepted as true memories by the patient, the therapist, other family members—and juries in courts of law. The result is that many families have been permanently torn apart and many people have been convicted of crimes that they almost certainly did not commit.

As stated, this controversy raged most fiercely in the 1980s and 1990s, beginning in the United States but spreading around the world to many other countries, including the United Kingdom. It still affects families today, as I know only too well from my experiences as a member of the Scientific and Professional Advisory Board of the British False Memory Society. It must be emphasized that childhood sexual abuse is far more common than many people realize, and the consequences can be devastating for the victims. All allegations of abuse should be taken seriously and investigated impartially. However, there are also many families in which the accusations of childhood sexual abuse are almost certainly solely the product of dubious forms of psychotherapy. The accusations are often never retracted, and families remain divided forever. There are a few happy endings, in which accusers eventually came to realize that the memories must be false and have the great courage to admit that they have been mistaken. Tragically, there are still therapists using such techniques in the mistaken belief that their clients' psychological problems could only be the result of such abuse and that healing requires that such memories be "recovered."

The one slender silver lining from these tragic cases is that they generated a huge amount of research into the nature of false memories.[41] It became clear that human beings are much more prone to false memories than anyone would have imagined. New experimental techniques were developed to reliably induce false memories in large numbers of people, and it was thus possible to investigate whether there is a particular type of personality that is most susceptible to false memories. It turns out that there is: as stated, individuals scoring highly on measures of absorption, dissociativity, and/or fantasy-proneness appear to be more susceptible to false memories compared to those with low scores.

Psychologists have used several techniques to investigate false memories, and the type of false memory implanted can range from the trivial (for example, misremembering a word as being on a list when in fact it was not) up to so-called *rich false memories* (detailed memories for entire events that never happened). A widely used example of the former is the DRM method, named after the first letter of the surnames of those who first used and popularized it (James Deese in 1959 and Henry L. Roediger III and Kathleen B. McDermott in 1995).[42] This technique involves the presentation of word lists. Within each list, each word is strongly associated with a critical "lure" word that is itself not included. For example, a list might consist of *bed*, *snore*, *dream*, *pillow*, *snooze*, and *doze* but not *sleep*. However, many people would erroneously claim the latter was presented. By counting the number of lure words that are reported when memory is tested across a series of such lists, a measure can be obtained of how susceptible to this type of false memory an individual is.

One technique used for implanting rich false memories for entire episodes involves repeatedly interviewing adult volunteers allegedly to obtain as much detail as possible about events that happened to them during their childhoods. Most of the events genuinely did occur, as testified by other family members, but one (for example, getting lost in a shopping mall) did not.[43] However, after repeated interviews and encouragement to try to recall the events, many people do report partial or full false memories for the event that never occurred, sometimes including additional details of what they think happened. Numerous well-controlled studies have shown that simply imagining events that did not happen increases the probability that they will subsequently be believed to have happened, an effect known as *imagination inflation*.[44]

Some studies use photographs to prompt childhood memories. However, the set of photographs include some that have been doctored to make it appear that a made-up event occurred (for example, going up in a hot-air balloon with a parent).[45] Once again, many participants elaborate on their "memories" of this day out that never happened.

Another approach came to be known as the *crashing memories* technique after it was first introduced by Hans Crombag and his colleagues in a study carried out in the Netherlands following the crash of an El Al Boeing 747 into a block of flats in Amsterdam in 1992.[46] Understandably, news of this disaster was widely reported, and it was headline news in the Netherlands for many days. As part of this study, participants were asked, "Did you see the television film of the moment the plane hit the apartment building?" Over half of the respondents reported they had seen this footage. However, no footage of the crash existed. This technique has been adapted to study false memories of a wide range of other high-profile news events that were not caught on camera, including Princess Diana's car crashing in the tunnel in Paris and the sinking of the Estonia ferry.[47] This technique and others have been used to identify a number of personality measures, such as absorption, dissociativity, and fantasy-proneness, that correlate with susceptibility to false memories.[48]

Absorption has already been mentioned several times in previous chapters. This is how Auke Tellegen and Gilbert Atkinson first defined absorption in 1974: "a disposition for having episodes of 'total attention' that fully engage one's representational (i.e., perceptual, enactive, imaginative, and ideational) resources. This kind of attentional functioning is believed to result in a heightened sense of the reality of the attentional object, imperviousness to distracting events, and an altered sense of reality in general, including an empathically altered sense of self."[49] Absorption has been shown to correlate with paranormal belief and reports of psychic or mystical experiences, as well as reports of alien contact.[50]

Dissociativity is as an individual's susceptibility to experiencing subjective separation between themselves and the world around them. Dissociation can be experienced in a variety of ways, including *derealization* (feeling that one's surroundings are unreal), *depersonalization* (feeling that one's consciousness is somehow disconnected from one's body), and *time distortion* (with time running either faster or slower than normal).

Almost everyone will have experienced at least mild dissociation at some point, perhaps as a result of stress, sleep deprivation, or the effects of mind-altering substances. It is only in extreme cases that psychiatric intervention may be advisable. Many individuals with moderately high levels of dissociativity function well in society but may appear to others to be a bit "spaced out" or, as my grandma would have put it, "away with the fairies." Several studies have reported correlations between dissociativity and paranormal belief and ostensibly paranormal experiences, including claims of alien contact.[51] Dissociativity in adults is associated with reports of past childhood trauma in the form of physical, sexual, or emotional abuse. It has been argued that the tendency to dissociate may develop as a psychological defense mechanism whereby a child learns to dissociate in order to psychologically distance themselves from the pain and harshness of their reality.[52]

Krissy Wilson and I used the crashing memories technique to directly test the notion that susceptibility to false memories might be associated with both dissociativity and paranormal belief/experience.[53] We asked our participants to provide details of where they were, who they were with, and what they were doing when they first saw various dramatic news footage such as the collapse of the Twin Towers on September 11 and Saddam Hussein's statue being toppled in Basra. Among the genuine examples was one that, although reported in the news, was not actually caught on camera: the terrorist bombing of a nightclub in Bali in 2002. As predicted, the 36 percent of respondents who claimed to remember this nonexistent footage scored more highly than those who did not on both dissociativity and paranormal belief/experience. Neil Dagnall and colleagues reported similar results in a subsequent study.[54]

The relationship between fantasy-proneness and paranormal belief and the tendency to report ostensibly paranormal experiences has been the subject of a considerable amount of research. Sheryl C. Wilson and Theodore X. Barber provided the first systematic description of this personality variable in the early 1980s.[55] Fantasy-prone individuals spend a great deal of their time daydreaming and have very vivid imaginations, claiming that they sometimes confuse fantasy and reality. They are extremely easy to hypnotize and often had imaginary friends during childhood. Notably, no fewer than thirteen of the twenty-seven female fantasizers in Wilson and Barber's sample had experienced false pregnancies. Very high levels of ostensibly paranormal experiences were reported by this group, including 72 percent who reported seeing apparitions, 88 percent who reported out-of-body experiences or waking dreams, and 92 percent who claimed to have psychic abilities. Two-thirds of the sample believed they could heal people by touching them.

Two distinct routes to developing high levels of fantasy-proneness have been identified. It may be that, as a child, the individual was strongly encouraged to engage in creative fantasy-based activities such as writing stories, acting, and so on. The second route is as a response to an aversive childhood environment involving abuse and/or social isolation and loneliness. This is a similar argument to that put forward to explain the development of dissociativity. In this case, it is argued that the child is able, to some extent, to distance themselves from the harshness of reality by mentally escaping into a fantasy world. Subsequent research by Steven Jay Lynn and Judith W. Rhue supported the general conclusions drawn by Wilson and Barber.[56] Fantasy-proneness has been repeatedly shown to correlate with paranormal belief and susceptibility to a range of ostensibly paranormal experiences.[57]

Opinion is divided as to whether those claiming alien contact are more fantasy-prone than the general population. This division of opinion appears to depend primarily on the approach: Those studies taking a biographical approach conclude that, as a group, abductees and contactees do indeed tend to have many of the classic characteristics of the fantasy-prone. Robert Bartholomew and colleagues analyzed descriptions of the lives of 152 such individuals, with the length of the descriptions ranging from a single paragraph to whole books.[58] In 132 of these cases they identified one or more characteristics of fantasy-proneness, including reports of high hypnotic susceptibility, out-of-body experiences, paranormal experiences (including apparitions), healing abilities, and physiological effects. The thirteen detailed cases presented by John Mack were similarly analyzed by Joe Nickell, leading him to conclude that all showed evidence of being fantasy-prone.[59]

In contrast, studies that compare groups using questionnaire measures of fantasy-proneness tend to find no significant difference between those who claim alien contact and those who do not. Of those studies that use the Inventory of Childhood Memories and Imaginings (ICMIC) to assess fantasy-proneness, neither Mark Rodeghier and colleagues nor Nicholas Spanos and colleagues found significant differences on this scale between those reporting UFO experiences and control groups, although the latter did find that scores correlated with the intensity of the experience.[60] Our own study using this questionnaire found a small but significant difference in scores between those claiming alien contact and a matched control group.[61] Peter Hough and Paul Rogers, using the Creative Experiences Questionnaire (CEQ) to assess fantasy-proneness, also reported no significant differences.[62]

Kenneth Ring and Christopher J. Rosing also reported no significant differences in fantasy-proneness between those reporting abductions and other UFO-related experiences compared to the general population using a questionnaire of their own design.[63] They did note, however, that as children, their sample of UFO experiencers, including abductees, were more sensitive to what they called "non-ordinary realities." That is, they claimed to see into "other realities" that were not perceived by those around them and to be aware of "non-physical beings." It might be suggested, in the absence of any proof that such other realities and beings exist, that their respondents were so fantasy-prone that they could not distinguish at all between perception and imagination.

It may be that those claiming alien contact really do not differ from others in terms of fantasy-proneness, but it is also possible that they are simply canny enough to recognize questionnaires that measure the degree to which they have what might be called "overactive imaginations." For example, the ICMIC includes items such as "When I was younger, I enjoyed fairytales" and "Now, I still live in a make-believe world some of the time," and the CEQ includes items such as "Many of my friends and/or relatives do not know that I have such detailed fantasies" and "I often confuse fantasies with real memories." For individuals who are motivated to convince others that the reported events really did take place, the implications of responding positively to such items would be all too clear. It is worth noting that those claiming alien contact report high levels of childhood trauma, which, as already stated, tend to be associated with heightened levels of both dissociativity and fantasy-proneness.[64]

Levels of paranormal belief and the tendency to report ostensibly paranormal experiences are typically extremely high among those reporting alien contact. This has been shown to be the case based on both anecdotal observations and systematic research using standardized measures of paranormal belief.[65] In other words, lots of weird stuff is often reported as having happened prior to any reported alien contact, although the frequency and intensity of such events may well increase following the alleged alien encounter.

To summarize, it appears that those reporting alien contact do indeed fit the psychological profile of people with a heightened susceptibility to false memories. They score higher than the general population on measures of absorption, dissociativity, and possibly fantasy-proneness, and they report high levels of paranormal belief and experience.

Two attempts, both using the aforementioned DRM technique, have been made to directly assess susceptibility to false memories in those claiming alien contact. The first was carried out by Susan Clancy and colleagues at Harvard University.[66] They compared three groups in terms of their scores on the DRM task. The first group reported that they had "recovered" memories of having been abducted by aliens; the second group believed they had been so abducted but had no memory of it; and the third group did not believe that they had ever been abducted by aliens. Those in the first group reported that initially they had no conscious memory of being abducted but had "recovered" those memories either while in therapy, as a result of techniques such as hypnotic regression, or else spontaneously after reading about abduction or watching TV programs or movies about it. Readers may be wondering, with respect to the second group, why anyone would ever believe that they had been abducted by aliens if they had no memory of any such event. The answer is that, as already mentioned, it is widely believed in ufological circles that the aliens are capable of wiping their victims' memories of the abduction. Certain telltale signs may be available, however, to indicate that an abduction has taken place. In this particular study, those signs include "insomnia, waking up in strange positions, unexplained marks on the body, preoccupation with science fiction." The scores on the DRM task indicated that the group with conscious memories of alien abduction had the highest susceptibility to false memories, and the group who did not believe that they had ever been abducted had the lowest. However, for reasons that are unclear, my own team found no significant differences in DRM scores between a group with conscious memories of alien contact and a well-matched control group (that happened to include my mum, who has never been abducted by aliens—as far as she knows).

Another reason to believe that claims of alien contact are often based on false memories is the widespread use of hypnotic regression to recover such memories. Thomas E. Bullard reported that hypnotic regression had been used in around 70 percent of the "well-investigated, high quality cases" in his sample of abductees, and both Budd Hopkins and John Mack were known to routinely use the technique, as do many other ufologists.[67] The idea that hypnosis can be used as some sort of magical key to unlock hidden or repressed memories is widely accepted but completely mistaken. In fact, the relationship between hypnosis and memory is poorly understood not only by the general public but also by many practitioners who actually employ it.

Daniel J. Simons and Christopher F. Chabris surveyed a large representative sample of US adults regarding their beliefs about how memory works and compared those responses with the views of an expert sample of professors, each with at least a decade's experience of memory research.[68] Whereas 54.6 percent of the general public agreed with the statement "Hypnosis is useful in helping witnesses accurately recall details of crimes," none of the experts agreed. The use of hypnotic regression in attempts to recover additional details of a witnessed event over and above those recalled in the nonhypnotic state runs the very real risk of confabulation. Gaps in memory may be filled in with whatever comes to mind, producing a detailed false memory that is then sincerely believed in.[69] Evidence produced as a result of hypnotic regression is rarely admissible in criminal trials.

Hypnotic regression is used not just to try to retrieve additional details of an event that was actually witnessed but also to recover allegedly "repressed" memories of traumatic events that have supposedly been banished from conscious awareness. In fact, most memory experts are dubious about the very concept of repression, pointing out that traumatic events are much more likely to be remembered than forgotten.[70]

People are sometimes convinced that hypnotic regression really is capable of mentally transporting someone back in time because they have witnessed the behavior of someone who has allegedly been regressed back to childhood. Such demonstrations are indeed superficially impressive. The regressed subject does not just describe what was happening on, say, the day of their seventh birthday, they actually appear to be reliving it. Their voices, vocabulary, mannerisms, and emotional responses all seem to be appropriate to the age to which they have been regressed. If asked to write something, their writing is childish, as are any drawings they produce. However, when their behavior is subject to detailed analysis by developmental psychologists, it turns out they are not really behaving in the way that a child of that age would behave. Instead, they are behaving in a way that most adults *think* a child of that age would behave.[71]

Michael Yapko collected data from psychotherapists regarding their beliefs about the relationship between memory and hypnosis.[72] His analysis revealed a worryingly high level of misconceptions:

> Survey data regarding hypnosis and suggestibility indicate that while psychotherapists largely view hypnosis favorably, they often do so on the basis of misinformation. A significant number of psychotherapists erroneously believe, for example, that memories obtained through hypnosis are more likely to be accurate than those simply recalled, and that hypnosis can be used to recover accurate memories even from as far back as birth. Such misinformed views can lead to misapplications of hypnosis when attempting to actively recover memories of presumably repressed episodes of abuse, possibly resulting in the recovery of suggested rather than actual memories.[73]

A minority of practitioners go even further, believing that hypnotic regression can be used to retrieve memories of life in the womb—and even earlier! (The latter will be discussed more fully in chapter 6.) James Ost and colleagues compared first-year undergraduate psychology students (during their first week at university), clinical psychologists, and hypnotherapists in terms of their actual knowledge of how memory works and their self-assessed knowledge of how memory works.[74] Hypnotherapists were found to have the poorest understanding of memory but scored highest in terms of their self-rating of their own knowledge.

The rich false memories resulting from the use of hypnotic regression are based on a blend of fantasy, expectation, and fragments of real memories, including memories of dreams, stories, and films. Sometimes the source of the images or conversations in the false memory can be traced directly to a film or television program. For example, it seems likely that Barney Hill's description of the aliens that abducted him and his wife was influenced by an episode of the TV sci-fi series *The Outer Limits*. Martin Kottmeyer points out that Barney's description of the aliens as having "wraparound" eyes through which they spoke telepathically is very similar to the aliens depicted in an episode of that series.[75] The episode was broadcast just ten days before the regression session in which Barney first described the aliens as possessing this unusual feature.

Perhaps the strongest evidence in support of the claim that reports of alien abduction are often based on false memories generated as a result of hypnotic regression is that produced by Alvin Lawson.[76] Lawson hypnotically regressed eight volunteers with little prior knowledge of UFOs, none of whom believed that they had ever been abducted by aliens, and asked them to simply imagine that they had been so abducted. The accounts of these so-called imaginary abductees proved to be remarkably similar to the accounts produced by those claiming that they had genuinely been abducted by aliens, down to the level of odd, minute details.

Hypnotic regression is not the only way in which false memories are produced. As already stated, false memories can also be implanted by repeatedly interviewing people regarding an alleged event in their past that never actually took place, especially if they are presented with deceptive information implying that it really did take place. Henry Otgaar and colleagues used this approach to implant false memories of being abducted by a UFO at the age of four in a number of younger children (aged seven to eight years) and older children (aged eleven to twelve years), the younger children being more likely to report the false memory.[77] One assumes that this study may have raised a few eyebrows on members of the ethics committee that approved it!

False memories can in fact develop without the involvement of either hypnosis or repeated (misleading) interviews. It is likely that, in many cases of individuals who have come to believe that perplexing events may have been the result of alien activity, false memories of alien contact are simply the result of imagining what such contact would be like. The formation of false memories is especially likely to happen in people with very vivid imaginations. Essentially, they end up mistaking their memory of something that they imagined for a memory of something that really did take place.

What might these perplexing events be that lead individuals either to seek the services of a practitioner of hypnotic regression or else to spend time imagining what it would be like to be abducted by aliens? Some of these events have already been referred to. The story of Betty and Barney Hill includes several examples: seeing a UFO, dreaming of aliens, finding marks on one's body, and the experience of "missing time." In all of these cases, there are plausible mundane explanations. UFO sightings can almost always be explained in normal terms if one makes the effort. Dreams of strange beings are common, and dreaming of aliens after believing one has just seen a UFO would not be surprising. As already explained, most of us could probably find unexplained scars on our bodies if we looked hard enough and, besides, highly fantasy-prone individuals are known to be susceptible to various psychosomatic effects.

An experience of "missing time" might reflect something as mundane as misreading a watch or clock, or it might reflect a psychological distortion of time perception, especially likely in those with high levels of dissociativity. One common, albeit underresearched, example of a time-distortion effect is referred to colloquially as *highway hypnosis*. This is the common experience that drivers have of driving on a long, monotonous road on "automatic pilot" and suddenly "coming to" and realizing that they do not recall the previous hours of driving.

One of the most common causes of the suspicion that one may have been abducted by aliens, even though one cannot remember it, is our old friend sleep paralysis, described at length in chapter 2. I am sure that when you read the description of Whitley Strieber's alien encounter earlier in this chapter, you immediately recognized it for what it was. Also, this is how Susan Clancy and colleagues describe the participants in their group who had "recovered" memories of alien abduction that they believed had been repressed: "They began to suspect that they had been abducted after a sleep episode characterized by awakening, full body paralysis, intense fear, and a feeling of presence. Several participants reported tactile or visual sensations (i.e., levitating, being touched, seeing shadowy figures)."[78] The researchers correctly identified these experiences as being due to sleep paralysis.

Many ufologists, such as Budd Hopkins, believe that the aliens have the ability to wipe their victims' minds of the memories of being abducted but a few telltale memories may survive. What is the nature of these telltale signs that the careless aliens fail to erase? The answer lies in a survey of around

6,000 US adults carried out by the Roper Organization at the request of Budd Hopkins, in collaboration with David Jacobs and Ron Westrum.[79] Hopkins and colleagues wanted to estimate how many people were being abducted by aliens, but they felt that simply asking people directly if they had ever been abducted would be futile given, as they believed, the aliens' ability to wipe memories. Instead, they adopted an indirect approach, including the following unusual experiences, among others, in their survey (the figures in brackets indicate what percentage of the respondents confirmed that they had had the experience described at least once in their lives):

Waking up paralyzed with a sense of a strange person or presence or something else in the room. (18%)
Experiencing a period of time of an hour or more in which you were apparently lost, but you could not remember why or where you had been. (13%)
Feeling that you were actually flying through the air although you didn't know how or why. (10%)
Seeing unusual lights or balls of light in a room without knowing what was causing them or where they came from. (8%)
Finding puzzling scars on your body and neither you nor anyone else remembering how you received them or where you got them. (8%)[80]

Hopkins and colleagues argued that if respondents answered affirmatively to four or five of the above items, they had probably been abducted by aliens and then had their memories wiped. However, it is quite clear that three of these items (the first, third, and fourth) describe common sensations experienced during sleep paralysis. They found that around one respondent in fifty reached their threshold of being a probable alien abductee, and then extrapolated to the adult population of the United States as whole—leading them to conclude that no fewer than 3.7 million American adults had probably been abducted by aliens! Much to my annoyance, this figure is frequently quoted in articles and programs about alien abduction. Even worse, it is frequently misquoted as "3.7 million American adults *believe* that they have been abducted by aliens" (my italics). No, they do not! They were never asked if they had been abducted by aliens!

As with other ostensibly paranormal experiences, there is no one-size-fits-all explanation for claims of alien contact and abduction. Having said that, the two-stage model presented in this chapter does provide a plausible explanation for most, if not all, sincere claims of this type. In the first stage, various types of unusual experience, such as seeing a UFO, "missing time," dreams featuring aliens, finding unexplained marks on their body, or episodes of sleep paralysis, may lead an individual to suspect that they have been the victim of alien abduction. This provides the motivation to "recover" the full memory of the alien encounter either by repeatedly imagining what such an encounter may have been like, based on reports from others, or by the use of hypnotic regression. Those with the appropriate psychological profile may well end up with detailed false memories of alien visitations that never actually happened.

Arthur C. Clarke is said to have quipped, "I'm sure the universe is full of intelligent life. It's just been too intelligent to come here."[81] In a more serious mood, however, he offered the following observation: "Two possibilities exist: either we are alone in the Universe or we are not. Both are equally terrifying."[82] Speaking for myself, I find the idea of intelligent life elsewhere in the universe considerably more appealing than the possibility that we may be all alone. I tend to agree with comedian and writer Ellen DeGeneres: "The only thing that scares me more than space aliens is the idea that there aren't any space aliens. We can't be the best that creation has to offer. I pray we're not all there is. If so, we're in big trouble."[83]

6 MANY HAPPY RETURNS?

During my time at Goldsmiths, heads of department did not occupy that stressful role on a permanent basis, presumably on the assumption that no sane person would put themselves forward for the position if that were the case. Instead, eligible staff members would each serve a three-year term in office, and then the next victim would have their turn. I had my turn from 1997 to 2000.

One afternoon during this period, as I was working in my office, the phone rang and a voice at the other end said, "What I am about to propose to you is going to sound very strange." It was indeed a rather unusual request. I was being invited to take part in a documentary investigating reincarnation claims among the Druse people in Lebanon. This would involve me spending a few weeks in Lebanon with a film crew, so I was unsure if my employer would be willing to allow me to go. Fortunately, senior management at Goldsmiths approved my request, and so I began to prepare for the adventure.

I was feeling both excited and nervous. Although I had read around the topic of reincarnation, I had never done any fieldwork investigation of the subject. I really did not know what to expect. I had my wife's blessing for the trip. Any doubts she had had evaporated the moment I told her the fee I would be getting. Even so, I did not relish the prospect of being away from her and my kids in the company of complete strangers for what struck me as a very long time. My feelings of guilt at leaving her at home with all the domestic responsibilities, supported only by our trusty nanny, Morag, were exacerbated when, a few days before I was due to depart, our three-year-old, Alice, broke her arm while playing.

As the taxi drove me to the airport in the early hours of the morning, it was still dark and, if my memory can be trusted, raining heavily. I regretted having agreed to take part, but it was too late to back out. Once I got to the airport and met the crew, most of my anxiety disappeared. They were a nice bunch of guys and immediately made me feel very welcome. And so began one of the most interesting investigations that I have ever taken part in.[1]

REINCARNATION: GENERAL CONSIDERATIONS

Before discussing the specific version of reincarnation espoused by the Druse, let us consider some general aspects of this widely believed concept.[2] Reincarnation is defined as follows in Michael Thalbourne's *Glossary of Terms Used in Parapsychology*: "A form of survival in which the human soul, or some aspects of self, is, after the death of the body, reborn into a new body, this process being repeated throughout many lives."[3]

Worldwide, a large proportion of the population believes in reincarnation. It is a central tenet of the Indian religions (Hinduism, Buddhism, Jainism, and Sikhism), which together account for over a fifth of the world's population, and large proportions of the population in Nordic countries and Western and Eastern Europe also endorse this belief.[4] Erlendur Haraldsson collated data assessing levels of belief in reincarnation across thirty-five countries. The average endorsement rate across five Nordic countries was found to be 22.6 percent, ranging from a low of 15 percent in Norway to a high of 41 percent in Iceland.[5] Among fifteen Western European countries, 22.2 percent expressed belief in reincarnation, with the Maltese having the lowest rate of endorsement (12 percent) and Switzerland having the highest (36 percent).[6] Both the United Kingdom and Portugal also had quite high rates of endorsement (29 percent in both cases). The average rate of endorsement was higher across fifteen Eastern European countries, at 27 percent.[7] The lowest endorsement rates in Eastern Europe were found to be in East Germany (12 percent) and the highest in Lithuania (44 percent).

By definition, all versions of belief in reincarnation hold that the soul is reborn into a new body following the physical death of the old body, but beyond that there is a lot of variation across cultures. Some cultures maintain that the soul is always reborn into a body of the same sex, and others believe that cross-sex reincarnation can occur. Some hold that humans are always reborn as humans, whereas some hold that one may be reborn as an animal or even an inanimate object. Some believe the soul enters the new body at the point of conception, others believe that it occurs at the point of birth.

Many different arguments have been put forward in support of the reality of reincarnation. One of the most common is that it accounts for apparent injustice in the world. Why is it that some are born into wealth and live long, healthy lives in luxury whereas others are born into extreme poverty and live short lives full of pain and misery? According to those who believe in the law of karma, the circumstances of our current lives reflect our behavior in previous lives. Therefore, if one lives a virtuous life this time around, one can expect to be rewarded with greater wealth and happiness in one's next incarnation. Similarly, if one lives an immoral life this time, one can expect to be punished with poverty and poor health in the next life. Some people find comfort in this belief insofar as it reassures them that, overall, the world is a just place. Personally, I find it an abhorrent idea, providing as it does an excuse to hold back from trying to improve the lives of those less fortunate than ourselves. After all, the argument goes, surely they deserve their current misery given their sins in previous lives?

Reincarnation is sometimes put forward as an explanation for puzzling phenomena such as the existence of child prodigies. Wolfgang Amadeus Mozart famously composed his first pieces before the age of five and performed all over Europe as a child. Mathematician, philosopher, and physicist Blaise Pascal completed a treatise on the subject of vibrating bodies at the tender age of nine, followed by his first proof at the age of eleven. Is it possible that these prodigies, along with many others, actually benefited from the transfer of skills obtained in previous lives?

It has also been argued that reincarnation explains the common anomalous experience of déjà vu. Most people have had the strange feeling that they have experienced something before—be it a place, an object, or an event—even though they know that they cannot have done so. Is it possible that this feeling of familiarity occurs because one really has had the experience before—in a past life?

We cannot claim to have definitive explanations for either the existence of child prodigies or the phenomenon of déjà vu, but promising lines of research within psychology and neuroscience provide valuable insights.[8] Mainstream scientists generally do not view reincarnation as a promising explanation for either phenomenon.

For some people, reincarnation provides an answer to a potential problem that arises with other versions of life after death: where does the soul go after the physical body dies? On reflection, though, the answer provided by reincarnation—that it simply inhabits a new body—raises more questions than it answers. Is there a gap between leaving a body at the point of death and entering a new one? If so, where is the soul during that period? What possible mechanism could allow a nonphysical soul to exit from one body while retaining certain memories, attitudes, and abilities, and then enter a new body and imprint those same memories, attitudes, and abilities into the physical brain of a developing embryo? No supporter of reincarnation has ever come close to answering these questions.

The strongest evidence in support of reincarnation comes from individuals who claim to remember aspects of their previous lives. Such memories fall into two general categories: those allegedly recovered through the use of hypnotic regression, and those that occur spontaneously, without any type of memory recovery technique. Memories of apparent past lives are the focus of the rest of this chapter.

PAST-LIFE REGRESSION

Past-life memories are usually sought through the use of hypnotic regression for one of two main reasons. The first is that of spiritual enlightenment. Many New Agers are keen to find out who they were in past lives as part of their ongoing spiritual journey. The use of hypnotic past-life regression is quite likely to recover rich and detailed memories of many fascinating former lives in which they were important historical figures (but rarely those accused of war crimes).

The second main reason for being hypnotically regressed into a past life is in a misguided attempt to deal with current psychological problems. A small minority of psychotherapists believe that psychological problems in this life may be caused by trauma experienced in a past life.[9] For example, they might argue that a phobia of dogs was caused by being torn to pieces by a pack of wolves in a past life or that a fear of flying is due to having been a fighter pilot in World War II who was shot down. Such claims are generally rejected by mental health experts as indicated by the results of a poll of over a hundred such experts who rated fifty-nine questionable treatments on a continuum from 1 (not at all discredited) to 5 (certainly discredited).[10] Past-life therapy was rated as 4.92 on this scale of shame with only crystal healing, orgone therapy, the use of pyramids for restoration of energy, and angel therapy being rated as more discredited.

It should come as no surprise to anyone who read the previous chapter that the available evidence strongly supports the claim that the apparent memories of past lives "recovered" by the use of hypnosis are, in fact, false memories based on a mix of expectation, fantasy, fragments of real memories, and suggestions from the hypnotist.[11] Past-life regression typically only works well on people scoring at the high end of hypnotic susceptibility. Most people either fail completely or "recover" only the most sketchy and vague "memories." But the highly susceptible individual may well experience detailed and vivid imagery, as well as strong emotional reactions, and may appear to be reliving their past-life experiences in the same way that those recovering memories of alien abduction appear to do. However, detailed analysis of the narratives produced reveals conclusively that they are typically describing the Hollywood version of the time and place they claim to be aware of, not the historically accurate version. Furthermore, they are typically unable to answer questions about that era that one might expect someone living at the time to be capable of answering, such as the name of the currency or the then current ruler and whether the country was at war.

The general public became aware of the idea of hypnotic past-life regression as a result of one particular case back in the 1950s: that of Bridey Murphy. In 1952, Morey Bernstein hypnotically regressed Colorado housewife Virginia Tighe on several occasions. While in the regressed state, Virginia spoke with an Irish accent. She claimed that her name was Bridey Murphy and that she lived in Cork in the early nineteenth century. She sang Irish songs and told detailed stories about her life in Ireland. This intriguing story was the subject of a bestselling book by Bernstein as well as newspaper and magazine articles (see plate 8) and a popular film.[12] The story also featured in two pop records and inspired a brief craze of "come as you were" parties. Unfortunately, subsequent investigations revealed that neither Bridey Murphy nor any of the people of whom she spoke had ever existed.[13] Furthermore, it turned out that in her younger days Virginia had been a talented amateur actor with a particular fondness for playing Irish characters. She had an Irish aunt and an Irish neighbor. The neighbor's name? Bridey Murphy Corkell.

There are a few other exceptional cases in which detailed and largely accurate historical accounts were given. I can still recall watching a documentary called *The Bloxham Tapes* on the BBC back in the 1970s and presented by one of the United Kingdom's most respected broadcasters of the time, Magnus Magnusson. The documentary investigated the work of Cardiff-based hypnotherapist Arnall Bloxham. Bloxham had made over 400 recordings of hypnotic regression sessions in which individuals had apparently been regressed into past lives. It was claimed that when these reports, sometimes packed with rich detail, were checked against the historical record, the only reasonable conclusion that could be drawn was that they provided incontrovertible proof of the reality of reincarnation. This was also the conclusion of a bestselling book written by the program's producer, Jeffrey Iverson.[14]

Probably the most impressive case from Bloxham's collection was a thirty-year-old Welsh housewife whom he referred to as Jane Evans. Jane had rich memories of no fewer than six past lives. Among them were memories of her life in York as Livonia, wife of the tutor to the family of the Roman governor Constantius during the Roman occupation of Britain. She gave the correct names of the governor's family members and others close to the family, as well as details of their daily lives.

She also recalled her life as Rebecca, the wife of a wealthy Jewish moneylender in 1189, also living in York. She described in detail how the Jewish community was persecuted by the Christians, culminating in her hiding in a crypt in a church where she was found and murdered. This revelation was particularly striking because it was not thought that any churches in

York had crypts. To his astonishment, Iverson was then contacted by one Professor Barrie Dobson and informed that a crypt, believed to predate the little-known massacre of the Jews, had been discovered in St. Mary's Church in the city. Understandably, this was taken as irrefutable proof that Jane had indeed once lived the life of Rebecca.

Jane gave another detailed and generally accurate account of her life as Alison, the Egyptian maid of Jacques Coeur, a wealthy French businessman and financier living in Bourges in 1450. She appeared to know a great deal about Coeur's life and business affairs and could even accurately describe his house down to the level of ornaments on his mantlepiece.

How could Jane possibly know so much about these lives and others unless she really was reincarnated, as Iverson claimed? The answers were revealed by subsequent investigations by Melvin Harris.[15] He hypothesized that the details recalled during the regression sessions may have come from works of historical fiction. Sure enough, he discovered that in 1947 a novel, *The Living Wood*, had been published about the life of Constantius. It contained all of the historically accurate details recounted by Jane—but, more tellingly, it also included fictional characters that the author had simply invented. These fictional characters also popped up in Jane's narrative.

It is not entirely clear where the details of Jane's Rebecca account came from, but Ian Wilson traced three individuals who claim that they recall a radio play from the 1950s telling the story of this historical massacre of the Jews in York. Professor Dobson had second thoughts about what had been originally described as a "crypt." It turned out to be a vault, built after the era of the Jewish massacre.

As for the life of Jacques Coeur, that was the subject of another historical novel: *The Moneyman* by Thomas B. Costain, published in 1948. The big giveaway this time was that Alison claimed that Coeur was unmarried, as indeed he was in the novel. But the historical record showed conclusively that he was not only married, he had five children. His house is one of the most photographed in France, appearing, for example, in Joan Evans's *Life in Medieval France*.[16]

The question naturally arises of whether Virginia Tighe, Jane Evans, and others who apparently recover memories of past lives are knowingly attempting to deceive others or whether they believe in the veracity of the narratives produced. Opinion is somewhat divided, but many commentators believe that their stories may be examples of *cryptomnesia* (literally, "hidden memories"). We know that sometimes we store information in memory but forget the source of that information. This is known as a *source attribution error*. Thus, the images that are perceived in the mind's eye during a hypnotic regression session may be based on books read, films and documentaries watched, and so on, combined with fantasy, expectation, and fragments of personal memory. The end result may be a detailed and vivid memory that feels as real to the individual as any of their memories of events that they have experienced firsthand.

This explanation of past-life memories is supported by the results of a series of studies by Nicholas Spanos and colleagues.[17] Results revealed that volunteers who reported memories of a past life in response to hypnotic suggestions had higher levels of fantasy-proneness and hypnotic susceptibility compared to those who did not, but the groups did not differ in terms of levels of psychopathological tendencies. In other words, the typical personality profile of those in the former group was the same as that found for those reporting alien abduction.

The influence of expectation on the content of the "recovered" memories was clearly demonstrated in these studies. Some volunteers were informed before the regression that the memories reported during such sessions often appear to be from identities of a different race and gender to those of the volunteers, and that they were often from exotic cultures. This was reflected in the content of the narratives produced compared to the narratives produced by volunteers who were not primed in this way. In another study, some volunteers were told before the session that child abuse was common in the past, and others were not primed in this way. Once again, the priming affected the content of the reported memories. Whether or not the volunteers accepted their apparent memories as genuine evidence for reincarnation depended primarily on their preexisting beliefs about reincarnation and whether they had been led to believe by the experimenters that reincarnation was a scientifically plausible concept.

Evidence supporting the idea that those reporting past-life memories are more susceptible to false memories was reported by Cynthia Meyersburg and colleagues.[18] Fifteen people reporting memories of past lives were compared to a control group in terms of their scores on the DRM task described in the previous chapter. Six of the former group had recovered their memories via hypnotic regression with the rest basing their claims on the experience of déjà vu, recurrent dreams, "flashbacks," and so on, emphasizing once more that hypnotic regression is not the only way to produce false memories. As expected, the past-life group scored higher in terms of propensity to false memories as assessed on this task. On the positive side, those reporting past-life memories report enhanced meaning in life and lower levels of fear of death.[19]

Within the field of reincarnation research, even those who are sympathetic to the possibility that reincarnation may be a genuine phenomenon have expressed strong skepticism toward accounts based on the use of hypnotic regression. For example, the world's leading researcher on cases of spontaneous past-life memories, the late Professor Ian Stevenson of the University of Virginia, referred to them as examples of "the psychotherapist's fallacy."[20] Although apparent past-life memories produced by hypnotic regression are of considerable interest in their own right for the light that they shed on the processes involved in the formation of false memories, there is little doubt that spontaneous past-life memories pose a greater challenge to skeptics. In the remainder of this chapter, I will describe my own thoughts and observations regarding such cases among the Druse in Lebanon.[21] Many, but not all, of the observations made would apply to spontaneous cases in general.[22]

REINCARNATION CLAIMS AMONG THE DRUSE

In the weeks before I flew to Lebanon, I spent many hours learning about the Druse and reading as much relevant literature as I could find regarding previous investigations of this type of claim. I could see why many people found such cases compelling. The ones that were written up in books and journals strongly supported the idea that children around the world, but especially in cultures where belief in reincarnation was prevalent, sometimes spontaneously described details of the life of someone who had died whom they had never met. It was claimed that, when checked, these details generally turned out to be correct and that, furthermore, there was no obvious way, other than reincarnation, in which the child could have learned them. It was not just autobiographical details that appeared to be passed on but also skills, preferences, fears, and so on.

I needed to learn as much about Druse culture as I could.[23] I learned that the Druse (also spelled *Druze*) are a religious sect found mainly in Lebanon, Syria, and Israel but also in many other countries around the world. In 1042, the sect broke away from the Ismaili doctrines of Islam and, from then on, no new members were allowed to join this closed religion. The Druse are not supposed to marry outside their own religion, and therefore their communities are tightly knit.

As stated previously, cultures vary with respect to the details of their reincarnation beliefs, and the Druse version is unique in holding that reincarnation occurs at the exact moment of death. The last breath of the old body is immediately followed by the first breath in a baby as it is being born. In other words, the soul enters the new body at the moment of birth not, as in many other versions of reincarnation, at the moment of conception. Furthermore, the Druse hold that humans always reincarnate as humans of the same sex, never as animals or inanimate objects. Druse will always reincarnate as Druse, Christians as Christians, and so on. I was reassured by a most eminent Druse religious leader during our visit that everyone reincarnates, even skeptics. However, not everyone will remember their past lives. This only happens if the previous life ended violently, an occurrence that was sadly all too frequent during the Lebanese civil war that ran from 1975 to 1990 and cost an estimated 120,000 lives.

It may seem strange to Western eyes, but the Druse adopt a very matter-of-fact attitude toward reincarnation. It is not seen as any kind of big deal, it is just something that happens to everyone. Their word for reincarnation, *taqamus*, literally translates as "changing one's shirt." One simply takes off the old body and puts on a new one.

The Druse believe that souls cannot be created or destroyed. They do not believe in karma, instead believing that the quality of one's current life is not related to one's preceding life. They maintain that each soul must experience a wide range of different lives, from the most unbearable misery to utmost happiness, in preparation for meeting God on Judgment Day.

Of course, one can only learn so much from books. Obviously, I would never learn as much about the Druse culture from books as someone who had grown up immersed in it. I hatched a cunning plan. I would make it a top priority when I reached Lebanon to find a member of the Druse community who did not believe in reincarnation—a Druse skeptic! If I could find such a person, they would surely be able to provide me with insights into the process by which these claims arose.

With my background in experimental psychology, I was keen to come up with a way to investigate reincarnation claims empirically, ideally to obtain results that I could analyze statistically. I came up with a second cunning plan. What we would need to find, I decided, was a Druse child who claimed to remember quite a lot about their previous life but had not yet made direct contact with their past-life family. We could then formulate questions to probe the child's memory. For example, we could show the child pictures of, say, five male adults and ask, "Which one of these men was your uncle in your past life?" We would need to ensure that the child had no way of knowing the correct answer by normal means and control for such things as family resemblance and so on, but I thought it should be possible to come up with a suitable series of questions. For the example question given, the child would have one chance in five of giving the right answer purely by guessing. Given a series of such questions, we could work out an exact overall probability that the child was simply guessing. If that probability was very low, it would strongly indicate that they really did possess accurate knowledge of the past-life family.

The idea for the documentary that I was taking part in was that I would investigate various Druse cases alongside Roy Stemmen, a firm believer in reincarnation and the editor of *Reincarnation International*. The program was made by Granite Productions and broadcast on Channel 4 in the series *To the Ends of the Earth*. We all benefited greatly from the background research done in advance of our arrival, particularly that done by producer Chris Ledger and Tima Khalil Majdalani, a journalist who worked with us as a translator. Our guide was Jad Al Younis, himself a Druse and also a strong believer in reincarnation. I was not surprised that Jad was a believer in reincarnation, but it soon became obvious that Cunning Plan No. 1 was doomed to fail—because every Druse I met believed in reincarnation!

One thing had already struck me about the specifically Druse version of reincarnation before I arrived in Lebanon: their belief that the instant of death of the past-life identity should correspond exactly to the time of birth of the new incarnation provided an obvious empirical test. In fact, this is hardly ever (well, okay, to my knowledge, never) found to be the case, the difference typically ranging from a few months to a few years. Some, including Ian Stevenson, have tried to explain this discrepancy as being partly due to poor record-keeping in Lebanon, but this is an unconvincing explanation.

The Druse themselves sometimes claim that the discrepancy occurs because the child is not actually remembering events from the life that immediately preceded their current one but the life before that. In other words, an individual died, their soul immediately entered a new body at the moment of birth, but this child died shortly afterward, having lived only a brief and unmemorable life. Its soul then entered the current body (at the moment of birth), and this child now recalled events not from the brief unmemorable life but from the one before that. Such an explanation is both implausible and untestable.

Another issue is their belief that the number of Druse souls has remained constant since their religion became closed in 1042, although no one knows what this number is. Clearly, the size of the population of Druse has not remained constant since the eleventh century, so how do the Druse account for this? The nonfalsifiable explanation favored by many Druse is that members of their sect are found in many locations, including such far-flung places as China—some even speculate that Druse might exist on other planets! In which case, the argument goes, any increase or decrease in, say, the size of the Druse population in Lebanon may be compensated for by decreases or increases in these remote locations. It is worth noting, however, that the past-life memories reported by Lebanese Druse children are typically those of someone who died within a fairly close distance, often a neighboring village, not some exotic foreign land or distant planet.

It is clear that the specific details of the Druse version of reincarnation cannot be based on empirical data but instead are culturally transmitted from one generation to the next. For many researchers in this area, however, it does not matter that the specific details of the Druse version cannot be objectively true. For them, the important thing is to find cases that show that reincarnation in some form really does sometimes take place, regardless of the specific details. Thus, it is to the investigation of individual cases that we now turn.

When Druse recount details of past lives, they often appear to follow a standard pattern. It is typically reported that, when the child is first learning to speak, they will sometimes come out with utterances that make little sense to their parents. As the child gets older and their language skills improve, it gradually becomes clear that they are referring to their memories of a previous life. For example, the child will insist that they have another family and will provide the names and other details of the members of this family. They may describe the house they used to live in and refer to their occupation in their past life. If the child was, say, a mechanic in their past life, they may show a precocious interest in and knowledge of engines. Disturbingly, they may even report memories of how they died. On the basis of the child's utterances, their parents will recognize their past-life identity.

As the child grows older, they often express the wish to visit their past-life family. At first, their biological parents may refuse such requests, but eventually the child's insistence will become so strong that a meeting is arranged. Members of the past-life family will not immediately accept the child as the genuine reincarnation of their deceased loved one. Instead, they will probe the child's knowledge by asking questions, taking great care not to give anything away. In fact, they may even try to trick the child. For example, they may point to a bedroom door in the family home and ask, "Do you remember your old bedroom?" The child may then reply, "That was not my room. *This* was my room," pointing to the door that was, in fact, the entrance to the deceased's room. Sometimes the proof that they are indeed the genuine reincarnation may be even more dramatic. For example, the child may recover some valuable object or document that had been hidden away, the whereabouts of which was not known to any living family member. Eventually, the sheer weight of evidence will convince both families that the child really is the reincarnation of the deceased loved one. Of course, no objective record typically exists of what took place when the child first made contact with the past-life family, so it is impossible to know how accurate such accounts are.

It is understandable that researchers who are focused on trying to establish whether genuine reincarnation occurs will concentrate on the most convincing cases that they can find. But it is also somewhat misleading. It is like trying to judge whether a psychic really does have the ability to foresee future events by only looking at predictions that appear to have been accurate. Would you be willing to conclude that such a psychic really did have paranormal ability if you realized that five successful predictions had been extracted from a set that also contained a thousand times as many unsuccessful predictions?

This was brought home to me when we visited a school in the Chouf mountains, a short drive away from Beirut. Of the 900 or so pupils, twenty-one reported that they had past-life memories. Clearly, the total number of Druse children reporting such memories across the whole of Lebanon and beyond must be very large indeed. Given that, how representative of the complete set are the impressive cases that are written up for publication? If our experiences are anything to go by, the answer to that is "not very."

We did not have the time and resources to interview all of children who claimed to remember past lives, but the most promising cases were selected for us. Even with these cases, the memories were often so vague that there was no possibility of checking their veracity. Once those cases had been filtered out, we were left with only a handful of cases where the reported memories included specific, checkable details.

Once we began to investigate these cases, it immediately became apparent that most of the details did not check out. For example, one child reported that in his past life he had been named Ramiz Haidar and had worked as a truck driver in the Pepsi depot in Beirut. However, when we visited the depot, no one remembered anyone of that name, including a supervisor who had worked there for two decades.

Mehdi Hibous recalled quite a lot of details of his alleged past life, some of which did check out. For example, he claimed that in his past life he was called Melhem Melerb and had been killed by a shell while working on his tractor. On checking, it turned out that there really had been a Melhem Melerb who had been killed in this manner. However, Mehdi also provided us with the names of his past-life wife and five children—and these names were all incorrect. The clincher came when we approached the past-life family directly to inform them that Mehdi claimed to be the reincarnation of their lost loved one. They patiently explained that this could not possibly be true, as they already knew the identity of the reincarnated Melhem Melerb. Clearly, if two boys are both claiming to be the reincarnation of the same person, at least one of them must be basing their claims on false memories.

It is understandable that researchers who are sympathetic toward the possibility that some cases of alleged reincarnation may be genuine will tend not to write up cases where the alleged past-life memories are too vague to check out or are proven to be inaccurate. But the result is that the published cases are in no way representative of cases in general. Those few cases that involve a reasonable number of specific past-life memories, most of which appear to check out, are the exception rather than the rule. Furthermore, it is noticeable that such cases are typically reported many years after the current family and the past-life family first made contact and the past-life family accepted the child as a genuine reincarnation. If we accept the accounts given of what did and did not happen all of those years ago, it would be hard to explain the events in nonparanormal terms. But is it wise to do so? A huge amount of psychological research into the unreliability of memory would suggest not, no matter how sincere the claimants are.

Much of the final documentary focused on the case of Rabih Abu-Dyab. Rabih was a very likable twelve-year-old who claimed to be the reincarnation of someone who was not only an international football player but also a successful pop singer. I must admit that my first reaction was essentially,

"Yeah, right!" After all, surely this would be the fantasy of many young boys? However, it turned out that my cynicism was misplaced. There really had been an individual who fitted that description. Saad Halawi was a national hero who lost his life in an explosion during the civil war. That meant, of course, that most people in Lebanon would know a lot about Halawi's life as he had been the focus of much media attention during his life and following his tragic death. Rabih's mother was a huge fan of Halawi before Rabih was even born. Rabih did not appear to know anything about Halawi's life that was not generally known by most Lebanese.

For one scene in the documentary we took Rabih to the trophy room at the stadium where his team played. If any location should prompt a flood of emotional memories, surely this was it. Rabih recognized the image of Saad Halawi in a large framed picture on the wall, but no specific memories of any matches he had played in appeared to come back to him. Even handling the ball that his past-life self had allegedly kicked to score a goal in a cup final, signed by all members of the team, did not do the trick.

I was beginning to realize just how naive my Cunning Plan No. 2 had been. I had arrived in Lebanon expecting to be faced by numerous children all providing copious detailed past-life memories, but the reality was far from this. To remind you, I had hoped that we could find a child who had not yet made contact with their past-life family. We could arrange a carefully controlled initial meeting and administer a questionnaire to allow us to probe the child's past-life memories. In this way, we could statistically assess the accuracy of those memories. I could now see that this plan was never going to work.

It still struck me that it was important to know what really happened when a child claiming past-life memories was first introduced to their past-life family. Were the family members careful not to give anything away, to possibly even attempt to trick the child, as they assessed the credibility of the claim? Unfortunately, no one, including veteran researchers like Ian Stevenson, has managed to record what really happens on such occasions.

The closest we got to being able to do so was by arranging a meeting between Rabih and his past-life sister. At that stage, some members of Rabih's alleged past-life family had already accepted him as a genuine reincarnation of Saad Halawi, and others remained unconvinced (and it is worth noting that Rabih was not the only child claiming to be the reincarnation of Halawi). Rabih's past-life sister was keen to meet him in order to make up her own mind. Some days after Rabih had said that he would welcome such a meeting, it was set up—but Rabih was simply told that he would have a visitor that day without being told her identity.

This part of the documentary was handled very sensitively. Emotions would be running high. How would Rabih react if his past-life sister was not convinced that he was the reincarnation of the brother whom she had loved so dearly? Only the people who were essential to the filming of the scene, such as the cameraman and the soundman, were allowed to be present, meaning that yours truly was one of those excluded. I only got to see what happened later.

Was Rabih's past-life sister careful to not give any clues regarding her identity? Quite the opposite. Short of actually having her name tattooed on her forehead, she could not have provided more clues. Under the circumstances, given that just a few days before Rabih had agreed to this meeting, he was extremely slow to figure out who his "surprise visitor" was—but he got there eventually. In a genuinely moving scene, just after this happened, Rabih was embraced by his past-life sister. She had accepted him purely on the basis of her emotional reaction toward him—he really was a very appealing child—but there had been absolutely no attempt to rigorously test Rabih's supposed memories of his past life. Obviously, this was only a single case, and we should be cautious in generalizing from a sample of $n = 1$, but I suspect that it may be typical of other such meetings.

Given that serious researchers into reincarnation claims generally accept that past-life memories produced by hypnotic regression are likely to be false, it is surprising that they rarely give this explanation much consideration when it comes to spontaneous past-life memories. How might false memories of a past life develop in such cases? Given the impossibility of tracking every interaction that a child has from the moment of birth to the point where they claim to have such memories, any explanation must be to some extent speculative, but I would suggest that the following is a plausible account for the cases I encountered in Lebanon.[24]

When a child is first learning to talk, they will sometimes come out with utterances that make little sense to those around them. In a culture where reincarnation is almost universally accepted, it would be natural for parents and others to wonder whether some of these utterances are the child's attempts to talk about past-life memories. This would be particularly likely to happen if the child says things that to Western ears would be interpreted as referring to an imaginary friend. If some of the utterances were taken as suggesting a specific deceased individual as a possible reincarnation candidate, it would be natural for the parents to probe further in an attempt to test their hypothesis. They may, for example, ask questions such as, "Was your mother's name X?" or "Did you live in the town of Y?" They may ask the child if they remember significant events in the life of the deceased, perhaps showing the child photographs in the process. No one would expect the child to have a perfect memory of the past life, so any errors in what they said would be disregarded.

The unintended consequence of such interactions may well be the formation of false memories on the part of the child. It should be kept in mind that among the Druse, all deaths are publicly announced in the area where the deceased had lived. There is also a high level of interconnectedness between families as a result of the prohibition against Druse marrying non-Druse. Thus, there are many possible pathways by which information regarding a deceased individual may reach the child or the child's carers. If the child is then introduced to the past-life family and accepted as a genuine reincarnation of their lost loved one, this process could continue with even richer sources of information about the life of the deceased.

If reincarnation is not possible, why has belief in reincarnation persisted for centuries among the Druse? The answer may well be that the belief itself confers certain advantages among believers.[25] It is reasonable to suggest that a belief in reincarnation can help the bereaved to cope with the pain of loss as well as reducing the fear of death itself. The Druse have a reputation for bravery in battle that has no doubt helped them to survive centuries of persecution. They are said to have a battle cry of "Tonight my mother's womb," reflecting their belief that if they are cut down, they will be immediately reborn.

In terms of social cohesion, it has already been mentioned that the prohibition against marrying non-Druse leads to a high level of interconnectedness within Druse society. This is strengthened further by the belief in reincarnation, as there are links between families based not only on biological relationships but also on reincarnation. In all cultures where reincarnation is said to occur, the past-life family is typically of a higher social class than the biological family of the claimant. Thus, belief in reincarnation can play a role in providing social support for the poor. If the past-life family accepts a claimant from a poorer background as a genuine reincarnation of their lost loved one, they are likely to use their influence to help take care of the claimant. Even if reincarnation is not real, there are real advantages to the Druse in living their lives as though it were real.

7 DYING TO KNOW THE TRUTH

The following is a fairly typical example of what has come to be known as a near-death experience (NDE). This is from a fifty-five-year-old man who found himself in great pain on regaining consciousness after cardiac bypass surgery:

> All of a sudden I was standing at the foot of the bed looking at my own body. I knew it was me in bed because there were certain features I could recognize easily. I felt no pain at all. I was a bit puzzled. I wasn't anxious or worried. I just didn't understand what I was doing at the foot of the bed. Then I was travelling. I was at the entrance of a tunnel. There was a light at the end of the tunnel just like sunlight. I had a feeling of comfort and all the pain was gone, and I had a desire to go towards the light. I seemed to float along in a [horizontal] position. When I reached the other end I was in a strange place. Everything was beautiful, yet it was more of a feeling rather than seeing. I could see other people. They ignored me completely and the next thing I heard a voice saying, "You must go back . . . it is not your time." Afterwards I seemed to think it was my father [speaking] but this may have been imagination. Then I was in my bed. There was no sensation of moving back. It was just as though I woke up and all the pain was back and [hospital staff] were working on me.[1]

Elsewhere, I presented the following general definition of a near-death experience:

> An apparently transcendental experience, often experienced by people who are, or believe themselves to be, near death. The components of the NDE can include feelings of peacefulness and bliss, an out-of-body experience (OBE),

travelling down a tunnel towards a white light, entering the light, meeting spiritual beings (such as religious figures or deceased loved ones), a life review, and a boundary of some sort where the decision is made that the individual is to carry on living.[2]

Raymond Moody deserves much of the credit for raising awareness of this fascinating phenomenon among both the general public and the scientific community with his bestselling book *Life After Life*, published in 1975.[3] Moody collected about 150 accounts of NDEs over a ten-year period, leading him to conclude that life after death is real.[4] In the subsequent decades, much more has been learned about NDEs, but opinion is still very much divided regarding their true nature.[5] Are they, as many researchers believe, proof that consciousness can be separated from the physical brain and may even survive bodily death? Or are they, as maintained by most mainstream neuroscientists, a rich hallucinatory experience generated by unusual activity in the dying brain?

Moody identified fifteen elements that commonly occur during NDEs, but he noted that none of these elements occur in all reports and that there is some variability in the order in which they occurred.[6] Kenneth Ring collected data from 102 individuals who had close shaves with death and found that 48 percent of them reported NDEs. He identified what he referred to as a "core experience" consisting of five stages that tended to occur in a specified order.[7]

The first stage, reported by 60 percent of Ring's sample, was a feeling of peace and well-being, sometimes followed by an out-of-body experience, during which the NDEr feels that their center of consciousness has left their physical body. This second stage, reported by 37 percent of Ring's sample, typically involves the sensation of viewing the scene from above. About half of those who reported an OBE in Ring's sample said that they could see their own physical bodies from this external vantage point. Around a quarter of the NDErs reported entering a transitional region of darkness, sometimes before the OBE, sometimes after it. A brilliant light, often interpreted as being a spiritual being, such as Jesus or God, was reported by 16 percent of Ring's sample. The NDErs felt drawn toward this benevolent light, which, although incredibly bright, did not hurt their eyes. Sometimes this stage would involve a panoramic life review in which key events from the NDEr's life would be replayed. Ring's fifth and final stage, reported by only 10 percent of his sample, involved passing through the light into another realm, such as a beautiful garden, where spiritual beings, including deceased relatives and friends, were encountered. Some kind of boundary, such as a gate or a river, is experienced here, and the decision is made, either by the NDEr or by a spiritual being, that the time is not yet right for permanent entry into this realm.

Having a near-death experience typically produces profound long-term changes in the individual. On the positive side, it is commonly found that the NDEr becomes more socially and environmentally aware, becomes more spiritual, and has a greatly reduced fear of death. However, they may be reluctant to talk about their experience, for fear of being ridiculed or dismissed as being "crazy." Hopefully, such ill-informed reactions are less common now that more people accept that NDEs are genuine experiences, whatever their true nature. Even if those around the NDEr are sympathetic, there may still be great frustration on the part of the NDEr insofar as they cannot find the words to describe this profound yet ineffable experience. Sometimes the generally positive personality changes that often follow an NDE can be difficult for those close to the NDEr to deal with. If a now-saintly NDEr announces to their partner that they have sold the family home and given the money to charity, this may well put strain on the relationship!

Although most NDEs are indeed extremely positive, occasionally they are extremely unpleasant. Bruce Greyson and Nancy Evans Bush identified three distinct types of distressing near-death experience.[8] The first type is phenomenologically similar to the prototypical positive NDE, but the person having the experience finds it unpleasant. For example, they may see themselves from above on an operating table, realize that this means they are dying, and be terrified by that thought. In some cases, this type of negative NDE can transform into the more usual blissful experience if the person stops struggling against it and just lets it happen. A second type of negative NDE involves hellish imagery, such as demons torturing their victims. For me, it is the third type of distressing NDE that sends the biggest shiver down my spine. This one involves the NDEr finding themselves alone in an infinite void, convinced that they will never ever escape. Not surprisingly, the aftereffects of such negative experiences are greatly *increased* fear of death and possibly even PTSD.

Attempts have been made to avoid the confusion potentially caused by different investigators adopting different definitions of the NDE. Typically, these attempts involve the use of standardized scales to measure the "depth" of the NDE. One of the most commonly used was developed by Bruce Greyson.[9] The scale consists of sixteen items in four groups (relating to cognitive, affective, paranormal, and transcendental features) with a maximum possible score of 32. A score of 7 or above is said to be indicative of a true NDE.

It is sometimes claimed that around 15 percent of the American population have had a near-death experience, but this figure is based on nothing more than a survey that asked people if they had ever had "a close call with death."[10] Almost getting knocked down by a speeding car as you step into the road is not a near-death experience in the sense we are discussing here. A much more interesting and pertinent question is to ask what percentage of people who come close to death have an experience of the type we are concerned with. For reasons that are unclear, even this question yields a very wide range of answers. It can be argued that the best data to use in answering this question are those that come from prospective studies, that is to say, studies that systematically assess the incidence of NDEs in preselected samples, such as all of those admitted to a cardiac unit within a particular period. In 2005, I reviewed the four studies of that type that had been done to date and argued, based on their methodologies and sample sizes, that the best estimate was 10–12 percent.[11] Since then, three additional prospective studies have been reported, but 10–12 percent still appears to be a reasonable estimate.[12]

When *The Lancet* published the large-scale prospective study by Pim van Lommel and colleagues in 2001, I was invited to write a commentary to accompany the article.[13] I suggested, among other things, that some reports of NDEs may in fact be based on false memories. My reason for doing so was based on the results of follow-up interviews carried out with a subsample of the original group some two years after the initial data collection. At this stage, four out of thirty-seven patients who had, despite coming close to death, initially not been classified as having had an NDE were deemed to have had one. I suggested that, in such cases, the patients may have learned more about what typically takes place during a near-death experience, imagined it happening to themselves, and, as a result of imagination inflation, ended up with false memories of having had such an experience. Although at the time I believe that this was a reasonable hypothesis, I would like to take this opportunity to state that I no longer believe it. Since that time, a number of studies have directly assessed the consistency of NDE reports over time and the memory characteristics of memories of NDEs.[14] Taken collectively, the results of such studies suggest that accounts of NDEs generally remain consistent across time and that memories for NDEs are typically reported as being more vivid, intense, and emotional than memories for other real emotional events and for imagined events.

So, if accounts of NDEs are not best explained as being based on false memories, what is the best explanation?

GLIMPSES OF THE AFTERLIFE–OR VISIONS OF A DYING BRAIN?

Different theoretical approaches to explaining NDEs can be conceptualized as belonging to three broad categories.[15] Spiritual theories (also sometimes referred to as *transcendental* or *survivalist* theories) maintain that the NDE is exactly what it appears to be to the person having the experience. That is to say, such approaches view NDEs as strong evidence that consciousness (or the soul, if you prefer) can become separated from the physical substrate of the brain and may even provide evidence for life after death. If such theories are correct, they would prove that consciousness can only be explained by adopting a dualist approach. A general problem with this type of theory is that it does not appear to generate any testable hypotheses.

The second general category of explanations are psychological in nature. One example is the idea that NDEs are based on memories of being born, as proposed by Stanislav Grof and Joan Halifax and, perhaps surprisingly, supported by none other than the late Carl Sagan.[16] The idea is that the oft-reported tunnel is a memory of the birth canal, the light at the end of the tunnel is the light in the delivery room, and the beings of light are the medical staff delivering the baby. This theory has been strongly criticized on the grounds that newborn babies simply cannot form autobiographical memories as their brains are not yet physically developed enough to do so. Furthermore, babies do not face forward as they move down the birth canal, and the experience would certainly not feel like gently floating along!

The final nail in the coffin for this theory was provided by Susan Blackmore.[17] She reasoned that if this theory were correct, the nature of one's birth should influence the type of NDE one experiences. However, her data showed that reports of tunnels are just as common among people who were born by means of Cesarean section, and thus never passed through the birth canal, as they were among those birthed vaginally.

Another, arguably somewhat more plausible, psychological theory is that proposed by Russell Noyes Jr. and Roy Kletti.[18] They argue that NDEs act as a psychological defense mechanism in which the individual experiences depersonalization, leading to detachment from stress and engagement in pleasurable fantasies. Typically of such psychological theories, this account provides a reasonably plausible account of some aspects of the NDE, such as the OBE, but not of others, such as the sense of "hyperreality" typically reported by NDErs.

The final broad category is that of the organic theories, which attempt to account for the various components of the NDE in terms of abnormal brain activity. Such approaches are sometimes subsumed under the general heading of the "dying-brain hypothesis," but this is somewhat misleading as NDEs can sometimes occur to people whose brains are definitely not on the verge of death. NDEs are often reported by people who merely *believe* that they are in mortal danger even though objectively this is not the case.

The remainder of this chapter will provide an overview of this approach and consider some of the challenges that have been directed toward it by those favoring spiritual theories. It must be emphasized that there are no clear boundaries between these three theoretical categories. They are employed merely as a matter of presentational convenience and often blur into each other. For example, a psychological theory must involve underlying neural mechanisms, and even spiritual theories may postulate that the spirit can only leave the body when the brain is in a particular state.

Investigating NDEs under tightly controlled conditions is rarely possible for both practical and ethical reasons. Most NDEs occur at unpredictable times, often when the individual is simply going about their daily life. In principle, one could deliberately bring individuals to the verge of death in the laboratory and then bring them back, all the time collecting data on their physiological condition, but such research would rightly never be allowed on ethical grounds. The closest one could get to this situation is when such data are being collected during a medical procedure, the execution of which happens to coincide with the patient having an NDE. Prospective studies of the type already described can also provide valuable physiological data. It turns out, however, that most of the components of NDEs also occur in non-NDE contexts in which it is possible to have better control over the conditions pertaining at the time.

We will now consider a number of physiological factors that have been put forward as explanations for different components of the NDE. It is likely that any comprehensive model of the NDE as a whole will require synthesizing these accounts along the lines proposed by Susan Blackmore.

The feelings of bliss that are commonly reported during NDEs may well be a result of endorphins. Endorphins are released as part of the body's natural response to stress and pain and are thus highly likely to be produced in many situations in which NDEs occur. They not only reduce pain but also produce a generally positive emotional state that may have an effect on the content of the hallucinations experienced during an NDE.

This possibility is supported by anecdotal evidence from a case described by I. R. Judson and E. Wiltshaw in 1983.[19] As described by Blackmore, this is an unusual case that turns from a typically blissful experience into a nightmarish one.[20] At one point during the blissful phase, the NDEr, a seventy-two-year-old cancer patient who had been found in a coma, finds himself standing on a high plateau. He is approached by kindly, compassionate beings, but suddenly he begins to panic as he realizes something is wrong. In his words, "Alarm deepens into panic as they close in upon me. I beg them to go away. To my horror, they lay hands upon me and try to pull me out of shape. The pain in unbearable . . . please, please let me alone. You are destroying me. O please, why do you do this to me?"[21]

As the beings continue to manipulate him, he feels a powerful vibration along with an overwhelming, abhorrent odor. His account continues, "I am lying on a bed, looking up into the eyes of the two beings who have been and are still 'manipulating' me. The scene has changed, moving from the sandy outdoor landscape with its great rock to this small room with its curtained door through which there are agitated comings and goings."

As he awoke, confused and frightened, in a hospital room, he repeated the words, "This is evil, this is evil!" as he struggled and tried to remove the cannula and drip from his arm. Why might this account support the idea that the positive tone of most NDEs may be a consequence of endorphin release? The reason is that the transformation from positive to negative experience appeared to occur at the point when the patient was injected with naloxone, a drug known to block the effects of morphine—and naturally produced endorphins.

Various other neurotransmitters have also been suggested as possibly playing a role in NDEs. For example, Melvyn Morse and colleagues argued that serotonin release plays a more important role than endorphin release, especially with respect to mystical hallucinations and OBEs.[22] Karl Jansen developed a model of NDEs based on similarities between the NDE and the effects of ketamine, such as seeing bright lights and moving through tunnels.[23] There are, however, important differences between the two types of experience, including the fact that ketamine-induced experiences are more likely to be frightening and to feel unreal.[24]

There is strong evidence to suggest that NDEs may result from hypoxia (reduced oxygen levels) in the brain. For example, it has long been known that fighter pilots sometimes pass out as they perform maneuvers that involve such extreme acceleration that blood is prevented from reaching the brain. This is known as a G-LOC episode (acceleration (+Gz)-induced loss of consciousness). The effects of such extreme levels of acceleration have been studied under controlled conditions, and the similarity between what is experienced and several components of the NDE is striking. James

Whinnery, basing his observations on almost a thousand such cases, summarizes the similarities as follows:

> The major characteristics of G-LOC experiences that are shared in common with NDEs include tunnel vision and bright lights, floating sensations, automatic movement, autoscopy, out-of-body experiences, not wanting to be disturbed, paralysis, vivid dreamlets of beautiful places, pleasurable sensations, psychological alterations of euphoria and dissociation, inclusion of friends and family, inclusion of prior memories and thoughts, the experience being very memorable (when it can be remembered), confabulation, and a strong urge to understand the experience.[25]

Hypoxia is often associated with hypercarbia (that is, increased levels of carbon dioxide), which is itself known to cause sensations of bright lights, OBEs, recall of past memories, and mystical insights.[26]

Arguably, the most comprehensive theory of near-death experiences is that proposed by Susan Blackmore.[27] As previously stated, Blackmore's model is essentially a synthesis of previous suggestions plus some novel explanations for particular NDE components. One example of the latter is her proposed explanation for the oft-reported tunnel experience. She argues that cortical disinhibition in the visual cortex leads to random neuronal firings in that part of the brain. As there are more cells devoted to the center of the visual field than to the periphery, this would be experienced as a bright light at the center of the visual field that would slowly get bigger as more cells started to fire. This would result in the subjective sensation of moving through a tunnel toward a bright light.

A great deal of evidence supports the idea that the temporal lobes, especially the temporoparietal areas of the brain, are involved in NDEs.[28] For example, some temporal lobe epileptics report experiences associated with seizures that are also reported during NDEs, including OBEs, visions of dead friends and relatives, and mystical feelings of oneness with the universe.[29]

It should be noted, however, that the alleged link between temporal lobe epilepsy and mystical experiences has been questioned by some commentators,[30] and at least one study, designed to directly "identify and characterize the mystical experience associated with seizure activity," concluded that "none of the patients' descriptions met the criteria for mystical experience."[31] There is much stronger evidence for a direct link between temporal lobe activity and OBEs, however, as we shall see.

So far, the OBE has only been referred to in passing, despite it being one of the most intriguing aspects of the NDE. After all, it is the OBE that offers the most potential for proving the paranormality of the NDE. All of the other components of the NDE—the feelings of bliss, the sensation of traveling along a tunnel toward a bright light, the entering of another realm where spiritual beings are encountered—are entirely subjective. No matter how real these experiences feel to the NDEr—and typically they are reported as being "realer than real"—there is absolutely no reason why an outside observer should accept that they are anything other than hallucinatory in nature. But if, during the out-of-body phase of a near-death experience, information was picked up from remote locations that could not possibly be accounted for in any other way, this would provide strong evidence that consciousness really can become separated from the physical body. Several classic cases have been presented that, it is claimed, prove exactly that.

DOES CONSCIOUSNESS REALLY LEAVE THE BODY DURING AN OUT-OF-BODY EXPERIENCE?

For many people, the ultimate proof that NDEs are genuinely paranormal rests on accounts that describe the accurate perception of events that were taking place at the time that the experience occurred, provided that no plausible alternative nonparanormal explanations are possible. In practice, very few, if any, such cases exist. Blackmore provides a concise summary of the nonparanormal factors that could plausibly account for accurate accounts of events that occurred during an OBE: "information available at the time, prior knowledge, fantasy or dreams, lucky guesses, and information from the remaining senses. Then there is selective memory for correct details, incorporation of details learned between the end of the NDE and giving an account of it, and the tendency to tell a good story."[32] With these factors in mind, let us consider a couple of the classic cases that have been put forward as providing the strongest evidence to date that NDEs really are paranormal.[33]

One of the first accounts alleged to provide irrefutable evidence of paranormal perception during an OBE was reported by Kimberley Clark in 1984.[34] Clark was a social worker in the critical care unit of a Seattle hospital when she met Maria, a migrant worker who had been brought in a few days earlier following a heart attack and had then suffered a cardiac arrest while in the hospital. Maria had then described what we would now recognize as a typical NDE during which she had had an OBE and watched the medical staff as they tried to resuscitate her. At one point, she felt that she was way up in the air outside the hospital. She noticed a tennis shoe on a third-floor ledge of the building. Intrigued by this odd but specific detail, Clark agreed to Maria's request to go and look for the shoe.

Clark claimed that from outside the hospital at ground level she could see very little, so instead she went in and out of patients' rooms to check what would be visible from those vantage points. According to her, she had to press her face against the screen to see the ledges at all as the windows were so narrow. Amazingly, however, she did find a tennis shoe as described by Maria! The fact that the shoe would not be visible from outside the hospital and could only just be seen from vantage points within the hospital appeared to rule out any possibility that Maria had at some point overheard someone commenting on having seen a tennis shoe in such an unusual location. Memory researchers are well aware of the phenomenon whereby the source of a piece of information may be forgotten although the information itself is remembered. But the possibility that Maria may have overheard this tidbit and then incorporated it into her NDE imagery appears to be ruled out, given the difficulty of seeing the shoe from either outside or inside the hospital.

As often happens with paranormal claims, however, this case may not have been as strong as it first appeared. For example, other researchers assessed Clark's claims regarding how difficult it would have been to see the tennis shoe by placing a shoe on a similar ledge in the same building.[35] They discovered that it was clearly visible from ground level outside the hospital and also from inside the hospital without needing to press one's face against the screen. It should also be kept in mind that the actual incident occurred in 1977, some seven years before Clark reported it. We do not even have a direct account of what happened from Maria herself. On reflection, this case is too weak to be taken as proof that the mind can leave the body.

The case that is often presented as being the strongest evidence available to support a survivalist account is that of Pam Reynolds, first reported by Michael Sabom.[36] Reynolds underwent an operation in 1991 to remove a giant basilar artery aneurysm. A complex procedure known as hypothermic cardiac arrest had to be employed, which involved decreasing body temperature to 60.8 degrees Fahrenheit, stopping breathing and heartbeat, and the draining of blood from the head. Once the aneurysm had been removed, Reynold's body was warmed up, normal circulation and heartbeat were restored, and wounds were closed. Following her gradual regaining of consciousness in the recovery room, Reynolds described a vivid NDE.

Reynolds reported that she had an OBE very early in the procedure and was able to watch herself being operated upon. Her description was very accurate. After her heart was stopped, she passed through a dark vortex into a realm of light where she was looked after by deceased relatives. These relatives helped her to return to her physical body, and she correctly reported that "Hotel California" was playing in the operating theater when she returned.

Does this case really defy conventional explanations, as often claimed? Not everyone thinks so. For example, G. M. Woerlee, an experienced anesthesiologist, examined the case in detail and presents a convincing argument that the explanation rests on the fact that sometimes, albeit rarely, patients actually regain consciousness as they are being operated on.[37] They are unable to move and (usually) feel no pain, but they may be fully aware of what is going on around them. Typically, the last sense to go when one is losing consciousness and the first to return when one is regaining it is the sense of hearing. Reynolds's eyes were taped shut, but if she regained consciousness, she would almost certainly be able to clearly hear what was going on around her (despite having small molded speakers in her ears).[38] In such circumstances, it is natural to mentally visualize the scene based on incoming auditory information and prior knowledge.

As Woerlee makes clear, Reynolds's vivid OBE occurred in the early phase of her operation, before the cardiac bypass equipment had even been connected. As with many other similar cases, the timing is crucial. Pim van Lommel and colleagues claim that "this patient proved to have had a very deep NDE, including an out-of-body experience, with subsequently verified observations *during the period of the flat EEG*" (my italics).[39] This is simply untrue.

Blackmore makes the point that this is not the only misrepresentation of which van Lommel and colleagues are guilty.[40] Elsewhere in the *Lancet* paper reporting their prospective study, they state, "During the pilot phase in one of the hospitals, a coronary-care-unit nurse reported a veridical out-of-body experience of a resuscitated patient."[41] On first reading it, Blackmore interpreted that sentence to mean that this allegedly "veridical" OBE had actually occurred during the pilot phase, as did I. Did you?

However, if you reread the sentence carefully, you will note that the sentence does not actually *explicitly* say that—although it clearly is the most obvious interpretation of the sentence. It actually says the NDE was *reported* during the pilot phase, not that it *happened* then. It turned out that the NDE in question had taken place in 1979, nine years before data collection had even begun in van Lommel's study. Space limitations preclude full discussion of this case, but it is worth noting that the nurse who reported the NDE was not actually interviewed about the full details until almost three decades later, several years after van Lommel described it as being an inexplicable example of a veridical OBE.[42] Many questions remain unanswered regarding exactly what happened all those years previously, and this case cannot be taken as more than an intriguing anecdote.

This was not the first time that Blackmore drew attention to misleading presentations of cases by NDE researchers. It has often been claimed, including by van Lommel and colleagues in their *Lancet* paper, that "blind people have described veridical perception during out-of-body experiences at the time of this experience."[43] One such case was vividly described by Larry Dossey. According to his account, patient Sarah's heart had stopped beating toward the end of a routine operation. Fortunately, defibrillation was successful, and she recovered. To the amazement of both Sarah and her surgical team, she then reported, in Dossey's words:

a clear, detailed memory of the frantic conversations of the surgeons and nurses during her cardiac arrest; the OR layout; the scribbles on the surgery schedule board in the hall outside; the color of the sheets covering the operating table; the hairstyle of the head scrub nurse; the names of the surgeons in the doctors' lounge down the corridor who were waiting for her case to be concluded; and even the trivial fact that her anesthesiologist that day was wearing unmatched socks. All this she knew even though she had been fully anesthetized and unconscious during the surgery and the cardiac arrest.[44]

And, as if this was not amazing enough, Sarah was congenitally blind!

Not surprisingly, Blackmore was intrigued by this case and wrote to Dossey asking for further details and hoping that she might be able to interview Sarah for herself. His reply was not quite what she expected. He candidly explained that he had simply made the case up for illustrative purposes!

If we exclude those rare accounts that turn out to be deliberate hoaxes or just concocted by NDE researchers for illustrative purposes, how are we to interpret near-death experiences?[45] Are they really glimpses of an afterlife? Or is it more plausible that they are vivid and life-transforming hallucinatory experiences? The accounts that receive the most attention from survivalists are, not surprisingly, those in which the descriptions of information obtained during the OBE appears to match what was actually happening at the time.

However, in many cases, what is perceived does not match what was actually happening, supporting the idea that at least some NDEs, and possibly all, are hallucinatory in nature. Keith Augustine provided numerous examples of NDE reports that did not correspond with reality, including cases involving encounters with still-living people, a feature that is more common among children than adults.[46] In one case, a woman observed her own cardiac bypass operation, including her heart pumping away, lying next to her outside her body. In fact, the heart is not removed from the body during such procedures. In another case, a child reported looking at her mother from above—but her mother had a flattened and distorted nose, "like a pig monster." Encounters with mythological beings, sentient plants, and talking insects have also been reported.

Survivalists might be tempted to argue that such experiences are not, in fact, true NDEs but this would clearly be a very arbitrary position to adopt. The most reasonable conclusion to draw is that at least some NDEs are indeed hallucinatory in nature. The burden of proof is therefore on those who argue that at least some are not hallucinatory but genuine examples of paranormal perception. One notable attempt to do exactly that was reported by Sam Parnia and colleagues in 2014.[47]

The AWARE (AWAreness during REsuscitation) project was an ambitious multinational study lasting four years and involving fifteen hospitals around the world (the final paper reporting the results had no fewer than thirty-one coauthors). The authors describe the aims of their study as follows: "The primary aim of this study was to examine the incidence of awareness and the broad range of mental experiences during resuscitation. The secondary aim was to investigate the feasibility of establishing a novel methodology to test the accuracy of reports of visual and auditory perception and awareness during CA [cardiac arrest]." It is this secondary aim that is the focus of our attention here.

Between fifty and one hundred shelves were installed on hospital wards where it was likely that cardiopulmonary resuscitation (CPR) would be necessary during the period of the study. Hidden targets, consisting of a single image of religious or national symbols, animals, people, or newspaper headlines, were placed on these shelves. The picture could only be seen from a vantage point above the picture, near the ceiling. The hope (if that is the right word) was that if anyone experienced an OBE during CPR, they might see the image and be able to describe it once they had recovered (figure 7.1).

Although there is undoubtedly something a little bit *Monty Pythonesque* about this approach to investigating the nature of consciousness during an NDE, I fully support it. An accurate description of an image that could only be seen from above near the ceiling would pose a serious challenge to skeptics such as myself. When funding was obtained for the AWARE project, I appeared with Sam Parnia to discuss the project on a *BBC Worldwide News* broadcast. I recall Sam saying words to the effect that if, by the time the study had been completed, no one had accurately reported one of these hidden images, that would prove that the NDE was hallucinatory. I disagreed, arguing that I could quite appreciate that if I

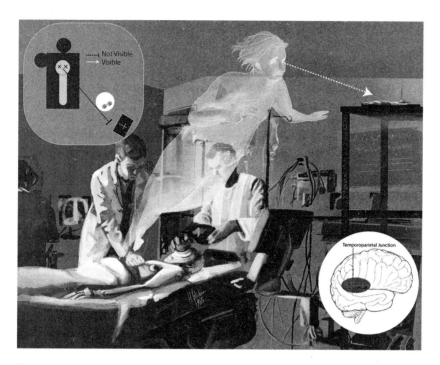

Figure 7.1

If someone having an out-of-body experience during a near-death experience could accurately describe a target stimulus that could only be seen from a vantage point near the ceiling (and all other possible means of obtaining that information had been ruled out), that would provide strong evidence that consciousness really can become separated from the physical brain. To date, this has not occurred. Evidence strongly supports the idea that out-of-body experiences are hallucinations caused by disruption to activity in the temporoparietal region of the brain. Illustration courtesy of Hawraa Wriden.

were to find my center of consciousness leaving my body during a cardiac arrest, my top priority might not be to pop over to take a peek at a hidden picture on a shelf!

A few years after this discussion, I appeared on an episode of the BBC's religious affairs program, *The Big Questions*, discussing the evidence for and against life after death. One of the other participants was Ken Spearpoint, who had been involved in the AWARE project. After the program, we traveled back to London on the train together, enjoying a friendly chat despite having been on opposing sides in the studio discussion. Now, I admit that

I cannot remember the exact words that Ken used during our conversation, but I do recall that I came away with the strong impression that he had said that one of the NDErs in the study had accurately reported one of the hidden images. I was stunned, anticipating that, once the results were published, I would have to either publicly acknowledge that this was indeed very strong evidence for paranormal perception or struggle to come up with plausible nonparanormal explanations for this finding. After all, no one likes to be publicly proved wrong.

It was therefore with some trepidation that I read the final report once it was published. The findings were certainly informative. Of 140 survivors of cardiac arrest, 101 were interviewed in detail regarding their experience as well as completing Greyson's NDE scale. Of these, 46 percent had memories of experiences during the cardiac arrest, 9 percent had NDEs, and 2 percent reported hearing and seeing events related to their resuscitation. One of the patients who reported experiencing such events was too ill to be interviewed in detail to allow for any checking of the details reported. The report from the other one, however, was a generally accurate description of what had taken place. So where was the report of an NDEr accurately reporting one of the hidden target pictures? There wasn't one! Despite the heroic efforts to collect the data in this study, only two cardiac arrest survivors reported memories related to their CPR—and neither of them were on wards with the hidden images at the time. I can only assume that Ken must have been referring to the one patient whose OBE memories of his CPR were generally accurate. But once again, that account can be explained on the assumption that the patient was actually conscious at times during the procedure.

ARE NEAR-DEATH EXPERIENCES A FUNDAMENTAL CHALLENGE TO NEUROSCIENCE?

Survivalists typically justify their belief that NDEs provide strong evidence that consciousness can become separated from the underlying neural substrate of the brain on two central arguments. The first is the claim that veridical perception can occur during the OBE phase, but we have already seen that such claims do not withstand close critical scrutiny. Instead, the OBE appears to be a compelling hallucinatory experience, as will be discussed further in the next section.

The second line of argument is that the findings from prospective studies simply cannot be explained in terms of our current understanding of neuroscience. In the words of Bruce Greyson:

> The paradoxical occurrence of heightened, lucid awareness and logical thought processes during a period of impaired cerebral perfusion raises particularly perplexing questions for our current understanding of consciousness and its relation to brain function . . . [A] clear sensorium and complex perceptual processes during a period of apparent clinical death challenge the concept that consciousness is localized exclusively in the brain.[48]

Similarly, Van Lommel and colleagues ask, "How could a clear consciousness outside one's body be experienced at the moment that the brain no longer functions during a period of clinical death with flat EEG?"[49] Parnia and Fenwick write,

> The occurrence of lucid, well-structured thought processes together with reasoning, attention and memory recall of specific events during a cardiac arrest (NDE) raise a number of interesting and perplexing questions regarding how such experiences could arise. These experiences appear to be occurring at a time when cerebral function can be described as at best as severely impaired, and at worse absent.[50]

These assertions are based on two underlying assumptions, both of which are open to question. The first is that is that the near-death experience occurs during a period of clinical death with flat EEG. However, as I have argued elsewhere, the possibility that the NDE occurs as the individual rapidly enters the period of flat EEG, or as they more slowly reemerge from it, cannot be ruled out.[51] Furthermore, recent research suggests an additional intriguing possibility. Lakhmir Chawla and colleagues recorded EEG in seven critically ill patients following the termination of life-support systems.[52] As would be expected, EEG activity faded as blood pressure dropped. However, in all cases, once blood pressure reached undetectable levels, there was an unexpected surge of electrical activity in the brain, approaching the levels normally associated with consciousness. This surge lasted for between thirty seconds and three minutes. Reasonably enough, the authors speculate that "since this increase in electrical activity occurred when there was no discernible blood pressure, patients who suffer 'near-death' experiences may be recalling the aggregate memory of the synaptic activity associated with this terminal but potentially reversible hypoxemia [low blood oxygen]."

Such a possibility is further supported by the results of a study by Jimo Borjigin and colleagues in which EEG was recorded in rats as they underwent experimentally induced cardiac arrest. [53] A transient surge of gamma oscillations (i.e., faster than 25 Hz) was recorded within thirty seconds after cardiac arrest. This gamma activity was global and highly coherent, and a large increase in gamma range connectivity between anterior and posterior areas of the brain was noted. This activity actually exceeded levels that occur during normal waking consciousness.

The second major assumption underlying the survivalists' argument that NDEs present a major challenge to neuroscience is the assumption that a flat EEG is always indicative of a totally inactive brain. Jason Braithwaite presents evidence from studies using a variety of methodologies to show that high levels of activity in deep structures within the brain may be undetectable in surface EEG.[54] It is reasonable to conclude that there is no good reason to accept the assertion that NDEs will never be explained by neuroscience.

THE PSYCHOLOGY OF THE OUT-OF-BODY EXPERIENCE

The one commonly reported component of NDEs that has been the focus of more research than any other is undoubtedly the OBE. One very important point is worth emphasizing here. The OBE, like other components of the NDE, often occurs outside of the NDE context. Indeed, there are some people who claim that by following various mental exercises they can induce an OBE more or less at will. Whereas the difficulties of studying OBEs within the NDE context are obvious, studying them outside of this context is relatively easy. The fact is that if even one such claimant could reliably demonstrate under controlled conditions their alleged ability to mentally leave their body and retrieve information from remote locations without the use of the known sensory channels, this would establish beyond doubt that paranormal perception is real. Despite thousands of anecdotal claims, we still await any such demonstration.

There is no good reason to assume that the psychology of OBEs experienced during NDEs is any different from that which applies to OBEs experienced in other contexts. Therefore, for the remainder of this chapter, I will outline the latest research on OBEs in general. In my opinion, the findings from OBE research are consistent with a general theory developed and refined by Susan Blackmore.[55] Blackmore's model relies on the generally accepted assumption that our sense of a unified self is a mental construction based on inputs from various sensory processes as well as top-down influences.

As discussed in chapter 3, perception of the outside world is a constructive process in which the input from vision, hearing, and other senses is integrated, guided by top-down influences, to produce a mental model of reality that is constantly updated as new sensory input is received. Under normal circumstances, our mental model of the outside world provides a very good approximation of reality, but the influence of top-down processes becomes stronger as the quality of the sensory input becomes more degraded.

In a similar manner, our sense of self and where we are in the world is also a mental construction based on input from vision, hearing, touch, proprioception (our sense of our own bodily position and muscular movement), balance, and so on. If asked to indicate where the sense of self is located, most people report that it is just behind the eyes, emphasizing the importance of visual input in this respect. The phrase "the eyes are the windows to the soul" primarily refers to the notion that we can read much about a person if we look into their eyes, but it also neatly encapsulates the idea of the mind (or consciousness or soul or whatever) looking out through the eyes at the outside world.

Blackmore argues that, at any moment, our mind will accept as real whatever corresponds to the content of the most stable mental model of the world (and our place in it) that is available at the time. Usually, the correspondence between our mental model and objective reality is high, as the model is constantly updated on the basis of new sensory input. However, under some circumstances, especially when the input is degraded (e.g., by drugs, anoxia, meditation, etc.), the best available model may be one based primarily on top-down influences such as memory, expectation, and imagination. The model of reality adopted may well also incorporate information from any remaining sensory input, both endogenous and exogenous.

The world of the OBE does indeed have much in common with the world of the imagination. OBErs report that they can fly, pass through solid walls, change size, and travel vast distances in the blink of an eye. The world that they inhabit is recognizably similar to the real world but also includes strange distortions of reality.

Several lines of evidence offer support for the idea that OBEs may be caused by disruption to the processes involved in integrating multimodal sensory information. One example of such disruption was referred to in chapter 2, where it was suggested that OBEs associated with episodes of sleep paralysis may be a consequence of the mismatch between activation in the vestibular system, responsible for coordinating eye and head movements and other proprioceptive feedback, and the absence of correlated visual input and feedback from actual head movements.

Many drugs that disrupt normal brain function are also associated with reports of OBEs, including ketamine, psychedelics such as ayahuasca, and cannabis, although interestingly cannabis users rarely experience an OBE while stoned. Studies have shown that brain damage, particularly in the region of the temporoparietal junction (TPJ), is often associated with experiencing OBEs.[56] The importance of the TPJ with respect to OBEs makes sense given that this part of the brain is known to be responsible for integrating information from various sensory modalities.

Other lines of evidence also demonstrate the involvement of the temporal lobes in OBEs. As long ago as the mid-1950s, classic experiments by Wilder Penfield showed that direct mild electrical stimulation of this area of the brain produced hallucinations, memory "flashbacks" (although these may in fact be false memories), and OBEs.[57] Such electrical stimulation is often carried out prior to brain surgery so that the neurosurgeon can identify the locations of important areas, such as those involved in language, in order to minimize damage to such areas.

In 2002, Swiss neurosurgeon Olaf Blanke and colleagues reported the results of electrically stimulating the cortex of a forty-three-year-old woman who had been suffering from epileptic seizures for eleven years.[58] Mild stimulation of the TPJ in the right cerebral hemisphere produced sensations of "sinking into the bed" or "falling from a height," as well as feeling that her arms and legs had been shortened. When the stimulation was increased, the patient reported, "I see myself lying in bed, from above, but I only see my legs and lower trunk." She then reported the sensation of floating six feet above the bed, close to the ceiling.

Similar results were reported from Belgium by Dirk De Ridder and colleagues.[59] In this case, the patient was a sixty-three-year-old man suffering from intractable tinnitus. Out-of-body experiences, during which the patient felt as if he was displaced to a location behind and to the left of himself, were repeatedly induced by stimulating the right TPJ, although in this case the patient did not report seeing his own body.

In one case it was possible for Lukas Heydrich and colleagues to record electroencephalographic activity in a ten-year-old epileptic boy as he actually had an out-of-body experience during a seizure.[60] The boy had a vivid sensation of being outside his body and floating near the ceiling, from which vantage point he could look down and see his mother but not himself. He reported that he had then been able to fly above the hospital and go way above the world, but he fully appreciated that the experience had been illusory. The focus of his epileptic activity was in the right temporal lobe, and a magnetic resonance image identified a lesion in the right angular gyrus.

The effects of drugs, brain damage, and direct cortical stimulation strongly support the idea that OBEs are a consequence of disruption of the neural processes responsible for integrating information from a range of sensory modalities, but can OBEs be induced in normally functioning brains? Recent research using virtual reality technology provides a positive answer to this question. But before we describe those studies, let us take a brief detour to describe a technique that produces an effect that is not as impressive as a full-blown OBE but is pretty striking nonetheless. I am referring to the so-called *rubber hand illusion*.[61]

The rubber hand illusion is a favorite demonstration in undergraduate psychology programs and at psychology department open days around the world. To induce the illusion, a volunteer must place, say, their left hand on a

Plate 1

Examples of *Skeptic Trumps* cards, courtesy of Crispian Jago and Neil Davies.

Plate 2
Did Jesus really turn water into wine? If he did, that is undeniably an example of psychokinesis in action and yet many people would not consider religious miracles as paranormal events. Note that, as is the case for many illustrations in this book, this is a drawing not a photograph.

Plate 3
A French postcard from around 1906 illustrating various symbols associated with good luck. Readers may be surprised to see the number 13, more commonly associated with bad luck, included but prior to the First World War the number was indeed viewed as portent of good fortune in France.

Plate 4
A fortune teller with crystal ball, Tarot cards, and a drained teacup.

Plate 5

The day I saw a large terracotta head of Queen Elizabeth II with flowers for hair on the back of a lorry really brought home to me the fact that virtually everyone carries a high-quality camera around in their pocket these days—and yet convincing photographs and videos of flying saucers remain elusive. Photograph by the author.

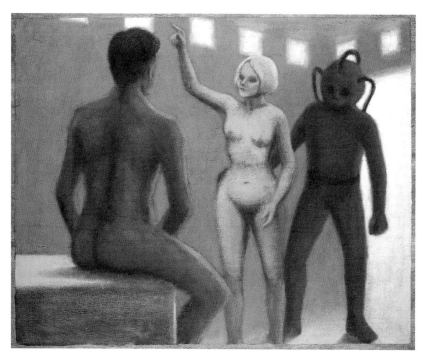

Plate 6
Artist's impression of Antonio Villas Boas with two of the aliens who abducted him. The one with blond hair is presumably the one with whom he had sexual intercourse (twice).

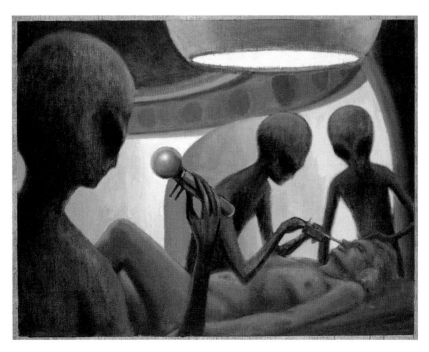

Plate 7
Artist's impression of aliens inserting an implant into an abductee's nose.

Plate 8
The cover of *Fate* magazine from November 1956 featuring the story of Bridey Murphy.

table beneath a screen that is positioned a few inches above the table's surface. They are thus unable to see their hand. However, a rubber hand is placed in full view on the upper surface of the screen, directly above the volunteer's real hand. Now the fun starts. The experimenter gently strokes the fingers of the rubber hand with a small brush while simultaneously making exactly the same brush strokes on the real hand. After a few minutes, many people report that they now feel that the rubber hand is their real hand. The illusion can be so strong that if the rubber hand is "threatened" with, say, a knife or a hammer, the volunteer reacts as if their real hand is in danger—as verified by simultaneously measuring their levels of physiological arousal at the time.

Clearly, there is a big difference between the distortion of one part of the body schema as induced by the rubber hand illusion and a full-blown OBE, but it does illustrate that distortions of the body schema can be induced by deliberately disrupting the processes that underlie multimodal sensory integration. In this case, it is clear that the visual input of the fingers of the rubber hand being stroked in synchrony with the tactile input from the real hand tricks the brain into constructing an inaccurate mental model of the real hand's position. This same basic principle has been extended by other investigators, using virtual reality technology, to induce OBE-like experiences in the laboratory.

Bigna Lenggenhager and colleagues used a similar approach to the rubber hand illusion but replaced the rubber hand with a virtual whole body.[62] Participants wore a virtual reality headset that fed them a live image of their own back being stroked with a stick as viewed from a vantage point some six feet behind where they were actually standing. When the visual input of the stroking was in synchrony with the tactile sensations of being stroked, participants felt as if the touch was on the virtual body rather than on their own backs. They also reported feeling as though their sense of self was outside of their real bodies and closer to the virtual body.

A similar, although subtly different, approach was taken by Henrik Ehrsson.[63] Participants in this study sat in a chair wearing a virtual reality headset that presented them with a view of their own back from six feet behind their actual position. Ehrsson stroked the participant's chest with a stick while simultaneously stroking the air just below the cameras in such a way that the stick came in and out view as it entered and dropped below the cameras' line of sight. Whereas in Lenggenhager and colleagues' study the sense of self had moved forward, with this method, after a couple of minutes of stroking, it was experienced as displaced farther back, toward the cameras.

In an interesting follow-up study, Lenggenhager and colleagues used a setup that allowed them to induce both types of experience in the same experimental context.[64] Male volunteers lay face-down on a table that had a hole cut out of it, thus allowing the experimenters to stroke either the volunteers' chests or their backs. Cameras were set up six feet above the volunteer, feeding an image of his back into a virtual reality headset. When the image of his back showed stroking that was in synchrony with the tactile stroking sensations, the volunteer reported that he felt as if he had moved downward, closer to the viewed body on the table. When his chest was stroked, the volunteer reported that it felt as though his sense of self had moved upward toward the camera. They also reported that they felt as if they were looking down on their body and that the stroking felt as if it were on a floating, imaginary body.

Valeria Petkova and Henrik Ehrsson used this same approach of disrupting the normal processes of multimodal sensory integration using virtual reality technology and synchronous stroking to induce an illusion of "swapping bodies."[65] Depending on the experimental setup, volunteers could be induced to feel not only as if their sense of self was located inside a mannequin but even that they were shaking hands with themselves!

Although it would be premature to claim that, at this stage of our knowledge, neuroscience can provide a comprehensive explanation of NDEs, real progress has been made in recent years especially with respect to improving our understanding of the nature of the OBE component. As our understanding of neuropsychological functioning in general continues to expand, aided by the use of increasingly sophisticated techniques for probing the neural substrate in living, working brains, so, too, we can anticipate that our understanding of the NDE will also grow. In contrast to spiritual accounts of near-death experiences, neuropsychological theories can be put to the test empirically—and that is how real understanding is obtained.

8 NO SUCH THING AS COINCIDENCE?

If mainstream science is right to reject the idea that paranormal forces exist, how are we to account for the fact that a sizable proportion of the general population claim to have had direct personal experiences of such forces? As we have seen in previous chapters, many of the more extreme claims can be accounted for in terms of unusual activity within the brain and the fact that hallucinatory experiences and false or distorted memories are much more commonplace than most people realize.

For many people, however, their general belief in the paranormal is often reinforced by more frequent, less spectacular, ostensibly paranormal events, such as knowing who is calling before picking up the phone, dreams that appear to come true, and acting on hunches that turn out to be correct. This chapter and the following one will consider the ways in which many of these experiences can be explained in terms of cognitive biases—that is, underlying biases in how we process information from the world around us that may lead us to believe that we have had a paranormal experience when, in fact, plausible nonparanormal explanations exist.

I first reviewed the research literature on this topic back in 1992. The tasks I set myself then were, "firstly, to demonstrate that certain situations are likely to be misinterpreted as described above and, secondly, that believers (sheep) may be more prone to such misinterpretation than disbelievers (goats)." Back then, there was not a great deal of relevant research to review. Based on what there was, I concluded, "It is clear that there is plenty of evidence to show that cognitive biases do exist which would lead to people misinterpreting certain situations as necessarily involving psi when in fact perfectly acceptable nonparanormal explanations are available." I went on to say, "The question of whether believers are more prone to such biases than goats cannot be answered with certainty, but the limited evidence available suggests that this is a real possibility."[1]

Fifteen years later, in collaboration with Krissy Wilson, I published an updated review of research in this area.[2] By this point, considerably more research had been carried out. Some of the cognitive biases identified in the early research did indeed appear to be stronger in believers than skeptics, but findings were mixed with respect to others. Now, some three decades after my initial review, there have been too many additional studies to comprehensively review in two chapters, but the overall picture remains the same.[3] This chapter will present an overview of research into the psychology of coincidences, and the chapter 9 will consider other cognitive biases of potential relevance.

WHAT ARE THE CHANCES?

As I was preparing to write this chapter, my sister-in-law, Jane Hatzimasouras, returned from a holiday on the Greek island of Spetses with a rather remarkable true story to tell. One day she was passing the time in a café with friends. One of her party returned from a swim with a ring that he had found on the seabed near a rock about 40 yards away from the main shoreline. A name and date were engraved on the inside of the ring: Athena Karagiannis and November 17, 1979.[4] It is traditional in Greece when a couple marry for the groom to have a ring engraved with his wife's name, along with the date of their wedding, and vice versa for the bride.

As the group excitedly mused on how the ring had come to be in its watery hiding place, Jane could not help but notice a woman who was not one of her group listening in intently. Eventually, this woman approached my sister-in-law's party and said, "I hope you don't mind me interrupting, but I could not help but overhear you discussing the ring that you have found. You see, my surname is Karagiannis and my mother's name is Athena—and about forty years ago, my dad lost his ring while swimming in the sea!"

The group then waited in anticipation as the stranger called her mother to check on the date of her mother's wedding day, only to be disappointed to learn that it had actually been March 8, 1980. It appeared that the ring was not the one that had been lost four decades earlier after all. But then the woman called her father to tell him what had happened, and he explained that he had had the rings engraved not with the wedding date but their engagement date—and that date was indeed November 17, 1979!

What were the chances that Mr. Karagiannis's daughter would just happen to be in the right place at the right time to overhear a conversation about a found ring that happened to be the very same ring that he had lost some four decades earlier? I think we can all agree that, even if we cannot work out the exact probability of this co-occurrence of events, it must be astronomically low. For some people, the probability would be so low that they would be strongly inclined to believe that some unseen force—perhaps fate?—had a role to play. Surely, something more than blind chance must have been at work here?

There is no denying that coincidences such as these have a definite emotional impact. They are the sort of thing that we cannot wait to tell our friends about, often beginning our anecdote with the words, "You'll never believe what happened to me!" Sometimes they end up being reported in newspapers and on TV and may be featured in books and academic articles. Here are a few of my personal favorites.

David Hand includes a lost-and-found wedding ring example of his own in his wonderful book *The Improbability Principle*.[5] In 1995, in Sweden, Lena Påhlsson lost her wedding ring, only to find it encircling a carrot she pulled up in the garden in 2011. However, Hand begins his book with an even more amazing coincidence:

> In the summer of 1972, the actor Anthony Hopkins was signed to play a leading role in a film based on George Feifer's novel *The Girl from Petrovka*, so he travelled to London to buy a copy of the book. Unfortunately, none of the main London bookshops had a copy. Then, on his way home, waiting for an underground train at Leicester Square station, he saw a discarded book lying on the seat next to him. It was a copy of *The Girl from Petrovka*.

As if that was not coincidence enough, more was to follow. Later, when he had a chance to meet the author, Hopkins told him about this strange occurrence. Feifer was interested. He said that in November 1971 he had lent a friend a copy of the book—a uniquely annotated copy in which he had made notes on turning the British English into American English ('labour' to 'labor' and so on) for the publication of an American version—but his friend had lost the copy in Bayswater, London. A quick check of the annotations in the copy Hopkins had found showed that it was the very same copy that Feifer's friend had mislaid.

Journalist and broadcaster Simon Hoggart recounted the following tale in his book, coauthored with Mike Hutchinson, *Bizarre Beliefs*:

> In 1994, Simon Hoggart's wife was traveling with a friend on a tube train. They were to visit an exhibition in Central London. She was talking about the actor, Richard E. Grant, who she did not know, but who lived in her neighbourhood. The friend said she could not place him; what did he look like? At that moment, Richard E. Grant got on to the train, and sat opposite them. He was holding a ticket for the same exhibition they were going to.[6]

Perhaps there is something about tube stations that facilitates the occurrence of amazing coincidences?

Research shows that we tend to be more impressed by coincidences that involve us personally than by the same coincidences happening to other people, so I will risk you being much less impressed than I was with the following coincidence that I experienced as a teenager.[7] At the time, I embraced a lot of New Age beliefs. Eager to learn more, I was reading the book *The Morning of Magicians* by Louis Pauwels and Jacques Bergier, a book crammed to breaking point with what I would now view as New Age nonsense, pseudoscience, and pseudohistory.[8] At the time, of course, I lapped it up. As I read it, I had David Bowie's *Hunky Dory* album playing in the background, specifically the track "Quicksand." I had listened to the album many times but did not understand all of the references in Bowie's poetic lyrics. Then, at the very moment that Bowie sang the line, "I'm closer to the Golden Dawn, immersed in Crowley's uniform of imagery," I read the words in Pauwels and Berger's book about the occult Golden Dawn society, one member of which was the infamous Aleister Crowley. Up until that point, I had never really heard of either. Needless to say, I saw this event as having tremendous mystical significance.

As with all claims of miraculous events, it is wise to be a little cautious in accepting all claims of amazing coincidences at face value. For example, Martin Plimmer and Brian King include the following account in their book *Beyond Coincidence*:

> In June 2001, ten-year-old Laura Buxton of Burton in Staffordshire was at a party where she wrote her name and address on a luggage label, attached it to a helium balloon and released it into a clear blue sky.
>
> The balloon floated 140 miles until finally coming to rest in the garden of another Laura Buxton, aged ten, in Pewsey, Wiltshire.[9]

A lovely, heartwarming little story, the only problem being that it isn't quite true. According to Snopes, checking back to the original story in the *Swindon Advertiser* reveals that the balloon did not land in (the second) Laura Buxton's garden but was found by farmer Andy Rivers in a field while out checking his cows.[10] Mr. Rivers knew his neighbors, the Buxtons, had a daughter called Laura so he "returned" the balloon to them. This was not, of course, the Laura Buxton who had launched the balloon (also, for fellow pedants out there, it turns out that the second Laura was nine, not ten, years old). This is still quite an interesting coincidence, but the balloon being found by someone who knew of a Laura Buxton is nowhere near as unlikely as the balloon being found by someone named Laura Buxton.

Despite the occasional exaggerated claim, there are numerous well-documented accounts of amazing coincidences. For example, people really do get hit by lightning more than once. In fact, Virginia park ranger Roy Sullivan was hit by lightning at least seven times (he claims he was also struck by lightning as a child, but we only have his word for it).[11] There are numerous examples of people getting very, very rich by winning lottery jackpots more than once. In June 1980, Maureen Wilcox found she had the winning numbers in both the Rhode Island Lottery and the Massachusetts Lottery.[12] Unfortunately for her, however, she had the winning numbers for the Rhode Island Lottery on the Massachusetts ticket and vice versa, so she did not win a cent—but, hey, it's still a pretty amazing coincidence!

Given our fascination with coincidences, it is not surprising that they have attracted the attention of both mathematicians and psychologists.[13] One of the reasons that we are more impressed with coincidences than perhaps we should be is the fact that we are all extremely poor intuitive statisticians.[14] In other words, in everyday life we frequently have to make decisions based on real-life probabilities, and we are simply not very good at doing so. Our poor intuitive understanding of probability can manifest itself in a variety of ways. One example is our frequent failure to appreciate what mathematicians refer to as *the law of truly large numbers*. This law states that with a large enough number of opportunities for an event to occur, even extremely unlikely events become probable.

The most obvious and familiar example of the law of truly large numbers in action is national lotteries. For example, the chances of winning the national lottery in the United Kingdom at the time of writing are around one in forty-five million—and yet we are not surprised when people regularly win the jackpot because, in this context, we fully appreciate that so many tickets are bought that it is indeed quite likely that at least one of them may match all six randomly drawn numbers (between 1 and 59).

In other contexts, the operation of the law does not appear to spring to mind as an adequate explanation for many people, such as when the same person wins the jackpot more than once. When newspapers report double winners, the probabilities are often reported in a way that, although technically correct, are not actually relevant to the question that should really be asked. Implicitly, the question that is usually being asked is "What is the probability that person X would win the jackpot in this particular lottery twice *if they only ever bought two tickets?*" It is immediately obvious with the emphasis added here that X may well have bought more than one ticket on each of the two days they bought their winning tickets. That would clearly increase the odds of them winning on each of those days. In fact, they probably bought multiple tickets on lots of other days as well, further improving their chances of winning the jackpot. But we can go even further. If we consider the fact that there would be a "truly large" number of people behaving in the same way for that particular lottery, it becomes even more likely that we will get a double winner. The final level comes when we consider that there are numerous national, international, and state lotteries in operation around the world. It becomes almost inevitable that multiple winners will crop up—and end up being reported in the media.

Consider the law of truly large numbers as it applies to my David Bowie/Golden Dawn example. I have no idea how many people had bought the *Hunky Dory* album by the time I was listening to it on that occasion, but I know it would be a "truly large number." A lot of those people would no doubt, like me, be playing it repeatedly. I suspect a smaller, but still very large, number of people read the cult classic *The Morning of Magicians*. Some of those people would, like teenage me, be in the habit of playing records while reading. It suddenly seems not quite so surprising that one of those people experienced the coincidence that I reported—it just happened to be me.

Sometimes the law of truly large numbers can operate in ways that are much less obvious than in the case of national lotteries. One nice example is the so-called Birthday Problem. The problem is usually stated as follows: How many randomly selected people would you need to have at a party to have a fifty-fifty chance that two of them share the same birthday (ignoring year)?[15] We will ignore leap years to keep things simple. If you have not come across this problem before, give it some thought before I reveal the correct answer. If you are very mathematically gifted, it is, of course, possible to work out the exact answer, but that is not a trivially easy task even then. We are more interested here in what you intuitively feel is approximately the right answer.

*** Short (and probably imaginary) musical interlude ***

Okay, here is the answer, presented in the form of an endnote (to minimize the chances of you just happening to glance at it accidentally while thinking about the problem).[16] That answer strikes most people as being far too low. It may be that, in some cases, people have misunderstood the problem as being "how many people would you need to have at a party to have a fifty-fifty chance that they share *your* birthday?" But even when it is made very clear that the problem is asking about the probability of any pair of people having the same birthday, people typically greatly overestimate the number required.

If you are still not convinced that the answer I presented is correct, you might like to consult, say, David Hand's book to see how the answer is calculated.[17] Or, if you are able to write simple computer programs, you could write a bit of code to generate random numbers between 1 and 365 (to correspond to each day of the year) repeatedly until a number is generated that has already been produced. You will see that the strings of numbers generated are less than or equal to twenty-three numbers long about half of the time. Or you may be convinced by visualizing someone throwing darts at random at a dart board divided into 365 sectors. Of the darts that actually hit the board, the first one could land in any sector and the second one would only have one chance in 365 of landing in the same sector. The third dart would have two chances in 365 of landing in the same sector as one of the first two—still very unlikely—but as more and more darts are randomly thrown, the chances of such an outcome would be improving all the time. By the time, twenty or more darts have been thrown, we might not be too surprised to see a dart land in an occupied sector.

What we typically fail to intuitively take into account when we first think about the Birthday Problem is the fact that the number of possible pairings increases exponentially as we add in each new person (or dart). When there are twenty-three people in the room, there are 253 possible pairings, so having at least one pair with the same birthday is not at all unlikely.

Another factor that has been identified as potentially resulting in what appear to be astronomically unlikely coincidences is that of *hidden causes*. A nice example of this, discussed by James Alcock, is the "perfect" bridge deal, in which each of four players is dealt thirteen cards of the same suit from a shuffled deck.[18] Now, it should be noted here that such an outcome is no more or less likely than any other prespecified bridge hand, but it is certainly more noticeable—and, like any other prespecified hand, it is exceedingly unlikely. There are, in fact, approximately 2.23 x 10^{27} (that's 223 followed by twenty-five zeros—indubitably, a truly large number!) possible distributions. To bring home just how mind-blowingly huge that number is, Alcock provides the following quotation from the McWhirter brothers: "Such an event having once occurred, it should not logically recur, even if the entire world population made up in fours and played 120 hands of bridge a day, for another 2,000,000,000,000 years."[19]

Given the extreme improbability of such bridge deals, it is therefore puzzling that cases of perfect bridge deals are reported every year. You may be tempted to concur with the McWhirters that such reports must due to either "rigged shuffling or hoaxing." Both possibilities would be good examples of hidden causes, but Alcock notes an even subtler possibility. New decks of cards are almost always organized in ascending order by suit. This means that if such a deck were subjected to two "perfect" shuffles (in which the deck was divided into two equal halves, which were then perfectly interleaved), the deal would always be a perfect bridge deal. It is therefore not that unlikely that this would indeed happen from time to time.

The more exact the coincidental match between the elements involved, the more impressed we are by the coincidence—but we are often almost as impressed by an almost exact match. For example, when giving public talks sometimes I illustrate the probabilities involved in the Birthday Problem by asking audience members to shout out, in turn, their birthdays until we reach someone whose birthday has already been called out.[20] As you will now expect, this point is reached about half the time with twenty-three or fewer members of the audience contributing. But people will often spontaneously comment when a previous audience member has shouted out a birthday either a day before or a day after their own. As another example, we would no doubt be very impressed when, as can happen, someone wins the jackpot twice in the same lottery—but we'd probably be almost as impressed if they won the jackpot twice in different lotteries or won the jackpot and a smaller, but still substantial prize, in the same lottery. Needless to say, our willingness to count close matches as being almost as impressive as exact matches will greatly increase the chances of such matches occurring.

COINCIDENCE AND OSTENSIBLY PARANORMAL EXPERIENCES

Why are coincidences relevant to ostensibly paranormal experiences? The answer is pretty obvious. In many cases, the most obvious nonparanormal counter-explanation is simple coincidence. At first sight, "mere coincidence" appears to be a very poor explanation for such events. Put yourself in the position of someone who has not heard from or even thought about a long-lost friend for several years who, just as they are thinking of that person, gets a call from them. Could that really be just a coincidence?

Not only does dismissing the event in that way feel wrong on an emotional level, it does not initially appear to provide an adequate explanation on an intellectual level either. After all, if that really was the first time that you thought about your friend in many years, surely the chances were literally several million to one against them just happening to call you at that precise moment? But a moment's reflection will reveal that coincidence may well be a more plausible explanation than any kind of spooky psychic link. Just consider how many people you think about in a single day. Most of the time when you think about them, either the phone does not ring or the phone rings and it is someone else calling. But, occasionally, simply based on the laws of probability, the person calling will be the person you were thinking about. In fact, it would be *really* spooky if this never happened.

Similarly, in the case of dreams that appear to come true, we tend not to think about the fact that pretty much everyone dreams pretty much every night. It would be amazing if no one ever had a dream that corresponded, purely by chance, to some unexpected event in real life sometime later. As we have seen, according to the law of truly large numbers, even very unlikely coincidences are likely to happen given enough opportunities.

American mathematician John Allen Paulos provides a nice illustration of how the law of truly large numbers applies to ostensibly precognitive dreams.[21] Let us suppose that you would only be willing to label a dream as "potentially precognitive" if details within the dream matched future events in real life with an estimated probability of occurring by chance of only one in 10,000. Surely, all of us would be pretty impressed if we were the one to have such a dream?

That means that on any particular night the chances are very high indeed that we would *not* have such a dream: 9,999 times out of 10,000, in fact. The chances of us not having such a dream over a period of a whole year are still very high. The actual probability can be calculated by multiplying 0.9999 by itself 365 times. That is about 0.964, meaning that 96.4 percent of people who dream every night will not have a "potentially precognitive" dream over a one-year period. But that means that 3.6 percent of those people *will* have such a dream—and that translates into millions of people worldwide.

The question that immediately springs to mind in such circumstances is "What are the chances that that particular coincidence would occur to me?" But to decide whether the event involves anything that cannot be explained purely in terms of probability theory, the question we should really be asking is "What are the chances that anyone at any time would experience any coincidence as unlikely as that?" That way, we can immediately see that anyone who experienced such an unlikely coincidence was bound to be blown away by it—it just happened to be you!

Another overlooked factor in such situations is that there is often more at work than pure chance even though we may not be aware of it. One such hidden factor in our hypothetical example of the phone call from the long-lost friend would be anything in the real world that would cause you both to start thinking of each other after a long period of not doing so. For example, suppose that you were both fans of a particular musician during your college days and frequently went to their concerts. Over the years, your musical tastes may have moved on, but you and your friend both happened to catch a minor news story about this (now aging) musician. That got you both reflecting on those carefree days at college and the many good times you had together. By the time your friend decides to track you down and give you a call, you might have both forgotten what prompted you into this train of thought in the first place—and you would be amazed that your long-lost friend just happened to call you "out of the blue" just as you were thinking about them.

The principle of "near enough is good enough" crops up in numerous paranormal contexts. For example, suppose it was not our friend who called us out of the blue just as we were thinking about him but his mother, father, brother, or sister, or a mutual friend, and so on. We would still be very impressed by the coincidence. Suppose we had a dream of being involved in a serious car accident as we were driving to work, and the very next day we had a minor bump driving to the shops. Again, we may be tempted to wonder if our dream provided us with a psychic glimpse into the future despite the fact that the match was close but not exact.

People do not expect paranormal sources of information to be 100 percent accurate. In fact, if they appear to be absolutely spot on, this may arouse suspicion. Nowhere is this better illustrated than in the way that true believers will often try very hard to find a way to interpret a clearly incorrect statement in a psychic reading as somehow "near enough" to be counted as accurate. For example, during a stage performance the medium may say, "I'm getting the name Elizabeth? Does that name mean something to someone in the audience?" In the unlikely event that no one replies positively (and, as we've seen, even highly unlikely events do occur), the medium may go on, "Elizabeth? Or Lizzy? Liz? Betty, perhaps?" If an audience member should timidly suggest, "Well, my dead grandmother's name was Lisa?" both medium and audience will happily accept that as "close enough."

A BRIEF DETOUR: POPULATION STEREOTYPES

There is a hidden cause behind a fun little demonstration of an ostensibly paranormal experience that I often include in public talks on anomalistic psychology, especially when I have a reasonably large audience. I explain to my audience that an important part of proper skepticism is to always be open to the possibility that you may be wrong. In that spirit, I tell the audience that I would like to do a little experiment with them to see just how psychic they are. I tell them that I am going to try to telepathically send a simple message from my mind to theirs. "I'm thinking of a number between one and ten," I say. "Not three, because that's too obvious. I want you to make a mental note of the first number that comes into your mind *now!*"

I then explain that, with such a large audience, we would expect around 10 percent of them to guess the number correctly just by chance, so we should only get ecstatically excited if considerably more than 10 percent of the audience get it right. I then, apparently somewhat nervously, ask, "Did anybody think of the number seven? If you did, raise your hand." With a large audience, I can, in fact, be very confident that around a third of them will put up their hand.

Feigning surprise, I will try another, slightly more complicated example. "This time I'm thinking of a two-digit number between one and fifty. They are both odd digits and they are not the same. So, it could be fifteen—one and five, both odd digits, not the same—but it could not be eleven—both odd digits but they are the same. What is the first number that fits that description that comes into your mind *now*?"

I then ask, as if expecting no one to have got it right this time, "Did *anyone* think of the number thirty-seven?" Once again, about a third of the audience will put up their hand. I will then add, "Did anyone think of thirty-five?" About a further quarter of the audience will raise their hand. "Sorry, that was my fault," I explain. "I thought of thirty-five first and then I changed my mind."

What is going here? Any audience members who believe in ESP may well think that it has just been demonstrated. More skeptical members may be at a loss to explain what they have seen (and possibly directly experienced). Is it possible that I had simply conspired with all those members of the audience who got it right by telling them in advance to raise their hands in response to my questions? That would seem unlikely. Was it just a coincidence that so many more people guessed correctly than would be expected on the basis of chance alone? Again, possible but extremely unlikely.

The actual explanation is a phenomenon that psychologists refer to as *population stereotypes*.[22] For most people faced with this task, when they are asked to make a mental note of the first number that comes into their head, they assume this is pretty much a random process. Therefore, they expect the frequencies of response to be more or less equal across the range of response options. In fact, this is not what happens. Responses tend to cluster in reliable and predictable ways, especially with large audiences.

In the first example, about a third of people will choose seven regardless of whatever number I may be thinking of (especially as I have ruled out three as a valid response, which otherwise would also be a popular choice). In the second example, about a third will pick thirty-seven and about a further quarter will choose thirty-five. Note that in neither example do the response rates approach 100 percent, but that is not a problem as people do not expect telepathy to be 100 percent reliable.

There are several other examples of population stereotypes that could be used to fool (at least some of) the unwary that you possess amazing telepathic powers. Tell them your telepathic target is two simple geometric forms, one inside the other. Around 60 percent will choose circle and triangle. Tell them you are thinking of a simple line drawing. Around 10–12 percent will draw a little house. It makes for a fun demonstration of the fact that not everything that looks like paranormal actually is. But would anyone ever seriously try to pass off such a demonstration as really involving telepathy?

The answer is yes. For example, in the mid-1990s Uri Geller took part in a TV program called *Beyond Belief*[3], presented by David Frost, in which, it was claimed, various paranormal phenomena would be demonstrated live for the millions of viewers at home. I was one of those viewers. Uri demonstrated his alleged telepathic powers by supposedly transmitting a message to viewers. Uri had chosen one of four symbols that were presented at the bottom of the screen in the following order: square, star, circle, cross. As the camera slowly zoomed in on his face, Uri said: "I'm visualizing the symbol in my mind . . . and you people at home, millions of you, I'm actually communicating now with millions of people, maybe eleven, twelve, thirteen million people. Here goes. I'm transmitting it to you. I'm visualizing it. Open up your minds. Try to see it. Try to feel it. I'm really strongly passing it to *you*! One more time . . . okay." By this point, the upper half of Uri's face pretty much filled the screen, with the four symbols still displayed across the bottom of the screen. Viewers were instructed to phone in using one of four different numbers to indicate their guess at Uri's choice of symbol. Over 70,000 viewers did so.

My days of believing that Uri really did possess amazing psychic abilities were long gone by this stage, and I was therefore watching from the perspective of an informed skeptic. It was pretty easy to come up with nonparanormal explanations for all of the demonstrations featured in the program. With respect to this particular demonstration, I was rather pleased with myself for not only stating in advance what Uri's telepathic target would be but for also correctly stating the order of popularity of the remaining symbols. It was very lucky for Uri that he chose to "telepathically" transmit the star symbol. It was by far the most popular choice of the viewers, with 47 percent of them indicating that this was the symbol that they had "received." The second most popular was the circle, with 32 percent of the "votes," followed by the cross (12 percent) and the square (10 percent). If the guesses were completely random, we would expect about 25 percent for each option, so 47 percent choosing the same symbol as Uri is a massively statistically significant outcome. The probability that almost half of the callers chose this symbol just by chance is astronomically low. So, was this really strong evidence of Uri's psychic powers?

Readers who are familiar with common techniques used to test for ESP will have recognized that the four symbols used in Uri's demonstration are taken from the five symbols used on Zener cards (the missing symbol is three wavy lines; figure 8.1). The cards are named after the person who designed them, perceptual psychologist Karl Zener. A full deck consists of twenty-five cards, five of each design. In a test of telepathy, a "sender" would take each card from a shuffled deck in turn and attempt to telepathically transmit the image on the card to a "receiver." The receiver would record their guess of which card the sender was looking at. By chance alone, we would expect around five of the receiver's guesses to be correct. If the receiver scores significantly more than five, this might be taken as evidence of ESP. However, it has been known for over eight decades that people are more likely to guess certain symbols compared to others. Back in 1939, Frederick Lund asked 596 people to each generate a random sequence of five symbols from the Zener set.[24] By far the most popular symbol was—you've guessed it—the star, accounting for 32 percent of the responses compared to the 20 percent that would be expected by chance alone. So, as I said, it really was lucky for Uri that he chose the star as his telepathic target (assuming that it was just luck).

Figure 8.1
The five symbols used on Zener cards.

PROBABILISTIC REASONING AND PARANORMAL BELIEF

As stated, we are all very poor at dealing with probabilities in everyday life. This is one aspect of what psychologists call *probabilistic reasoning*, but this term covers more than just our (lack of) understanding of probabilities per se. Other aspects of probabilistic reasoning that typically cause us problems include the *base rate fallacy* as well as our inadequate understanding of the true nature of randomness.[25]

The base rate fallacy (also known as *base rate neglect*) can be demonstrated in many different contexts. In the words of Maya Bar-Hillel, "The base-rate fallacy is people's tendency to ignore base rates in favor of, e.g., individuating information (when such is available), rather than integrate the two. This tendency has important implications for understanding judgment phenomena in many clinical, legal, and social-psychological settings."[26]

A classic demonstration of this tendency is the cabs problem:

> . . . where people are told that 85% of the cabs in a city are blue and 15% are green. They also are told that there was a hit and run accident at night and that a witness, who has been shown to be able to correctly identify the color of the cab at night 80% of the time, identified the cab as being green. People are then asked how likely it is that the cab is green, and many report an 80% likelihood. However, this conclusion does not consider the base rate of the two colours of cabs within the city. . . . Specifically, it is important to consider both the individual descriptive information about the witness's accuracy and the base rate of green cabs. Based on both, the correct probability of the cab being green is 41% because, even though the witness is 80% accurate and identified the cab as green, the witness is still wrong 20% of the time; thus, due to the low base rate of green cabs (15%), it is more likely that the cab was blue than green.[27]

Our generally poor appreciation of the true nature of randomness is vividly demonstrated by the fact that, when asked to do so, people are incapable of generating true random sequences without the aid of external physical devices such as dice.[28] When trying to generate truly random sequences, people show a strong tendency to avoid consecutive repetition of the same element. In fact, of course, each element in a random sequence is entirely independent of all other elements in the sequence, including those that come immediately before or after it. Thus, in a random sequence of digits from zero to nine, a six is just as likely to be followed by a six as any other digit. Runs of the same digit are to be expected.

The same repetition-avoidance bias occurs when we try to judge whether some external process, such as a roulette wheel, is truly random or not. One notorious example of this is known as the *gambler's fallacy*. This is the tendency that some people have to assume that a black number, as opposed to a red number, is more likely to occur on the wheel after a long run of red numbers. In fact, of course, black and red are still equally likely.

Given our problems with everyday probabilistic reasoning, the question naturally arises of whether believers in the paranormal may be somewhat worse than nonbelievers in this respect, thus making them more likely to seek a paranormal explanation for various experiences where in fact no such explanation is required. Several studies have addressed this possibility, although a comprehensive review of this literature is beyond the scope of the current chapter.[29]

A very wide range of tasks are used in these studies, along with different measures of paranormal belief, conditions of testing, and types of participant. Many studies have indeed obtained results supporting the idea that believers in the paranormal show stronger deficits in probabilistic reasoning compared to nonbelievers.[30] Others have found such effects for some probabilistic reasoning tasks but not others,[31] and at least one failed to find any such effects.[32] Taken as a whole, it appears that, in general, believers in the paranormal are somewhat poorer at certain aspects of probabilistic reasoning than nonbelievers, and this is probably one factor that contributes to the believers' tendency to interpret more experiences, such as coincidences, as involving paranormal forces. However, it may well be that misperceptions of chance per se are less important than other aspects of probabilistic reasoning, such as poor appreciation of the true nature of randomness and an increased propensity to find connections between separate events.

The latter has been said to be a factor in explaining another type of reasoning error often associated with paranormal belief: the *conjunction fallacy*, first described by Amos Tversky and Daniel Kahneman.[33] Consider the following:

> Linda is 31 years old, single, outspoken, and very bright. She majored in philosophy. As a student, she was deeply concerned with issues of discrimination and social justice, and also participated in anti-nuclear demonstrations.
> Which is more probable?
> 1. Linda is a bank teller.
> 2. Linda is a bank teller and is active in the feminist movement.

You may find that you, in line with 85 percent to 90 percent of undergraduate students, feel that the second option is more probable. But this is logically impossible. If you think about it for a second, you will realize that the set of bank tellers who are also active feminists is a subset of the set of all bank tellers; therefore, the second option cannot be more probable. Tversky and Kahneman argued that people are misled because they rely on the *representativeness heuristic* in making their decision. Because the description of Linda corresponds to a stereotypical description of a feminist, this has an undue influence on our thinking, causing us to commit a logical error.

The idea that believers in the paranormal might have an increased propensity to find connections between separate events lends itself naturally to the possibility that they may be more susceptible to the conjunction fallacy insofar as they may tend to overestimate the probability of conjunctive (co-occurring) events compared to component (singular) events. This hypothesis has been supported across a number of studies, particularly those carried out by Paul Rogers and colleagues.[34] It should be noted that the strength of the effect varies with such factors as the content of the hypothetical events (e.g., paranormal versus nonparanormal, confirmatory versus nonconfirmatory) and the measure used to measure paranormal belief. However, Neil Dagnall and colleagues argue that overall misperception of the nature of randomness is more strongly associated with paranormal belief than is susceptibility to the conjunction fallacy.[35]

Before we leave the topic of coincidences, it is important to point out that our tendency to pay particular attention to the co-occurrence of matching events has been hugely important to our evolutionary history. Although we may sometimes be tempted to infer a causal connection between two co-occurring events when no such connection exists, it is clearly the case that sometimes there really is a causal connection.

This important point has been emphasized by Mark Johansen and Magda Osman.[36] It would be a big mistake to dismiss all reports of apparent coincidences as being due to nothing more than chance. Johansen and Osman define coincidences as "surprising pattern repetitions that are observed to be unlikely by chance but are nonetheless ascribed to chance since the search for causal mechanisms has not produced anything more plausible than mere chance."[37]

They argue that the experience of coincidences should not be viewed as an example of human irrationality but instead as a fundamental consequence of rational cognition. We can only describe the relationship between two events as truly coincidental if we have first excluded potential causal links. Indeed, sometimes scientific breakthroughs, such as the discovery of penicillin by Sir Alexander Fleming in 1928, are based on a serendipitous observation that, fortunately for humanity, was not dismissed as "just a coincidence."

As will be discussed further in the next chapter, many of the cognitive biases that are associated with paranormal belief have evolved because, on balance, they conferred more advantages than disadvantages upon our species in increasing our chances of surviving long enough to pass on our genes to the next generation. We may sometimes be misled into thinking that there is a causal relationship between two events when in fact there isn't—but this is outweighed by the advantages of being able to detect such relationships when they really are there to be detected.

9 TRICKS OF THE MIND

As described in the previous chapter, our poor appreciation of probability sometimes leads us to draw faulty conclusions and see significance in events where we should not. This chapter will consider other cognitive biases that may underlie our tendency to misinterpret events as involving paranormal forces when, in fact, mundane explanations will suffice.

SUBJECTIVE VALIDATION

Back in the 1970s, Russell Targ and Harold Puthoff published sensational results in the prestigious journal *Nature* in support of the idea that it was possible to accurately perceive information at remote locations without the use of the known sensory channels.[1] They were reporting the findings from investigations using the technique of *remote viewing*. A typical remote viewing study involves one or more individuals visiting a randomly selected location, such as a park, a bridge, a shopping center, or a library, at a pre-specified time and attempting to telepathically transmit information about that location to an individual acting as a "receiver" back at base.[2] The receiver describes any impressions going through their mind at the time and may also make drawings to record any visual images. An experimenter, who is also blind to the target location, may ask the receiver to clarify anything ambiguous or unclear. A number of such trials are carried out. Transcripts of the receiver's impressions are then issued to independent judges, who visit the randomly selected sites and assess the degree of match between each transcript and each site. The receiver themselves may also sometimes be involved in such judging.

Targ and Puthoff claimed that the results they had obtained from the study reported in *Nature* were highly statistically significant. Across nine locations, their independent judge had correctly matched transcripts to locations on no fewer than seven out of nine trials. Intrigued by these impressive results, David Marks and Richard Kammann attempted to replicate the study. Their results offered no support whatsoever for the original claim of paranormal information transfer. Furthermore, after persistent detective work, they discovered that there were major flaws in the original study that fully explained the apparently significant results reported by Targ and Puthoff without any need to invoke psychic powers. The debate regarding whether remote viewing studies really do provide strong evidence for extrasensory perception rumbles on, but I want to focus here on an interesting phenomenon noted by Marks and Kammann in the course of carrying out their attempted replication.

Over a series of thirty-five experiments, Marks and Kammann noted strong similarities between transcripts and locations on many occasions, as did their independent judges and the receivers themselves. This initially gave all concerned the very strong impression that some kind of psychic communication had taken place. The only problem was that the transcripts that appeared to match were rarely the ones generated when the senders were visiting that particular location. Any correspondences noted were purely the result of coincidental matches between transcripts and locations. Marks and Kammann coined the term *subjective validation* to refer to this phenomenon, which they described as occurring "when two unrelated events are perceived to be related because a belief, expectancy, or hypothesis demands or requires a relationship."[3]

In the case of remote viewing studies, the transcripts (and accompanying drawings, if any) contain a large number of elements. Similarly, looking around at any randomly selected location, one might view a large number of distinct elements that provide potential matches to those in the transcripts. It is very likely that some matching elements will be found. Subjective validation comes into play in many other paranormal contexts. When a psychic or other type of diviner gives a reading for a client, there is huge potential to find matches between the many elements in a reading, on the one hand, and the rich tapestry of a human life, on the other. As we shall see in chapter 11, subjective validation is an important factor in explaining why dreams sometimes appear to match future events.

SYLLOGISTIC REASONING

Probabilistic reasoning is not the only type of reasoning that human beings find difficult. We also often struggle with certain types of deductive reasoning, including *syllogistic reasoning*.[4] A syllogism consists of two statements, known as premises, and a third, the conclusion. Performance on syllogistic reasoning tasks is typically assessed by presenting participants with a set of syllogisms and asking them to judge whether the given conclusions are valid or invalid, assuming the premises to be true.

For example, one type of syllogism is known as a *categorical syllogism*. A typical example is:

All men are mortal.
Aristotle is a man.
Therefore, Aristotle is mortal.

The conclusion drawn here is valid, meaning that it follows logically from the two premises. If the first two statements are true, then the third statement must also be true.

Note that in syllogistic reasoning tasks one is not being asked whether the premises or conclusion are *true*, simply whether the conclusion follows logically from the premises *if* the premises are true. For example, the following syllogism is valid even though we might disagree with the conclusion:

Everything that reduces unemployment is good.
War reduces unemployment.
Therefore, war is good.

We tend to be influenced in our judgments about the validity of syllogisms by the degree to which we believe the conclusion, an effect known as *belief bias*. Thus, for example, we are more likely to judge invalid reasoning as valid if we agree with the conclusion given and vice versa.

Another type of syllogism is known as a conditional syllogism. An example would be:

If Alice works hard, she will pass her exams.
Alice works hard.
Therefore, Alice will pass her exams.

It is a trivially easy task to confirm that it follows from the first two premises here that the conclusion must be true. Sometimes, however, it is not so easy to judge the validity of conditional syllogisms. A common error is what logicians refer to as *affirmation of the consequent*. Here is an example:

If vaccinations are harmful, the authorities would deny it.
The authorities do deny it.
Therefore, vaccinations are harmful.

This logical error is very common among conspiracy theory believers where a lack of evidence supporting a conspiracy is often interpreted as evidence of a cover-up.

Readers may have spotted that this is the error made by those claiming that 3.7 million Americans have been abducted by aliens, as described in chapter 5. Even if people really were being abducted by aliens, and even if these abductions really did cause people to experience paralysis, a sense of presence, and so on, the conclusions drawn from the responses to the Roper poll would still not be justified. Here is their argument presented in typical syllogistic form:

If 3.7 million Americans were abducted by aliens, they would report paralysis, a sense of presence, and so on.
3.7 million Americans did report paralysis, a sense of presence, and so on.
Therefore, 3.7 million Americans were abducted by aliens.

The conclusion would only logically follow in this case, as with all similar conditional syllogisms, if the *if* in the first premise meant *if and only if*. In other words, if there are other possible reasons why the second premise might be true, they cannot logically be ruled out. In this case, there is a very plausible alternative explanation for reports of paralysis, a sense of presence, and so on—our old friend sleep paralysis.

Might it be the case that believers in the paranormal are somewhat poorer at deductive reasoning than nonbelievers, and therefore somewhat more prone to drawing erroneous conclusions from the available evidence? An early study by Michael Wierzbicki supported this hypothesis, but a subsequent replication attempt by Harvey Irwin failed to find any difference between paranormal believers and nonbelievers in terms of their reasoning ability.[5] Irwin speculated that the different pattern of results obtained in the two studies might be due to an experimenter effect.

Both studies used student samples. Irwin proposed that the brighter students in Wierzbicki's study may have been aware that Wierzbicki was likely to be rather skeptical regarding paranormal claims, in line with the general attitude of most psychologists. These students may therefore have tended to be reluctant to admit to any pro-paranormal beliefs they held, thus producing a negative correlation between reasoning ability and (professed) level of belief. In contrast, Irwin's own students would know him to be more sympathetic toward paranormal claims and thus would feel free to express their true attitudes toward the paranormal. As a consequence, Irwin claimed, no artifactual correlation would result. Although this suggested explanation is certainly plausible, Maxwell Roberts and Paul Seager highlighted other possible reasons for Irwin's failure to find any correlation between reasoning ability and paranormal belief, claiming that Irwin's reasoning task differed in important ways from that used by Wierzbicki.[6] In their own study, Roberts and Seager did indeed replicate the negative correlation reported by Wierzbicki.

One study of the relationship between paranormal belief and syllogistic reasoning, by Caroline Watt and Richard Wiseman, did indeed find experimenter effects.[7] Watt and Wiseman followed identical procedures and yet, whereas Watt obtained a large negative correlation between the two variables, Wiseman failed to find any correlation. Unfortunately for Irwin's hypothesis, however, these results are the opposite of what that hypothesis would have predicted. It is Caroline Watt who is known to be more sympathetic toward paranormal claims whereas Richard Wiseman is a well-known and vocal skeptic—another case of nice theory, shame about the data!

NONCONSCIOUS PROCESSING: NOT KNOWING WHAT WE KNOW

It is generally accepted within psychology that we are not consciously aware of most mental processes. Typically, we are aware of the products of mental processing rather than the processing itself. As a simple demonstration, mentally answer the following question: What was the name of your first school? The answer, no doubt, simply automatically popped into your head, but you had no conscious access to the processes involved in reading and understanding the question and then retrieving the correct piece of information from all the information stored in your long-term memory. The same is true for other psychological processes. For example, during waking consciousness we feel as though we have immediate direct access to the visual and auditory world around us, whereas in fact that mental model of the world and our place in it is the result of complex interactions between bottom-up and top-down processing, as described in chapter 3.

The distinction between conscious and nonconscious processing is at the heart of an interesting concept that was first proposed by Michael Thalbourne and Peter Delin in 1994 as a possible explanation for the fact that measures of paranormal belief and experience are found to consistently correlate with a range of other psychological variables, including creativity, susceptibility to mystical experiences, and certain psychopathological tendencies.[8] Their hypothesis was that a single common factor might underlie all of these intercorrelations. They labeled this factor *transliminality*, which they initially defined as "the extent to which the contents of some preconscious (or 'unconscious' or 'subliminal') region of the mind are able to cross the threshold into consciousness (in its sense of 'awareness')." Subsequently, they defined it as "a hypersensitivity to psychological material originating in (a) the unconscious, and/or (b) the external environment. 'Psychological material' is taken to cover ideation, imagery, affect and perception, and thus a rather broad concept."[9]

A useful way to think about transliminality is to imagine that the mind contains both conscious and unconscious areas and that they are separated by a semipermeable membrane through which material must pass in order to cross the threshold from unconsciousness into conscious awareness. The idea is that people differ in terms of the permeability of this membrane. Thus, a highly transliminal individual may have conscious access to material that would not normally be accessible to a low transliminal person. Thalbourne, a believer in the reality of psi, originally developed this concept as a possible explanation for differences between individuals in terms of their psychic abilities. If ESP signals are received initially by the nonconscious part of the mind, he reasoned, such differences may be explained by inherent differences in transliminality.

Although I personally do not believe that psi exists, I too found the concept of transliminality of interest. A number of commentators had already suggested that certain experiences that a person might interpret in paranormal terms might in fact be explicable in terms of nonconscious processing.[10] For example, in some of the early studies of ESP by J. B. Rhine using Zener cards (figure 8.1), as described in chapter 8, it is said that some of the cards were so heavily embossed that some participants could just about make out what the symbol was from the back of the card.[11] One might expect participants would let the experimenter know that they were not actually trying to use ESP but instead could see the symbol using normal vision. Is it possible that these participants were not consciously aware that they had processed this information even though it was influencing their guesses on each trial? Furthermore, might it be the case that highly transliminal individuals would be more likely to score above the level expected by chance by this means compared to low transliminals?

Such a possibility is supported by the results of a study that I carried out in collaboration with Susan Crawley as part of her PhD research.[12] Participants were asked to take part in a computerized ESP task. On each trial, the computer would randomly select one of the five Zener card symbols, and then the image of a swirly pattern, like those often seen on the back of playing cards, would appear on the screen. Participants would then enter their guess. What participants were not told was that on half the trials the correct answer had been subliminally presented to them by flashing it on the screen so briefly before the swirly pattern appeared that it was below the threshold of conscious awareness. Participants also completed a questionnaire designed by Thalbourne to assess transliminality. As we had expected, the scores on this scale correlated with the number of correct guesses on the computerized task—but only on the subliminally primed trials. On the unprimed trials, there was no correlation. Although no evidence for actual ESP was found, it would not be surprising if our high transliminal participants seriously entertained the idea that they might be psychic. After all, how else could they explain their above-average performance on the task?

THINKING, FAST AND SLOW

If the numerous cognitive biases identified in recent decades are as ubiquitous as they appear to be, they may provide some sort of advantage in terms of our evolutionary history. At first glance, this may appear to be somewhat surprising. Surely any kind of cognitive bias that leads us to misinterpret, misperceive, or misremember events would be a *disadvantage* in evolutionary terms and thus would have been eliminated by natural selection? The solution to this puzzle is provided by an appreciation of what has become known as System 1 and System 2 thinking.[13]

It is now widely accepted that human thought operates in two distinct modes, as reflected in the title of Daniel Kahneman's bestseller, *Thinking, Fast and Slow*.[14] System 1 thinking corresponds to what we might loosely refer to as *intuition*. Such thinking is fast, nonconscious, effortless, emotional, and automatic and involves no sense of voluntary control. A typical example of System 1 thinking is meeting someone for the first time and taking an instant liking or disliking to them. By contrast, System 2 thinking is slow, conscious, effortful, and unemotional and feels under voluntary control. Such thinking corresponds loosely to what might be called *rational thought*.

A typical example of System 2 thinking is attempting to solve a mental puzzle. Many of the cognitive biases discussed in this book are based on System 1 thinking. A great deal of evidence has been produced showing that System 1 influences our attitudes, beliefs, and behavior to a much greater degree than we might assume.

In an evolutionary context, one of the most important differences between System 1 and System 2 thinking is speed. Our brains have evolved to maximize the chances of us staying alive long enough to pass on our genes to the next generation. Our ancestors lived in a world full of dangers from enemies and predators. Relying primarily on System 1 thinking, which was both fast and usually delivered the right answer, was a better bet than relying primarily on System 2 thinking, which, although likely to deliver the right answer a bit more often, was much slower.

It simply made sense for our ancestors to rely on the quick and dirty rules of thumb of System 1, technically known as *heuristics*, in their day-to-day interactions with the world around them rather than on System 2. For example, a rustling in the bushes may or may not have been a predator. By relying on System 1 thinking and assuming it was indeed a threat, our ancestors were more likely to stay alive long enough to procreate. If they were correct and responded appropriately in terms of either fight or flight, they maximized their chances of staying alive. If they were incorrect, the cost was minimal. To make the opposite error of assuming that there was no threat or simply taking too long to make any decision at all, they maximized their chances of becoming lunch. Reliance on System 2 thinking was clearly not the better option.

Evolutionary pressures such as these were likely to result in cognitive systems with a bias toward what scientists refer to as false positives. In science, these are also known as a *Type I errors*. For example, a medical researcher may wrongly conclude that a new form of treatment is effective, perhaps because the results came from a clinical trial that was affected by confounding variables. *Type II errors*, in contrast, are false negatives, for example, a medical scientist wrongly concluding that a treatment was ineffective, perhaps as a result of basing their conclusion on unreliable data.

Our success as a species depends on our ability to spot meaningful patterns and recognize cause-and-effect relationships in our environment. Our ancestors would have used such skills in hunting, avoiding predators, and developing agriculture, for example. The problem is that sometimes, in line with our general bias toward Type I errors, we may think we see meaningful patterns or cause-and-effect relationships when in fact there are none, as previously discussed with respect to pareidolia, our poor appreciation of the true nature of randomness, and subjective validation.

Michael Shermer refers to our tendency to see meaningful patterns where there are none as *patternicity*.[15] Others have used the term *apophenia* to describe this tendency, although it should be noted that when Klaus Conrad first coined the latter term in 1958, it was intended not only to refer to the tendency to see meaningful patterns in randomness and to make connections between unrelated things but also to perceive deep personal significance in such occurrences, as may occur in schizophrenia.[16] Thus, although we may all notice a pattern resembling a face in a stain on a church wall, only some of us will see this as more than an accidental configuration of marks. These days, *apophenia* is used more loosely than Conrad intended and, in some contexts, is used interchangeably with *patternicity* and *pareidolia*.

Humans also appear to have an in-built tendency to assume that when something happens in their vicinity, it happens because someone or something made it happen for some specific reason. Again, this would make sense in terms of human evolution, as it would lead us to routinely assess any event occurring nearby in terms of the degree of threat that it might pose, assuming it was caused by some sentient agent with specific intentions toward us.

Justin Barrett has plausibly argued that belief in gods was a natural consequence of this tendency.[17] He hypothesizes that humans possess a mental module that he refers to as a *hyperactive agency detective device* (HADD) that provided an evolutionary advantage in ensuring that we would be more inclined to make Type I errors in ambiguous but potentially threatening situations and thus more likely to survive. We would also be naturally inclined to assume that some sentient being or beings caused thunder and lightning, natural disasters, and crop failures, not to mention disease and injuries, perhaps as punishment for our wrongdoing. Such reasoning clearly can be applied to belief in a range of supernatural beings, not just gods.

Our brains evolved over millennia, and the pace of evolutionary change is slow. Our brains are therefore pretty much indistinguishable from those of our recent evolutionary ancestors. When considered in this context, it is not at all surprising that much of our thinking is based on the heuristics that characterize System 1 as opposed to the more rational System 2 with all of the advantages and disadvantages that follow from that.

10 SKEPTICAL INQUIRIES

The main focus of anomalistic psychology is on trying to produce and empirically test nonparanormal explanations for ostensibly paranormal phenomena, as the previous chapters illustrate. It must be always borne in mind, however, that the notion that paranormal forces do not exist is itself not proven. Indeed, logically it is one of those negative statements that can never be proven. Even if paranormal forces have not been shown to exist after well over a century of systematic research, it is always possible that proof of their existence will be found at some point in the future.

The idea that psi does not exist is simply a working hypothesis guiding research in anomalistic psychology, which, by its very nature, could never be proven. Having said that, the more ostensibly paranormal phenomena that anomalistic psychology can explain, the less need there is to entertain paranormal hypotheses. In light of the fact that we recognize the theoretical possibility that psi just might exist, however unlikely we may judge this to be, members of the Anomalistic Psychology Research Unit at Goldsmiths (APRU) put considerable time and effort into directly testing psychic claims. This chapter and the next describe some of these tests, most of which were carried out as part of television documentaries and have never appeared in print before.

BRITAIN'S PSYCHIC CHALLENGE

Britain's Psychic Challenge is a TV series that was broadcast in the United Kingdom by Channel 5 and presented by Trisha Goddard. The main series was broadcast in 2006, but this was preceded in December 2005 by a special one-off program to introduce the series. The idea was that the series would be a kind of *X Factor* for psychics. Each week, a group of psychics would be set various tasks to perform using their claimed psychic abilities, and at the end of each program, the worst-performing psychic would leave the competition. Eight psychics were selected from almost 2,000 applicants to take part in the series. In the final program, the three remaining psychics competed against each other, and ultimately one was declared the overall winner—in some people's eyes, this was to be officially declared Britain's top psychic!

The psychics' performances were discussed and judged by a "panel of skeptics" consisting of myself, Philip Escoffey, and Jackie Malton. Philip is a professional mentalist and conjurer.[1] Jackie is a former senior police officer and now works as a TV script consultant. Her police career provided the inspiration for the fictional DCI Jane Tennison in the hugely successful TV series *Prime Suspect* written by Lynda La Plante and starring Helen Mirren. I enjoyed working with both Philip and Jackie, but I think it is fair to say that Jackie turned out to be not much of a skeptic as the series progressed.

Needless to say, taking part in this series raised concerns on my part about being a token skeptic whose appearance might lend some legitimacy to a venture that perhaps did not deserve it. After all, I knew in advance that at the end of the series I would be partly responsible for declaring one of the contestants to be Britain's best psychic. In the end, though, I decided to do the program for a number of reasons. First, because I reasoned that if I did not take on the role, someone else would; second, because I believe that it is important for the views of an informed skeptic to be included in such programs; third, because I needed the money; and finally, because I anticipated, correctly as it turned out, that it would be fun.

From the outset, it was clear that this program was going to present Philip and me with something of a dilemma. Just like *Haunted Homes*, its intended target audience would be believers, hungry for evidence of amazing abilities that defied conventional explanation. If the tests had been designed by Philip and me, combining his knowledge of the deceptive art of mentalism with my expertise in parapsychological research, we can be pretty certain that very few of the psychics would have passed any of the tests. At least, that would seem to be a fair conclusion based on my many years of carrying out such tests under well-controlled conditions. It is unlikely that the show would have retained its intended audience if, week after week, the psychics had all failed every test set for them. So, as you might have guessed, the tests were not designed by us. In general, they were deliberately designed to ensure that, even on the basis of guesswork alone, there would be some successes, and the conditions were far from properly controlled.

It should be noted that the tests devised for the program were also problematic from the psychics' point of view insofar as they were not designed around the specific claims of those particular psychics. If a psychic claimant insists that they are able to divine the contents of sealed envelopes, it is clearly inappropriate to test them by, say, asking them to psychically determine which of several women are pregnant. When members of the APRU put psychic claimants to the test, great care is always taken to design the test around the specific claims being made. If this is not done, failure on the test tells us nothing. In fact, we routinely get claimants to sign a document in advance stating that the test is a fair test of their alleged abilities. When the claimants on *Britain's Psychic Challenge* failed particular tests, as they often did, they were usually justified in simply shrugging and pointing out that they never claimed to have the specific psychic abilities required to pass that particular test.

Over the six episodes broadcast in 2006, a total of twenty different tests were set for the psychics. Given that there were eight psychics in the first episode that were whittled down to three in the final episode, this means that there were 118 opportunities for the psychics to demonstrate their powers. Unfortunately, this rarely happened.

Given the large number of tests set for the psychics, space limitations preclude a detailed analysis of each and every one of them, but a few general observations are in order. Some of the tests did allow a definite success or failure result. This would obviously be a very good thing if the test itself was well designed and well controlled, but sadly this was virtually never the case. To give an example, one of the tests set for six psychics in the one-off introductory program was to see if they could identify in which of fifty cars in a parking lot was a volunteer hidden in the trunk. If this test had been designed and carried out properly, each psychic would have only a one in fifty chance of being right on the basis of guesswork alone. Should we therefore be impressed that three of the six psychics actually identified the correct vehicle?

The short answer is no. There were a number of serious flaws with this test. For example, the volunteer was allowed to choose which trunk he wanted to hide in. The fifty cars varied considerably in size, color, position, and so on. There would be an understandable desire to choose a car with a reasonably spacious trunk that was not parked in a position that would make it difficult to get in and out of. In an ideal world, the test would involve fifty identical cars, but at the very least fifty similar cars should have been used, all parked in such a way that the trunks were easily accessible. Furthermore, the car that the volunteer was to hide in should have been chosen at random, and all of the other cars should have had weights equivalent to the weight of the volunteer in order to ensure that one car did not stand out because its back end was lower than its front end in comparison to others.

Another major flaw was one that undermined the validity of many of the other tests in the series. No attempt was made to use a double-blind methodology. In any test involving selecting the correct target from a number of potential targets, it is not enough that the person being tested has no prior knowledge of the right answer; it is essential that no one in the vicinity of the test has such knowledge. In the case of the "body in the trunk" test, it was clear that the camera crew, as well as Jackie Malton, who was supervising the test, knew where the volunteer was hidden. This raises the very real possibility of the crew giving unintentional cues as to the target's location. For example, a camera operator may well zoom in when the psychic is near the correct car. On at least one trial, Jackie was standing right next to the target car as she encouraged the psychic to make a choice!

Given how poorly controlled most of the tests were, the psychics still managed to perform remarkably poorly on most of them. For example, when asked to identify the two women who were pregnant from a group of ten, none of the eight psychics identified both women correctly, and five got neither of them right. Their total score of three was the same as that obtained by a group of eight, presumably nonpsychic, student volunteers and did not exceed what would be expected on the basis of chance alone.

Many of the other tests were even worse insofar as it was virtually impossible to determine just how successful or unsuccessful the psychics had been. These were tests where the possibility of cold reading, either intentional or unintentional, had simply not been taken into account. Once again, a good example was featured in the one-off introductory program. The program was broadcast from Knebworth House, an English country house in Hertfordshire. The residents of the house are screenwriter Henry Lytton-Cobbold and his family. Two of Henry's ancestors had been killed in tragic circumstances: Antony Lytton-Cobbold died in a plane crash in the 1930s, and his brother, John, was killed during a tank battle in the Second World War. The three psychics who were successful in the "body in the trunk" challenge were handed two envelopes, each containing a photograph of one of the brothers. Their task was to glean information about the two individuals using their psychic ability.

Often in programs that include psychic readings, the reading is heavily edited to include only those parts of the reading that could be deemed, sometimes with a little imagination and interpretation, to be hits. In this case, even careful editing could not produce this impression. This test, along with numerous other similar tests throughout the series, did provide the viewers with nice illustrations of cold reading in action, though. For example, consider this exchange between Henry Lytton-Cobbold and one of the psychics, Amanda Hart:

> **AH:** What I'm picking up so far is a boat. It's got no sails, but it's got lots of masts. There's lots of—all I can see is lots of masts and crisscrossing. It's quite like—I've got a little boat.
> **HL-C:** Is it definitely a boat, might it be an airplane?
> **AH:** It's not that clear. It's like a man with a helmet but it's made of, almost, leather—and goggles.

This is a very clear example of the sitter trying hard to fit the utterances of the psychic to what they know to be the facts and, in the process, steering the psychic in the right direction. Elsewhere in this and other readings throughout the series, all of the standard tricks of cold reading, whether intentional or unintentional, can be seen in action.

DOWSING WITH DAWKINS

In 2006, I was contacted by a researcher working on a two-part documentary series titled *The Enemies of Reason*, which was to be presented by the eminent evolutionary biologist and author Professor Richard Dawkins. The researcher wanted to know if we were carrying out any research projects that might be suitable for inclusion in the program. As it happened, we were planning on a double-blind test of dowsing at the British Association Festival of Science 2006 that was to be held in Norwich that year. Following some discussion, it was decided that our test was indeed suitable for inclusion in the program, which was broadcast in the United Kingdom in 2007 on Channel 4.

Dowsing, as defined by Ray Hyman, is "the practice of finding hidden objects or substances with the aid of a forked twig, metal rods, or some other device that indicates the location of the desired target by its movement."[2] One example of dowsing is the alleged ability to locate a hidden object using a forked rod. A forked stick (referred to as the *dowsing rod*, *dowsing stick*, or *divining rod*) is the most commonly used device, but dowsing rods can be made of almost any other material, including iron and steel.[3] Although the rod may vary in material and shape, it is almost always forked and very light in weight.[4]

The standard method is to hold the dowsing rod in front of the body, arms outstretched with palms facing upward and one fork in each hand, so that the elbows are inward against the body.[5] Dowsing can also be performed using L-shaped metal rods, one held in each hand (see figure 10.1). The rods are held in such a way that the long arms of the L point forward in parallel with each other. The belief is that as the dowser approaches the target object or substance, the dowsing rods will cross. Normally, dowsing is conducted outdoors in open spaces, but it can also be practiced indoors. Some dowsers even claim to be able to search for an object not in the immediate vicinity by dowsing over a map. Map dowsers typically use pendulums rather than dowsing rods.

The origin of dowsing is debatable, as is its original purpose. However, most researchers now generally concur that dowsing began, and then became prominent, in sixteenth-century Europe, where it was used by German and

Figure 10.1
The author trying his hand at dowsing (without success). Photograph courtesy of Valerie Heap.

Figure 10.2
Woodcut by Georgius Agricola, dated 1556, illustrating the use of dowsing to locate deposits.

English miners to locate underground metal (figure 10.1). Originally used to locate both metal and underground water, dowsing is now used to locate almost anything, including lost objects, missing persons, precious metals, and the flow of electrical current and is even used as a decision-making tool. Some paranormal investigators claim to be able to communicate with spirits using dowsing rods.

Even amongst dowsers themselves, there is considerable debate regarding the question of how dowsing might work. Some practitioners are convinced that when the dowsing rods move it is because some external force has moved them. Others accept that the movement is caused by muscular activity on the part of the dowser, even if those muscular movements are unconscious, but they may still believe that the fact that the rods moved precisely when they did is because the dowser had obtained knowledge of the target location by psychic means.

Not surprisingly, there are many dowsers who embrace a host of other New Age beliefs. But interestingly there is a sizeable minority who very much want to distance themselves from what they would deem to be such 'woo woo' ideas. These are people who very much see themselves as rationalists and strong supporters of science. I suspect they have come to believe in dowsing on the simple basis of trying it out—and being surprised to experience what appeared to them to be the movement of their chosen dowsing device of its own accord. Rather than asking whether there might be a psychological explanation for this phenomenon, they immediately assume that there is some real physical force at work, perhaps electromagnetism, that could potentially be measured objectively.

The big question is, of course, does dowsing actually work? Anecdotal evidence and field studies are often cited by those arguing in support of the validity of dowsing.[6] In contrast, whenever dowsing is tested under properly controlled conditions it appears to be no more effective than pure guesswork.[7] Critics of dowsing point out that the apparently positive results obtained in field studies may be based upon the dowsers, either consciously or unconsciously, making use of sensory cues available in the environment. For example, if dowsing is being used to locate underground water, the lie of the land and patterns of vegetation may supply such cues.

We wanted to keep our test as simple and straightforward as possible. Different dowsers, as stated, make very different claims. Although some dowsers insist that dowsing can only be used to locate flowing water, many others claim that dowsing can be used to locate pretty much anything. Our test investigated whether dowsing could be used to identify which of six plastic boxes contained a bottle of water, the others all containing bottles of sand. This test was certainly not rocket science, as the saying goes, but even so it took a lot of setting up and I am grateful to the Society for Psychical Research for providing a small grant to pay for materials and to my trusty voluntary research assistant, Mark Williams, for obtaining all of the required equipment. Thanks are also due to Elaine Beattie and Rosie Bunton-Stasyshyn for their assistance in running the formal double-blind test.

We had recruited eight experienced dowsers for our test, four male and four female, aged between 47 and 76. The number of years of dowsing experience ranged between ten and fifty-six years. All participants considered themselves to be amateurs with the exception of one who was semi-professional. Seven of the participants believed that the forces responsible for moving the rods were grounded in natural science (e.g., conductivity, electromagnetic fields, energy fields, gravity), while one participant believed that God is responsible for moving the rods.

The test was conducted outdoors in a tented area with an audience present and the whole test was recorded not only by us but also by the film crew working with Richard Dawkins.[8] Each dowser was allowed to walk around the area to establish clear spots for the location of containers and dead spots where their rods moved indicating 'interference.' Any dead spots located were clearly marked and containers were not subsequently laid on these areas.

The dowsers all used their own dowsing rods (or, in one case, their own pendulum). They were asked to check the conditions in order to confirm that their dowsing rods were responding as they normally did. Each dowser was shown a bottle of water and a bottle of sand and asked to hold their rods above each bottle. As the dowsers would expect, the rods crossed above the water bottle but not above the bottle containing sand. They were then asked to repeat this procedure with the bottles inside the plastic boxes with the lids closed. Once again, when the rods were above the box that the dowser knew contained a bottle of water, the rods crossed. Over the other box, there was no movement of the rods. Having satisfied themselves that the conditions of the test were fair, each participant was led away from the test site to a holding area and required to sign a consent form agreeing to take part in the formal test.

Each dowser was then required to take part in six test trials. In accordance with the rules of the tests, all were informed that any one of them scoring four or more hits out of the six trials would be deemed to have passed the test. There was just less than one chance in a hundred that any individual dowser would obtain a score this high or higher purely on the basis of guesswork alone. For each block of trials, the test area was closed off completely from everyone except Rosie. She determined which randomly determined container would hold the bottle of water for each trial, placed the bottles of water and sand as required, and then removed herself completely from the test site. There was thus no opportunity for any information to be conveyed to the dowsers or to anyone else regarding the location of the hidden water bottles. This is what is known as a *double-blind* testing procedure where no one present during testing knows the correct answer until testing has been completed. This ensures that no one can give any clues to the dowser, either intentionally or unintentionally.

When it was their turn to be tested, each dowser entered the test site and moved up and down the rows of containers, dowsing each one. They were allowed to take as much time as they liked. When they felt they had identified the location of the bottle of water, I placed a marker next to the chosen container. At the end of each individual trial for each dowser, they watched as the containers were opened and they could see how well (or badly) they had done. Of our eight dowsers, two failed to identify any of the containers containing a bottle of water, four got one right, and two got two right. The overall average number of hits was one out of six, exactly the outcome expected on the basis of pure guesswork.

Despite not passing the test, however, all of our dowsers remained confident in their ability to dowse successfully. Their lack of success on our test did not appear to shake their belief at all. They offered a range of reasons as to why they felt their attempts to locate the water were unsuccessful. Four of them suggested that their inability to locate the water was due to it being situated above ground and suggested that they would have been more successful dowsing for moving water below ground level. Two suggested that the plastic containers were interfering with the electrical signals transmitted from the water. One participant felt that their lack of success was due to the fact that God did not want them to find the water on this occasion. However, this dowser decided some time later that he had simply not been asking God the right question. Another dowser felt that there were too many dowsers present and hence the signals from the water could not be detected.

This type of *post hoc* reasoning is common when dowsers fail a controlled scientific test. Our dowsers appeared to be genuinely surprised at their results but nevertheless remained steadfast in their conviction that dowsing works.

The dowsers all seemed to forget that in the pre-test checks, the dowsing rods had shown exactly the expected responses when held over a bottle of water and a bottle of sand, whether in full view or inside a container. Of course, the difference between these checks and the formal double-blind test trials was that the dowsers knew the locations of the water and the sand in the pre-test checks and the rods behaved in accordance with their expectations. The explanation of the movement of the rods was thus due to our old friend, the ideomotor effect, as described in Chapter 4.

Despite the lack of good quality evidence from well controlled studies, many people are still absolutely convinced that dowsing does indeed work, including such no-nonsense people as those who run water companies in the UK. Science blogger Sally Le Page was shocked to discover that no less than ten out of twelve major water companies in the UK confirmed that, in 2017, they were still making use of dowsing to locate leaks in water pipes.[9] Her findings were reported on in the *Guardian* by Matthew Weaver, followed a week later by an influx of angry letters from supporters of dowsing defending the practice.[10] The Reverend Martin J. Smith of Wilmslow, Cheshire, was typical in insisting that the reason the water companies and others use dowsing is simple: "it works, it is replicable and it is independently verifiable." The good reverend and the other angry correspondents are almost certainly mistaken as the results of numerous well-controlled tests of dowsing, including our own, convincingly demonstrate.

THE BABY MIND READER

In 2006, the future was looking very bright for psychic Derek Ogilvie.[11] In April that year, he had published a book with the title, *The Baby Mind Reader: Amazing Psychic Stories from the Man Who Can Read Babies' Minds*.[12] A couple of months later, his own series launched on Channel 5 in the UK.

Derek's unique selling point as a psychic was his claim that he had an amazing ability to read the minds of very young children, even including pre-verbal babies. Whereas you might assume that babies would rarely be thinking about much other than "I'm hungry" or "Whoops, I've just pooed myself!," it turns out, according to Derek, that they are all too aware of many other aspects of family life including marital problems, employment issues, even issues relating to the family car or home. Who knew?

In 2007, Derek agreed to take part in a further documentary to be broadcast on Channel Five, this time a one-off in the *Extraordinary People* series. The title of this programme was *The Million Dollar Mind Reader*, referring to the fact that Derek was to be tested by not only members of the APRU but also by the one and only James Randi. If Derek passed the test set by Randi, he would receive a cool one million dollars (if he passed our test, he would receive a nice cup of tea and a biscuit—which he could also have even if he did not pass the test).[13]

In his 2006 series, Derek would typically give his reading for babies and young children in the presence of their parents. The parents often expressed amazement at the accuracy of his readings. The problem here will by now no doubt be obvious to the reader. No attempt was made to rule out the possibility that Derek was using, either intentionally or unintentionally, standard cold reading techniques, let alone hot reading. The correct way to test Derek was pretty obvious. He should do readings for a number of children of whom he had no prior knowledge whatsoever without the parents (or anyone else who knew them) being present at the time. Once all readings had been transcribed, the parents would then be invited back and asked to choose the reading that they felt had been done for their child. If Derek could really do what he claimed he could do, there should be one reading that stood out as containing lots of specific and accurate information about that particular family's circumstances. I was told that Derek referred to our test, as he was being driven down to Goldsmiths to be tested, as "a piece of piss," a charming British expression which translates as "easy peasy."

With the assistance of the television production company that was making the documentary, Krissy Wilson and I set up the test as described. On the day of testing, each of six children (aged between 15 months and 2.5 years) was brought, one by one, into the testing room by a registered child minder who remained throughout. Typically, Derek would walk around as he did his readings, sometimes gesticulating wildly. Krissy and I watched proceedings from behind a one-way mirror. Of course, we had no idea as the readings were being given for the children whether they were accurate or not but something interesting did happen at one point. Derek appeared to forget that he was supposed to be giving a reading for the child and started giving a reading for the child minder. She, for one, was impressed with that reading—but, of course, cold reading was one obvious explanation. To pass our test, four or more parents would have to pick the correct reading for their child from the six that they would be given. Purely on the basis of chance alone, such a result would occur less than one time in a hundred.

It was a long day's filming but the real hard work was just about to begin for one poor researcher who had the task of preparing transcripts of Derek's readings. It was originally felt that the most efficient approach would be for the transcripts to consist of a summary of the most important points made in each reading but Derek was very unhappy with this idea. When he saw the original summaries, he felt that too much information had been edited out. The actual transcripts used for the judging task thus ended up being rather lengthy, averaging over 500 words each. Certain themes tended to recur in the readings including health issues of the child and parents, relationship and emotional issues, and problems relating to the state of the family home and car.

The level of alleged knowledge regarding the family car was sometimes incredibly detailed, as this extract from the reading given for a 2.5-year-old girl demonstrates:

> The child was telling me about a car. The following could be associated with one or two cars. One car has or had a wheel or tyre problem on the front driver's side; this is associated with a dark colour car. There is also a scratch or sticker on this car on the lower part of the windscreen on the driver's side. There has also been problems with that car with the rear, driver's side, tyre and a brake problem. On the inside of the passenger seat there is a mark or tear beside the handbrake. A door handle doesn't seem to be working all that well either. There is also a noisy exhaust. The child mentioned that there is an issue with getting a car into first gear with a parent.

So, how well did Derek do? By chance, we would expect one reading out of six to be chosen correctly—and that was exactly the result obtained. Derek was extremely upset at this outcome, proclaiming through his tears, "Well, that's it then! My career is over!" I tried to reassure him that the results of our test would not have any effect on his fans whatsoever.

The next day, Derek set off for Florida to be tested by Randi, in the hope that he would be returning home one million dollars wealthier. Sadly for him, he completely failed that test as well. One might have hoped that his complete failure on two fair tests of his ability would be enough for viewers to draw their own conclusions but the programme-makers apparently felt the need to provide at least a crumb of comfort for the true believers.

In the final part of the programme, Derek had his EEG recorded by Dr Gerald Gluck while attempting a psychic link with another toddler. Dr Gluck opined that Derek's EEG was very unusual, leaving the more gullible viewer with the distinct impression that maybe he was psychic after all. Even if Dr Gluck's interpretation of Derek's EEG was accurate, such a conclusion would be, of course, completely unjustified. The EEG recorded was simply a record of Derek's brain activity when he was *trying* to be psychic, not when he was actually *being* psychic. We can be pretty sure that if the reading produced while Derek's EEG was being recorded had been at all accurate, the narrator of the documentary would have made a point of saying so. It is also worth noting that Dr Gluck was himself a strong believer in various New Age ideas, even proclaiming himself to be an "energy healer."

PUTTING THE CLAIMS TO THE TEST

In 2008, Mrs Patricia Putt contacted the James Randi Educational Foundation (JREF) asking if she could be tested with a view to claiming Randi's million-dollar prize. In common with other mediums, Pat believed that she could communicate with the spirits of the dead and, in this way, glean information about complete strangers without having to speak to them or even see them. Pat believed herself to be the reincarnation of an ancient Egyptian by the name of Ankhara, a discovery that she had made as a result of hypnotic regression. She had worked for many years as a professional psychic.

Before Pat could try for the million dollars in Florida, JREF required her to pass a preliminary test in the UK. Richard Wiseman and I were asked to carry out this preliminary test and we agreed to do so. Once again, the basic idea behind the test was very simple. Pat would be asked to produce ten readings for ten complete strangers using only her psychic powers. Those ten readings would then be presented in random order to each of the individuals for whom the readings had been done and they would be asked to pick the reading that they believed was done specifically for them. If Pat could really do what she claimed she could do, there should be one reading in the collection that stood out from the rest for each volunteer participant as containing lots of accurate, specific details. If five or more of the volunteers chose the correct reading, Pat would be deemed to have passed the preliminary test.

An initial protocol had already been worked out between Pat and members of JREF before Richard and I were contacted and a further round of tweaking the protocol took place after we were contacted. Wherever possible, all reasonable requests from claimants in tests such as these are granted, provided that doing so does not compromise experimental control. For example, Pat asked if the volunteer sitters could read a short pre-specified passage of text as she believed that "the Spirit enters and makes contact through the sound of the sitter's voice." Her request was granted. As always, it was crucially important that the claimant was happy with the agreed test conditions. Pat signed a statement to this effect prior to being tested.

Although the basic idea behind this test was very simple, there were many methodological considerations that needed to be taken into account that might not be so obvious. For example, as noted elsewhere:

> Mrs Putt agreed not to include in her readings anything that might give an indication of the position of the reading in the series (e.g. "Feeling more confident with this one" would indicate that this could not possibly be the reading for the first volunteer). She also agreed not to make any reference to events that she might overhear outside the testing area (e.g. had there been the sound of children playing during one reading and reference was made to "happy children" in the reading itself). She agreed that all of the participants could be selected from the same ethnic group (Caucasian), be of the same

gender (female), and within a restricted age range (18–30). This is because a person's voice gives away much information regarding such factors.[14]

The actual test took place in May 2009 (with the invaluable assistance of volunteer helpers Panka Juhasz, James Munroe, Suzanne Barbieri, and Fabio Tartarini). Each volunteer sitter was led into the testing room by myself and seated on a chair facing the wall. Only then was Pat brought in by Richard and seated at a desk at the opposite end of the room, where she quietly wrote her readings down. When she had finished her reading, Richard led her away and the next trial began. One aspect of the protocol agreed between Pat and JREF before Richard and I got involved did give the proceedings a slightly surreal appearance. In order to ensure that Pat could not pick up any clues about the volunteer sitters from their appearance, each one wore a ski mask, wraparound sunglasses, white socks, and an oversized graduation gown.

Once all of the readings had been completed, all of our volunteer sitters returned and were handed a booklet containing the complete set of ten readings, each in a different random order. They read the set of transcripts carefully and selected the one that they felt was most applicable to themselves. Sadly for her, Pat did not get a score of five or more correct. By chance, one would expect one correct hit out of ten purely on the basis of guesswork but Pat did not even manage that. Not one of the sitters chose the actual reading that had been done for them.

Initially, Pat's reaction to her failure on the test appeared to be refreshingly different from what we had expected. Instead of coming up with excuses for her failure and deciding that the test was not fair after all, she simply declared herself "gobsmacked" and accepted the result. Unfortunately, within a day or two she had changed her mind. As she wrote in an email to JREF, "With them [the volunteers] being bound from head to foot like black mummies, they themselves felt tied so were not really free to link with Spirit making my work a great deal more difficult."[15] In fact, no one was "bound" and Pat did not speak to the volunteers at any point. Any knowledge of how they felt during the test must, one assumes, have been obtained psychically.

In addition to claiming that no one could be expected to get any hits in this test because of the conditions of testing, Pat also went on to argue that, in fact, she had hits on every trial. In a comment on Richard's blog, she wrote, "with hindsight I realised that every girl had accepted each and every message that I had written down not one had been discarded, not one thrown away each and every one of the ten girls had gone away with something to me this makes a total of 10 out of 10."[16] Of course, this entirely misses the point that each sitter was required by the protocol to choose one reading from the ten as the one most applicable to themselves even if they felt that none of them was particularly applicable!

Despite feeling, in hindsight, that this test had not been fair after all, Pat did volunteer to be tested again in 2012. Along with Kim Whitton, a spiritual medium and healer with 15 years' experience, Pat had responded to a challenge issued by myself, science writer Simon Singh, and Michael Marshall of the Merseyside Skeptics Society. We presented our Halloween Challenge as an opportunity for psychics to prove to the world that they really did possess powers that defied explanation in terms of conventional science.[17] After all, if they could really do so, this would be an amazing breakthrough for science—perhaps a Nobel Prize or two was within reach? Unfortunately, all of the British high-earning celebrity psychics (including Sally Morgan, Colin Fry, Gordon Smith, and Derek Acorah) whom we invited to take part flatly refused to accept the challenge, but Pat and Kim were willing to step up to the plate.

This test took place at Goldsmiths on October 21, 2012 and was similar in design to the test that Pat had failed in 2009. Once again, our psychics would do readings for a number of complete strangers without seeing them or communicating with them in any way. Our volunteer sitters would then be required to rate each reading for accuracy on a scale from 1 to 10 and to choose the one reading that they felt best described them. Because we were testing two psychics on the same day, we limited the number of volunteer sitters to five. To be deemed to have officially passed the test, all five sitters would have to choose the reading that had been done specifically for them.[18]

There were a few changes made to the protocol used in the 2009 test. Rather than having our sitters wear the strange outfits worn on the previous occasion, we simply had them sit behind a specially constructed screen. During their reading, they were asked to think about the sort of issues that they might expect a psychic to tell them about. As usual, our psychics signed a statement in advance of the test confirming that they believed it was a fair test of their claimed ability. In addition, they were asked after each reading to indicate their level of confidence that the sitter would be able to recognize herself in that reading. On a 7-point scale (where 7 = totally confident), Kim gave an average rating across all sitters of 5.2 and Pat's average was 5.8.

Here is one of the readings produced by Kim (presented with the sitter's permission):

> Affectionate, touchy-feely sort of person. Can't sleep well at night, all sorts of thoughts running through my head. Badminton, tennis. Wants children. Not yet, too young. Glasgow is an important place. Singing and dancing. Would like to develop performance skills. Collecting old photos of places I've been to. Individual likes and dislikes—laws and government, not good. Learn to get along together better would be a good start. Can't get him out of my head. I wish we could be together. Sometimes pain in leg. Like things with salt not sugar. Dutch ancestry? Optimistic and open-minded. Want to get university degree. Living in London enjoyed very much. September is an important month. Help in canteen sometimes. Brother in Holland at home. I can get into paints and crayons. Birthday November. I enjoy winter months. Have many friends here in London. Want to go home soon to see family. Wants to go to South America.

Psychics often have their own unique style, as we can see in this reading where Kim frequently adopts a first-person perspective as though actually inside the sitter's mind. The reading above was rated as 8 out of 10 by the sitter, who gave ratings of 3 or less for the other four readings. She acknowledged that some of it was inaccurate (for example, the Dutch ancestry) and some of the comments were pretty safe bets (for example, living in London), but this looked like a pretty impressive hit.

Unfortunately, it was Kim's only hit, perhaps suggesting that even the hits in this reading were based on nothing more than lucky guesses. Overall, Kim's target readings were on average rated as 3.2 out of 10 by the sitters, compared to 2.4 for the nontarget readings. The difference was not statistically significant. None of our sitters chose the correct reading from those done by Pat. In fact, for Pat the nontarget readings had a higher mean accuracy rating (4.2) than the target readings (3.2). Once again, the results of our test had failed to produce any compelling evidence that psychics really do have the amazing abilities that they claim to have.

Skeptics are often accused of being closed-minded and afraid of accepting the evidence that their critics believe provides overwhelming proof that paranormal powers are real. Those making such criticisms have typically devoted little or no time to directly testing paranormal claims for themselves. Given the considerable amount of time, effort, and resources that I have put into directly testing paranormal claims for myself, I hope the reader understands why I find it a little bit annoying when such criticism is directed my way.

11 DREAMS OF THE FUTURE

In chapter 8 we discussed the application of the law of truly large numbers as one possible explanation for reports of apparently precognitive dreams. Essentially, this law states that given a sufficiently large number of opportunities for a specified extremely unlikely (but not impossible) event to happen, it is virtually inevitable that it will happen. Given the extremely large number of dreams that are recalled every single night, it is inevitable that sometimes a dream will bear a striking correspondence to a future event purely on the basis of chance alone. Such coincidences, when considered in isolation, may indeed sometimes be extremely unlikely, but they should not be considered in isolation.

Chance is not the only relevant factor here. One of the most commonly reported examples of precognitive dreams is dreaming of the death of a loved one only to discover upon awakening that that person did indeed die in the night. But it is likely in many such cases that the dreamer was already aware of, say, the fact that the loved one was seriously ill before they had the dream. Their anxiety about their loved one's health caused them to have a distressing dream of the death, but the death itself was pretty likely to happen under the circumstances.

In chapter 8 we also discussed the principle of "near enough is good enough," and this again is relevant with respect to apparently precognitive dreams. Suppose your loved one did not die that night but a week later. You might still feel that your dream provided you with a psychic glimpse into the future. The same would apply if your loved did not actually die on the night of your dream but came very close to death, perhaps only surviving as a result of medical intervention. Furthermore, many people believe that dreams represent future events not in their literal form but in symbolic form, as indicated by several prophetic dreams reported in the Bible that required interpretation in order to be properly understood.

Although such factors may account for many apparently precognitive dreams, can they fully account for those exceptional individuals who claim to have such dreams on a regular basis? This is a very difficult question to answer. Consideration of the law of very large numbers would lead us to expect that quite a lot of people would experience one or two such dreams in their lifetimes, and even that a few individuals would experience several purely on the basis of chance alone. But if someone claims to have such dreams very regularly, it is not unreasonable to suggest that it might stretch the explanation in terms of coincidental matches beyond breaking point. The rest of this chapter describes investigations of two individuals making exactly this claim.

THE MAN WHO PAINTS THE FUTURE

In 2002, I was approached by a television company making a documentary about David Mandell, a sixty-nine-year-old artist living in Sudbury Hill in northwest London, for the *Extraordinary People* series on Channel 5 in the United Kingdom. What was extraordinary about David was that he claimed that his dreams often foretold future events, particularly disasters and terrorist attacks. Being an artist, whenever he had one of his precognitive dreams he would try to fix an image of the dream in his mind as he woke up. He would then get up and paint or draw that image, sometimes with a few accompanying notes for clarity or extra detail. He would then wait, sometimes for days, sometimes for weeks, sometimes for years, for the events that he had dreamed about to occur. He never knew when this was likely to happen, but often, he claimed, he would see a news report in the media of the very event that he had witnessed in his dream, the level of correspondence between the dream and the event being so high that it could not possibly be explained, as far as David was concerned, as being a mere coincidence.

David was aware of one obvious objection to his claims. How could people know that he did not just paint his pictures after a newsworthy event had taken place and then simply *claim* that he had painted them in advance? In a charmingly amateurish attempt to rebut that accusation, David would often take his newly painted picture into his local bank and have his photograph taken as he stood in front of the date and time indicator on the wall. Of course, such photographs could have been faked even back then if David was nothing more than a clever fraudster, but for what it is worth, I felt then and continue to believe that David was totally sincere in his claims (while fully accepting that that is exactly what a clever con artist would want me to feel!).

We realized early on that it would not be possible, in the time available, to carry out any definitive test of whether David really was able to psychically foretell the future through his dreams. The main problem was that David himself had no idea when the events he had dreamed about would actually occur in the real world. So we could not simply take a sample of his dreams and then wait to see if the events occurred within some prespecified time frame. Instead, we took an indirect approach.

The most obvious nonparanormal explanation of the correspondences between David's dreams and future events was that they were indeed simply coincidental. After all, the aftermath of one earthquake is likely to bear a resemblance to the aftermath of many other earthquakes—or, indeed, to a dream of the aftermath of an earthquake. But some of David's dreams were much more specific than this. Perhaps his most striking prediction was the collapse of the Twin Towers on September 11, 2001. David had two dreams about this historic and terrible event—one of which occurred *exactly* five years before it happened.

Skeptics might object that what David painted in this notable picture does not correspond exactly to what happened on that unforgettable day. Although his painting is undoubtedly a depiction of the New York skyline, the tower on the left in his picture is depicted as toppling over into the tower on the right, whereas in fact both buildings collapsed vertically downward independently of each other. Furthermore, David himself had originally thought that the collapse of the buildings was the result of an earthquake, not of a terrorist attack. For all that, it is undeniable that the pictures do portray the collapse of the Twin Towers.

Although the collapse of the Twin Towers was arguably the most impressive of the predictions based on David's dreams, several other major news events were said to be depicted among the hundreds of other paintings and sketches that David had in his collection, including the collision on the river Thames of the pleasure steamer *Marchioness* with the dredger *Bowbelle* in 1989, resulting in the deaths of fifty-one people; an earthquake in San Francisco in 1989; a mortar attack by the IRA on Heathrow Airport in 1994; the *Braer* oil tanker running aground off Shetland in Scotland in 1993; the sarin gas attack on the Tokyo subway in 1995, carried out by members of the Aum Shinrikyo cult, killing thirteen people; the suicide in prison of serial killer Fred West in 1995; riots in Oxford Street, London, in 1994; fifteen people injured in a mass stabbing in Rackham's department store in Birmingham, UK, in 1994; a train crash in Watford, UK, in 1996 that killed one person and injured sixty-nine others; the death of Princess Diana in Paris in 1997; the Dunblane massacre in Scotland in 1996, in which sixteen pupils and one teacher at a primary school were killed; and Concorde crashing into a nearby hotel in Paris shortly after takeoff in the year 2000, killing all 109 people on board as well as four people in the hotel.

The task that my PhD student Louie Savva and I set ourselves was to see if we could find other reports in the news archives that matched David's paintings at least as well as, if not better than, those chosen by David himself.[1] Forty of the pictures that David felt were good matches to subsequent real events reported in the news were selected from his collection of two hundred or so precognitive pictures. These were presented to thirty volunteers, each one accompanied by a brief description of the event that David felt he had predicted as well as a brief description of an alternative news event that we felt also matched the picture to some extent. Where available, copies of actual reports in newspapers were also provided, as David felt that some of the photographs that accompanied the articles contained details that clearly matched his pictures. Our volunteers were asked to take as long as they needed to carefully study each picture and then rate the degree to which they felt it matched the news events chosen by David as well as the alternative news events chosen by us on a 7-point scale (where 1 = no match at all and 7 = a perfect match).

To give the reader a clearer idea of what was involved in this task, here is a more detailed description of just one of the forty trials—but please note that not all of the news stories featured in this study were as horrendous as this particular example. The picture itself was not particularly gruesome. It appeared to show a hand hammering a chisel into something and a lot of red paint. The note written on the picture itself was considerably more disturbing:

> Dream SATURDAY MORNING 11 MARCH 89 [dates were edited out of the pictures used in the study] of three women's faces covered with canvas type cloth with slashed slits for the eyes in it. Then a man with a large cold chisel and hammer smashed their faces in hammering the cold chisel several times in different parts of the faces through the cloth—it was awful dream!

David believed that this dream picture was a description of a particularly vicious crime that occurred some years later. Here is the description of that crime used in our study:

> 10th July 1996: The three female members of the Russell family were attacked with a claw hammer in a country lane near Chillenden, near Canterbury, Kent. Only Josie, aged 9, survived. Both Josie and her mother were blindfolded and suffered severe blows to the head. Meagan, aged 6, was hit at least seven times. The words *hammer* and *hammering* appear in the text.

A copy of the report of the crime from the *Times* accompanied the picture.

Our suggested alternative news event, selected from the archives, was described as follows:

> 27th September 1986: Donna Jester (aged 37), her blind cousin Dalpha (aged 64), and Laura Lee Owens (aged 20) were discovered dead in Lancaster, Texas. All three died as a result of numerous chopping wounds to their heads and faces, which were inflicted with a hatchet. The text refers to *three women*.

For the forty pictures used in this study, our volunteers gave significantly higher correspondence ratings between seven of the pictures and David's choice of matching news event compared to our chosen alternative news stories. Or, to put it another way, there was no significant difference between the correspondence ratings of David's choice and our suggested alternative for 82.5 percent of the pictures. Having said that, a comparison of the average ratings for each participant across the full set of forty pictures did reveal a small but highly significant difference (just over half a point on our 7-point scale) in favor of David's choices.

What is one to make of these results? Given that these were, in David's judgment, the best examples of correspondences between his dreams and subsequent events, believers in precognitive dreams might be disappointed that our alternative news events were judged to correspond at least as well to the pictures as David's choices in the vast majority of cases. It should also be borne in mind that, whereas David had the luxury of selecting news events as they occurred over many years, our selections from the archive had to be made within a couple of weeks in order to meet the filming schedule of the program makers. Given more time, we may well have been able to come up with even better matches. I suspect that believers in the paranormal will, with some justification, be more impressed by the fact that David's choices were rated as better matches than ours for seven out of forty of the pictures and better matches overall. Personally, I'd be happy to class this one as a tie.

When going through the papers that I had stored relating to this study, I came across copies of eight pictures with brief descriptions, all with the words "Event has not occurred yet" written on them. I could not help but wonder if any of the foretold events had occurred in the two decades since David gave me these pictures. Here are the notes accompanying those pictures:

- CIVIL WAR NORTHERN GREECE (dream occurred around May 2002): This painting depicts a civil war in Northern Greece. Mr. Mandell thinks it is occurring on the border. The image shows hotels burning, and "military men," including a "little Hitler type Commisar telling us hotels were on fire on hills." Notes also state: "the Salonica threat."
- CHILD KILLERS (dream occurred 11 June 2002): The painting depicts a "gang of 5 or 6 children who are child killers" led by "a girl in a blue spotted dress."
- PLANE WITH PLASTIC STRIPS (dream occurred 28 April 2002): Mr Mandell envisioned this event happening at "Heathrow or City Airport."

He dreamt about a "plane with plastic type strips trailing behind the tailplane" which cause "very violent movement sideways of tail of airplane." The plane is "trying to crash land over water near to airport."

- ROYAL LONDON HOSPITAL—HELICOPTER (vision occurred 19 February 1995): This vision concerns the helicopter that operates from the roof of the Royal London Hospital, unable to make a safe landing (for further details see letter enclosed, form [sic] Mr Mandell to the hospital). Notes on painting: "Descent too fast to stop or slow down"; "Top of building hiding landing ramp."
- EARTHQUAKE IN LONDON (dream occurred 28 April 1992): See notes on copy of drawing for details. [Notes on and accompanying this drawing indicate that the dream was in two parts. In the first part, buildings viewed through a window were seen "moving sideways past each other" and then swinging back "to return to their original positions." In the second part, the building that Mr Mandell was in lifted up and down several times. "It was just like being on a seaside rollercoaster."]
- BEATLE SHOT (dream occurred 30 November 2001): The drawing shows a member of the Beatles being shot by a left-handed gunman.
- PLANE HITTING DARK GLASS BUILDING (dream occurred around May 2002): This painting shows a plane exploding, that has flown into the side of dark plate glass building that looked "like new towers side of Canary Wharf" [by the River Thames]. Mr Mandell's notes state that "tower collapsed just like NY tower did."
- PADDINGTON BRIDGE CRASH (dream occurred 11 January 2002): This painting depicts a "bad accident" at "perhaps Paddington," caused by a "collapsed steel girder bridge." The word 'bomb' is also noted on the painting. Mr Mandell shows "devastation around bridge."

As far as I have been able to ascertain, none of these predicted events have taken place in the two decades or so that have elapsed between David giving me copies of these pictures and the time of writing. It may be, of course, that events will take place in the future that do indeed bear a striking resemblance to some or all of these dreams (although the earthquake dream sounds especially unlikely). If that is the case, it would certainly provide some support for the claim that dreams can sometimes provide a psychic glimpse into the future. But I am struck by the fact that David's hit rate appears to have plummeted over the last two decades compared to the years preceding our test. One can only speculate regarding the possible reasons for this decline.

THE MAN WHO DREAMS THE FUTURE

The reason that we could not directly test David Mandell's claim that his dreams matched future events to a greater extent than could be explained by coincidence and other normal factors was, as stated, simply that David himself had no idea whether his dreams would come true within days, months, or decades. If he had been able to give specific dates for his predictions coming true, or at least a reasonably narrow range of dates, directly testing his claims would have been much easier. Fortunately for us, the opportunity to test someone who claimed to be able to do precisely that presented itself a few years later. That person was Chris Robinson, the self-styled "dream detective." Our test of Chris's precognitive ability was featured in (yet another) program in Channel 5's *Extraordinary People* series, first broadcast in the United Kingdom in 2007.

Chris believes that through his dreams he has foreseen the future on dozens of occasions, a skill that he acquired following a near-death experience. Like David Mandell, he claims to be particularly gifted when it comes to predicting terrorist attacks (such as the attack on the Twin Towers on September 11, 2001, and the bombings in London on July 7, 2005) and disasters (such as that at the Chernobyl nuclear power plant in 1986). He also claims to be able to solve crimes and find missing persons through information received in his dreams, information that he claims is sometimes provided to him by the spirits of the dead.

Chris generates his allegedly precognitive dreams by thinking about the specific topic that he would like to dream about before he goes to sleep. On occasion, he may even write down a specific question before going to sleep or have a relevant object, such as an item of clothing belonging to a missing person, beside his bed. Upon awakening, he will record details of his dream in a dream diary. According to Chris, sometimes his dreams are a literal representation of a future event, as though he was directly witnessing it as it happened, and sometimes the events are represented symbolically. However, the symbols often have consistent meanings such that dreams of dogs always represent terrorists, snow always represents imminent danger, meat always represents carnage, and so on.

The advantage that Chris has over other psychics from the point of view of testing is that he claims to be able to "dream to order," so to speak. Thus, if Chris knows that on a prespecified day he is going to be taken to a mystery location, he claims that his dreams prior to being taken there will contain specific information corresponding to that location.

Parapsychologist Gary Schwartz was convinced that Chris's claimed ability was real on the basis of the results of a ten-night study in which Chris's dreams from the previous night bore a striking correspondence to a series of mystery locations that he was taken to. The degree of correspondence between the dreams and the locations was such that Schwartz did not believe that it could be explained purely on the basis of coincidence. The problem, as the reader has no doubt already realized, is that this approach makes absolutely no allowance for the possibility of subjective validation as discussed in chapter 9. With this technique, we have simply no way of knowing for each trial if the level of match between aspects of the target location and elements of Chris's (very detailed) dream description are higher than we might find for an alternative possible location. Fortunately, this methodological problem was relatively easy to solve.

For our test of Chris's claimed ability, we tweaked the approach taken in the Schwartz study in small but important ways. As in that study, Chris was informed that he would be taken to a mystery location on a particular day and asked to record his dreams, as usual, in his dream diary. However, whether his dream was deemed to match the location was not decided purely on the basis of looking around the location to see if there were any details that appeared to match the dream. Instead, a pool of six possible target locations was assembled in advance by Lesley Katon, the producer and director of the documentary. Lesley made an effort to choose six very different locations, and neither I nor Chris had any knowledge of them. On the specified day, Chris and Lesley would discuss the dreams that Chris had recorded and Lesley would decide which of the six locations was the best match to the contents of those dreams. It was only then that the actual location was decided on—by the roll of a die. Chris was then taken to the location, whether or not it was the location deemed to be the best match to his dream, as the logic of the methodology required this to be the case.

It was extremely important in the interests of fairness that the independent judge choosing which of the six locations was the best match to Chris's dream was someone whom he trusted and who wanted Chris to succeed. After all, a mean skeptic who wanted Chris to fail the test could simply have chosen the worst match even if Chris really did have the ability he claimed to have. For that reason, Chris was allowed to choose who the independent judge would be. The program maker herself was the obvious choice. Lesley would clearly have a much more interesting documentary on her hands if Chris could pass a test set by a skeptical psychologist rather than if he failed.

Three trials as described above were set. The probability of Chris being successful on each trial on the basis of chance alone was one in six. Therefore the chances of him being successful on all three trials on that basis was one in 216 (that is, 6 × 6 × 6). If Chris could pull it off, that would be a pretty impressive performance. So, how did he do?

Here is part of what Chris had to say about one of his dreams on Day 1: "Paint, white, sheets painted all white. In the dream, it wasn't my house, it was a person I knew when I went to school, who was an artist. He could paint and draw anything. So 'painting' is like a key word if you like. You have got to choose from your pool of places which one you think that little lot matches the best."

Lesley chose location number four from the pool of possible targets as being the best match. This was the private home of a painter called the House of Dreams, an apparently excellent match to aspects of Chris's dream. Unfortunately, when the die was rolled to determine the actual target location, it came up with number two, corresponding to the ICEBAR, a cocktail bar in London. Given the logic of the experimental design, based as it was on the fundamental claim that Chris would dream about where he was actually taken, we then had to take Chris to the ICEBAR even though we already knew that, from his point of view, this trial was a failure. To make matters even more surreal, Chris had insisted that he should be taken to all target locations blindfolded. He reasoned that if this precaution were not taken, he might dream about the route to the target location not the target location itself. Interestingly, very few Londoners even batted an eyelid as I led this blindfolded man along the street from the taxi to the bar.

Chris's reaction when his blindfold was removed was to exclaim, "There's a lot of white." Clearly, subjective validation was at work. Any thoughts of "painting" being the key word were forgotten as Chris insisted that he was "surrounded by white." In fact, the walls of the ICEBAR consisted of pure, transparent ice, not white snow, with the gray walls behind them clearly visible. However much Chris might have wanted to explain it away, this trial was a clear miss.

Regarding one of his dreams for the second trial, Chris had this to say: "This was an observer dream so I am watching other people. It's not as if I'm part of it. And I'm rolling or unrolling something. That could be, if you're in a printing works, it could be rolls of paper. Somebody's now made a pie, so it could've been pastry. And we then put this pie in the oven. And there was a joke about not putting the oven on for too long so you don't burn it. And I see these faces of people go past me."

Finding a match was trickier this time, but Lesley finally chose target location number one, St. Bride's Crypt in Fleet Street, on the basis of this famous street's historic links with the printing industry. Unfortunately for Chris, when the die was rolled, it came up with number five, corresponding to a city farm. This time, I not only had to lead Chris blindfolded into the farm, knowing full well that this was another failed trial, but help him up onto a horse while still blindfolded. The things I do for science! When the blindfold was removed, Chris simply said, "There's a surprise," and made no attempt to try to claim this as another hit.

For the third and final trial, here were some of Chris's comments on his dream: "In the dream, I'm in a sort of a room and there are people that are sort of people." Pointing to his dream diary, he continued, "And there in big letters I've put 'DEAD.' It's in this room. There was actually glass cups and glasses. Now, cups is always the same. Cups, if I see them, means dead people. And if we don't go where there are dead people, I don't know what that was all about!"

St. Bride's Crypt was still in the target pool as, although Lesley had chosen it as the best match on the previous trial, it had not actually been used. This time it seemed an even better match so Lesley chose it again—and this time when the die was rolled, it did indeed come up with number one. So Chris scored one hit out of three trials, very far from a statistically significant result. In commenting on his failure, he explained that the information he receives in his dreams is "bandwidth limited," allowing him to only see the "silhouette" of the future event, not the details. One cannot help wondering why this limitation only seems to occur when Chris is tested under properly controlled conditions, not when subjective validation is given free rein.

Despite the general lack of empirical support from well-controlled studies for the claims of Chris and other similar seers, there continues to be a demand for their services. One source of such demand is from desperate people who are searching for a missing loved one, including the parents of forty-three-year-old Marcy Randolph. Marcy had set off on a sightseeing flight from Deer Valley Airport, fifteen miles north of Phoenix, Arizona, in a plane piloted by fifty-four-year-old William Westover on September 24, 2006. It was believed that the pair were intending to fly to Sedona Airport and then return to Deer Valley on the same day. Sadly, radar contact with what was believed to be that plane was lost about nine miles southwest of Sedona.

It was arranged for Chris to fly to Arizona in the hope that his dreams might provide vital clues in the search for the missing plane, which was now presumed to have crashed somewhere in the area. Previous searches for the wreckage in the rugged and extreme terrain had proved unsuccessful. On the basis of information in Chris's dreams, it was decided to focus the search in an area around Snake Canyon and Diablo Canyon. Unfortunately, this search also failed to discover the crashed plane, but the viewer was left with the strong impression that it might indeed be somewhere in that area.

In April 2009, however, long after the documentary had been broadcast, the wreckage of the plane was found, along with the skeletal remains of its occupants. It turned out that on the day the plane went missing, a couple of hikers had reported seeing a small fire in remote Loy Canyon northwest of Sedona. They reported the fire to the authorities, but neither they nor the authorities to whom they reported the fire were aware that a plane had gone missing at that time, so no one connected the two events. It was not until over two years later that someone fortuitously happened to come across the report, thus sparking further searches in that area by volunteers including Phil Randolph, Marcy's father, who had never given up in his tireless efforts to locate the missing plane. These volunteers narrowed the search area down further by matching aerial photographs to a photograph taken by the hikers, but it was ultimately the hikers themselves who found the physical wreckage after returning to the area. The wreckage and the remains of the occupants of the plane were found many miles away from the area searched as a result of information provided by Chris.

I had known Chris Robinson for many years prior to submitting him to the test just described, our paths having crossed many times on live TV shows and elsewhere. On one memorable occasion in April 1993, both Chris and I took part in a late-night live discussion program. At the time, it was very common for regional TV channels in the United Kingdom to broadcast such programs on Friday nights, presumably aimed at people coming in from the pub after a few drinks. I do not remember the details of most of the program (I did a lot of them and they all became a bit of a blur in my memory). But I do recall that Chris, in the final seconds of the program, was challenged to give the name of the winner in the following day's Grand National race held annually at Aintree Racecourse near Liverpool. Chris was not expecting the question, and although he did offer a name that I cannot now recall, he told me immediately after the program that he had simply felt under pressure to say something and had little, if any, confidence in his answer. This seemed reasonable to me, as Chris had not been expecting to be asked the question and had not sought the information using his usual dream diary technique.

The TV company had provided overnight accommodation for us in a local hotel. When Chris came down for breakfast, he was very excited, as he had now sought the information in his dreams regarding who would win that day's race. Although he did tell me—and it was indeed a different horse than the one named the night before—I am afraid I cannot now remember the actual name he gave me. What I found fascinating, however, was Chris showing me his dream diary and explaining how he uses it to predict future events. Chris's dreams are usually long and detailed, providing many potential elements that could match future events (the brief descriptions provided above from our documentary are but a fraction of the details provided by Chris for each dream). Given that sometimes the dreams represent events literally, sometimes symbolically, and sometimes a mix of the two, the scope for subjective validation is enormous.

Some readers may be thinking that it is unfortunate that I can no longer remember the name of the horse that Chris had predicted would win the Grand National on April 3, 1993. I can assure you that, at the time the race was run, I had the name of that horse at the forefront of my mind, and, as you might imagine, I watched the race on TV with considerable interest. As some readers may recall, 1993 was the first and to date only occasion upon which this famous steeplechase was declared void. This was because thirty of the thirty-nine runners set off despite there being a false start. If Chris had told me over toast and marmalade that that was what he expected to happen in the 147th running of this race, I would have been most impressed! But he didn't. He told me who the nonexistent winner would be.

The technique used in our test of Chris's claims in the 2007 documentary was in fact based on a previous test designed in collaboration with Chris, parapsychologist Keith Hearne, James Randi, Richard Wiseman, and others a few years before. Keith Hearne is convinced that dream precognition is a genuine phenomenon and fully expected Chris to pass the test. He was very familiar with Chris and his technique for interpreting his dreams. Finalizing the protocol took a great deal of effort, with many email exchanges, not least because many of Chris's supporters not only did not trust Randi (or me or Richard) but actively despised him.

Prior to us agreeing to test Chris's claims, his more extreme supporters would often declare that the "closed-minded" skeptics would never dare to actually test him, as proof of his psychic powers would show the world what fools they were. As soon as we agreed to carry out such a test, those same people pleaded with Chris for him to refuse to take part on the grounds that any test would be rigged by the evil skeptics so as to ensure that Chris failed. This type of paranoid distrust, especially that directed at Randi, sometimes reached hilarious heights. When one of Chris's supporters learned that we intended to use a coin to make a random choice on each trial, he sounded the alarm, pointing out that magicians have various techniques for influencing the fall of a coin. That may be so, but the concern was misplaced for two reasons. First, due the experimental design used, the person tossing the coin would not know at that point whether heads or tails would lead to success or failure. Second, Randi would actually be on the other side of the Atlantic, not in Chris's hometown of Dunstable in the United Kingdom when the test took place. Now, Randi was indeed a great magician, but influencing the fall of a coin on the other side of the Atlantic would have been a big ask even for him!

Although the basic idea was the same as that for our 2007 test described above, one major difference was that there were only two potential target locations on each trial, not six (hence we could decide the actual target for each trial by a coin toss). This meant that on each trial, there was a 50 percent probability of the correct location being chosen purely on the basis of guesswork. The plan was to have ten such trials. If the correct location was chosen across all ten trials, this would strongly indicate that more than mere chance was at work, as this outcome would only be expected to occur one time in every 1,024. Keith Hearne himself was the person chosen by Chris to decide which location was the better match to Chris's dream on each trial. He was a perfect choice, having worked with and been impressed by Chris for many years prior to the test. It was also agreed that the study would continue until all ten trials had been run, even if Chris failed on any trials, as this would allow the protocol to be refined with a view to carrying out further studies in the future.

It had been agreed that a TV company would film the first trial so that it could be included in a documentary. It struck me as being a bit silly to only include the first trial in a documentary, given that there would be a 50 percent chance of success purely on the basis of chance. This would be like trying to assess whether someone could psychically influence the toss of a coin on the basis of a single coin toss. But I supposed that it would at least give the viewing public an idea of how paranormal claims can be put to the test. They would get some indication of the amount of time, effort, and careful planning that is required. Of course, the TV company also went to a great deal of effort and expense to film the proceedings, requiring as it did a director and film crew for a full day.

In the event, I need not have worried about a single trial being presented as if it was any kind of meaningful test of Chris's alleged ability. Without going into details, suffice it to say that he failed on the first trial, an outcome that surprised and upset Keith Hearne greatly—and none of that footage was ever included in any documentary. Had he been successful on that single trial, I am sure it would have been, given the time and resources that had been required to film the trial. Sadly, Keith Hearne appeared to lose interest in the study at this point, and no further trials were ever carried out.

12 LESSONS FOR (THE) LIVING

Anomalistic psychology is not just a fascinating topic in its own right, it is also a relatively painless way of picking up important critical thinking skills that can be applied in all aspects of life, including those which may appear to be a million miles away from the paranormal.

Among the more obvious applications of such skills is the assessment of controversial scientific claims. Newspapers and other media constantly bombard us with such claims, and many people understandably feel unable to decide what to believe. One of the most important current areas of apparent controversy is whether global warming is really happening and, if it is, the degree to which it is caused by human activity. The very future of life on our planet may depend on addressing such issues correctly. The coverage of this topic provides a perfect example of how the media often do not simply report on genuine scientific controversies but sometimes play a major role in generating what we might call *pseudo-controversies*. When the vast majority of the scientific experts in an area speak with one voice but are opposed by a very small but vocal minority, we would probably be wise to listen to the majority and, in this context, adopt the precautionary principle with respect to looking after our planet.

On a more personal level, critical thinking skills are certainly required if we are to make wise choices when it comes to consumer behavior. Many of the cognitive biases, particularly those related to reasoning, that lead us to draw faulty conclusions when considering the paranormal also appear to skew our judgment when it comes to buying, selling, investing, and (especially) gambling. Models of classical economics assume that people make such decisions based on an entirely rational analysis of costs and benefits. Unfortunately, this central assumption is just plain wrong, as shown by numerous ingenious and compelling experimental investigations.

Medical decision-making is often literally a matter of life and death, and yet here too our judgments and behavior often depart radically from anything that could be described as rational. Where our own health and the health of our loved ones are concerned, the stakes are at their highest, and we find it hard to deal with the inevitable inherent uncertainties involved in such situations. Emotions run high, and this too will cloud our judgments. This has been demonstrated all too often during the COVID-19 pandemic. And yet it is precisely in such areas that we really do need to do our absolute best to make the correct choices regarding medical treatments and healthy lifestyles. The popularity of complementary and alternative medicines is undeniable proof of our human irrationality where health is concerned. Once again, the media is often at fault, making exaggerated claims regarding miracle cures, on the one hand, and health risks, on the other, and generating yet more pseudo-controversies.

On occasion, insights from anomalistic psychology are directly relevant to evaluating controversial claims in areas that appear to be completely unrelated to the topics covered in this book. For example, chapters 5 and 6 discussed the use of hypnotic regression to allegedly recover blocked memories of alien abduction and past lives, respectively. In both cases, the evidence very strongly suggests that such "recovered memories" are, in fact, false memories. However, the same method is used by some therapists in attempts to recover allegedly repressed memories of childhood sexual abuse, including extreme claims of ritualized satanic abuse. I suspect that most people would be rather skeptical regarding reports of alien abduction or past lives based on hypnotic regression (although clearly a substantial minority are not). But many of those people who would dismiss such reports may well be inclined to accept reports of recovered memories of childhood sexual abuse obtained in this way even in the complete absence of any independent evidence that such abuse ever took place.

Why might many people feel it is more reasonable to believe that "recovered memories" of childhood sexual abuse are likely to be true even though they would reject memories of aliens and past lives? For one thing, any reasonable person would know that childhood sexual abuse really does occur throughout society on a scale that is much higher than was once recognized—and it really can have devastating psychological consequences for the victims. The evidence that alien abduction and reincarnation really take place is thinner on the ground, to put it mildly.

A second factor is the widespread acceptance by both the general public and professionals of psychoanalytic notions such as repression despite the fact that most memory experts are very dubious regarding this very concept.[1] In general, those suffering traumatic events are much more likely to remember than to forget them. The lesson from anomalistic psychology is that, in the absence of independent evidence, one should treat all reports of recovered memories with a great deal of caution.

Elsewhere I have discussed two examples of contexts in which some knowledge of the ideomotor effect, so familiar to students of anomalistic psychology, may well have helped to prevent needless tragedies.[2] As readers will recall, the ideomotor effect is a phenomenon whereby one's own beliefs and expectations can result in unconscious muscular movements. It is the explanation for several ostensibly paranormal phenomena as previously discussed, including table-tilting, the Ouija board, and dowsing.

The first example is that of the ADE-651, a device that was claimed to be able to detect not only explosives and weapons but also human bodies, contraband ivory, truffles, drugs, and even banknotes. It was apparently able to do so even if the target object was several miles away, underground, or underwater. Not surprisingly, this amazing piece of technology did not come cheap, costing up to $60,000 per unit, but if it provided an effective means to prevent the loss of hundreds of lives in terrorist attacks, it would clearly be a price worth paying. The only problem was that it didn't. It was a piece of junk that cost a few dollars to produce and was no more effective than a chocolate teapot What is more, the British company that produced the device, Advanced Technical Security & Communications (ATSC), was fully aware of this fact.

In the first decade of this century, these useless devices were sold to twenty countries in the Middle East, including Afghanistan and Iraq. The

Iraqi government alone spent around $80 million on them. Each device consisted of a handheld unit into which was mounted a swiveling antenna. It was claimed that the device was powered by static electricity alone and could be charged up by the operator holding the device while walking or shuffling their feet for a few moments prior to use. The inventor of the ADE-651 and founder of ATSC, James McCormick, claimed that the device worked on a similar principle to that of dowsing. That should have set alarm bells ringing very loudly given that, as discussed, dowsing has never been shown to be effective when tested under properly controlled conditions. James Randi publicly offered $1 million to anyone who could demonstrate that the device really worked. Tellingly, no one from ATSC ever responded to the challenge. It is clear that McCormick was fully aware that the devices did not work. It is reported that on one occasion, when challenged about the effectiveness of the device, McCormick replied that it did "exactly what it's meant to . . . it makes money."

In all probability, this cynical scam cost many hundreds of innocent lives. In January 2010, the BBC's *Newsnight* program broadcast an exposé of the device, and in April 2013 McCormick was convicted of fraud and sentenced to ten years in prison.[3] In 2018 his sentence was extended by two years following his refusal to meet a shortfall of $2.5 million in repayments to recompense organizations defrauded by him.

A second example of tragic consequences resulting from a lack of knowledge of the ideomotor effect is the discredited pseudoscience of *facilitated communication* (FC). FC, which is sometimes referred as *progressive kinesthetic feedback* and *supported typing*, is a technique that attempts to allow people with severe communication difficulties to express their thoughts and feelings with the assistance of a facilitator. The facilitator steadies the disabled individual's arm or hand enough for them to operate a keyboard or other device. Proponents of this technique believe that the communication problems of such individuals, perhaps resulting from extremely severe autism or cerebral palsy, are primarily due to motor difficulties, not impaired intellect. Thus, it is claimed, by reducing the impact of such motor impairments, the person can express their inner voice. Needless to say, such a positive message was music to the ears of the impaired individual's loved ones. The parents of many severely impaired children were now convinced that their beloved offspring were no longer prisoners of silence. Sadly, it turned out to be too good to be true.

This technique first became popular in the mid-1970s thanks to the efforts of its inventor, Australian teacher Rosemary Crossley. Sociologist Douglas Biklen then introduced the technique into the United States, and from there it spread around the world. Although media coverage was initially overwhelmingly positive and uncritical, some were skeptical from the outset. The critics pointed to the fact that none of the disabled individuals now expressing themselves so eloquently had ever had any formal training in reading and writing. Furthermore, messages were often being produced when the individual was not even looking at the keyboard. The reader has probably guessed by now that the true source of the messages was not, in fact, the disabled individual at all. It was the facilitator, albeit without any conscious awareness of this fact on their part. They were, in effect, using the disabled individual as a kind of human Ouija board.

This unwelcome suspicion was confirmed by the results of dozens of properly controlled double-blind tests.[4] These tests established that the correct answer to questions was only ever produced if the facilitator knew the answer. In situations where the disabled person was shown a picture of a test object, such as a ball, but the facilitator was led to believe that a different picture had been presented, say of a teddy, the answer produced was always the one that the facilitator thought was the correct answer, not the one that was actually correct.

Some may argue that even if facilitators, teachers, and loved ones were mistaken in believing that FC was a valid method of communication with the severely disabled, it was still worth doing as it brought joy into the lives of carers and did no actual harm. Sadly, even this is not true. There have been literally dozens of allegations of sexual and physical abuse based on messages generated using facilitated communication. In other cases, disabled people were sexually abused by facilitators who believed that consent for such activity had been given via FC.

THE LIMITATIONS OF SCIENCE

As the examples above demonstrate, the scientific method often provides a powerful technique to help us to differentiate between true and false claims. However, one of the most profound lessons I have learned from my decades of working in psychology and parapsychology is the need to be aware of the weaknesses as well as the strengths of the scientific method. I became particularly aware of the former when reflecting on my experience of attempting to publish a set of failed replications of a controversial parapsychological study some years ago.

This series of events began in 2011 with the publication of a set of nine experiments in the prestigious *Journal of Personality and Social Psychology* (*JPSP*) by Professor Daryl Bem of Cornell University.[5] Bem is a well-respected social psychologist who has made many important contributions to the field. What is unusual about him is that, in contrast to most psychologists, he is a strong believer in psi. On first reading, Bem's series of studies, involving as they did over a thousand participants, appeared to present strong evidence that people could somehow sense future events before they happened; in other words, that precognition is real. What is more, Bem had published his results in a mainstream psychology journal, knowing full well that had they been published in a parapsychology journal, they would have been ignored by the world's science reporters. As it was, they were reported and discussed around the world, much to the delight of most parapsychologists.

There was a theme running through the nine experiments, all but one of which reported statistically significant results supporting the existence of psi. Bem had adapted several standard psychological techniques by "time-reversing" them. In many experimental psychology studies, a manipulation at time T1 will have a predictable effect on performance at time T2. According to Bem, even if performance is measured *before* the manipulation takes place, it will still be affected by that manipulation. Rather than describe all nine of Bem's experiments, I will illustrate the idea of time reversal by describing the technique employed in Experiment 9, the experiment with the largest effect size.

Experiment 9 investigated what Bem called the *retroactive facilitation of recall*. If you were to look at a list of words only once and then have your memory for those words tested, you would not be surprised if your performance was poorer than if you had been allowed to look at the words several times. After all, there is nothing controversial about the idea that rehearsal improves memory performance. What is rather more controversial is the notion that rehearsal will improve memory performance even if the rehearsal does not take place until *after* memory has been tested. What Bem did in this experiment was to present a set of forty-eight words, one at a time, on a computer screen to each participant. They were then allowed five minutes to write down every word that they could remember. The computer then randomly selected half of the original words and presented them again for extra processing (i.e., rehearsal). Results suggested that the words that were randomly selected for additional processing were remembered better than the other words—even though this additional rehearsal did not take place until *after* memory had been tested. This appears to be a claim worthy of *Alice in Wonderland*, and yet this is precisely the claim that appeared to be supported by the results of Bem's Experiment 9.

As you might expect, the experimental techniques and statistical analyses employed in this series of studies were criticized by skeptics from the outset, but the feeling generally was that, although the criticisms were generally valid, each of the flaws identified was not in itself a major concern.[6] The ultimate test of whether the effects reported by Bem were real was if they would replicate. After all, the mantra is often heard that "replication is the cornerstone of science." To his great credit, Bem was indeed keen for other researchers to replicate his findings, even offering to make his software freely available to any researchers who wanted to carry out such replications. Was it possible that Bem had finally discovered the Holy Grail of parapsychology—a robust and replicable paranormal effect?

In collaboration with Stuart Ritchie and Richard Wiseman, I decided to take him up on this kind offer. Given that all three of us were skeptical about actually replicating Bem's psi effects, the reader may be wondering why we chose to do this. The honest answer is, we had an ulterior motive. We thought this would be a relatively quick and easy way to get a paper published in a top psychology journal, the *Journal of Personality and Social Psychology*. After all, the world's science media had widely covered the original controversial findings. If, as we expected, the effects failed to replicate, this would surely be worthy of publishing in the self-same journal, especially given Bem's explicit encouragement for other researchers to attempt replications.

We decided to each carry out an independent replication of Bem's Experiment 9, the study of the retroactive facilitation of recall that had produced the largest effect size of all in his experimental series. We consciously decided to replicate Bem's methodology as closely as we could, a task made much easier by the fact that we were using the same software as him. There was a good reason for this. There is a tendency for parapsychologists to dismiss the results of failed replications if there was any deviation from the exact methodology used in the original study that produced results in favor of the psi hypothesis. Of course, such deviations are not seen as being a problem if the previous results are apparently successfully replicated!

Much to our complete lack of surprise, none of our three studies replicated the results of Bem's Experiment 9. We wrote up our findings and submitted our paper to the *JPSP*. In response, we received a polite reply from the editor rejecting our paper without even sending it out for peer review. His reason? The *JPSP* simply did not publish replications. In light of the huge amount of publicity that Bem's paper had received, we argued that it was important that failures to replicate these effects should be published, but the editor still refused to send our paper out for peer review. We submitted our paper to two more high-impact journals, *Science Brevia* and *Psychological Science*, and received exactly the same response. Note that we were not arguing that any of these journals should automatically publish our paper, simply that they should send it out for peer review in the standard way.

Our initial failure to get a journal to send our paper out for peer review caused a considerable amount of comment in the media, as it revealed in a very stark manner the problem of *publication bias*.[7] If journals simply refuse to publish replications—especially failed replications—the research that they do publish will in no way be representative of the field as a whole. Instead, it will strongly overrepresent significant, positive findings of novel, often counterintuitive effects. Such effects may well attract considerable media attention, but, given their novelty, it is unclear at that early stage whether they would replicate at all.

We were pleased that when we submitted our paper to the *British Journal of Psychology*, it was finally sent out for peer review. One of the referees was very positive about our work and felt that the paper should be published pretty much as it was. The other referee was far less positive and most certainly did not feel the paper should be published as it was. On the basis of the comments made by the second referee, we suspected that we knew who it was. We thought it was probably a certain Professor Daryl Bem of Cornell University—the very man whose results we had failed to replicate. When we asked Bem if he was indeed the second referee, he confirmed that he was. We pointed out to the editor that this appeared to be a rather glaring conflict of interest and requested that a third referee be sought to decide the issue. Our request was turned down.

Almost on the verge of giving up hope of ever getting our paper published, we decided to submit it to the open access journal *PLoS One*. Much to our relief, the paper was sent out for peer review and, following minor revisions, was finally published[8]. This was the first paper I ever published in an open access journal, and I was very glad I did. Because it was open access, anyone could download the paper free of charge—and it turned out that lots of people were interested in our results. At one point, the paper was receiving over a thousand views a day. To date, it has received over 50,000 views.

Our timing was fortuitous. Bem's paper had appeared at a time when many researchers were starting to express concern regarding replicability issues within psychology. It had long been recognized that parapsychology's biggest challenge was to find techniques that could be used to reliably demonstrate psi effects under controlled conditions, but the magnitude of this problem within mainstream psychology was only just beginning to be appreciated. Concerns were being expressed that even some of the effects that had been featured in standard psychology textbooks for decades may not actually be real. This *replication crisis* became the subject of a great deal of discussion and debate within the field.[9]

Even prior to the replication crisis, researchers accepted that some of the effects reported in the literature, and even in their own research output, must in fact be spurious. This follows from the reliance on the use of *p*-values in evaluating the results of experiments. Without going into details, these days the numerical results of an experiment are typically fed into a computer running appropriate software to produce a test statistic and its associated *p*-value. The *p*-value is the probability that you would get that particular value of the test statistic or one even more extreme if, in fact, the effect you are hypothesizing is not actually present. If the *p*-value is low, the researcher will reject the *null hypothesis* and conclude that the results are probably due to a real effect. The arbitrary *p*-value conventionally taken to indicate statistically significant results within psychology is $p < .05$. Taken at face value, this implies that spuriously significant results will arise purely on the basis of chance once in every twenty statistical tests.

In fact, however, the situation is much worse than this, and false positive results are actually present in the research literature at a much higher rate than the 5 percent *p*-value implies. This is because of what are known as *questionable research practices* (QRPs).[10] We are not referring here to blatant fraud involving the manipulation or even complete fabrication of data. Such fraud does indeed occur in all areas of science, but it is probably quite rare. Any researcher found guilty of fraud will face very serious consequences, including losing their job and their reputation. QRPs, in contrast, do not at first appear to be such heinous crimes. The term is used to describe the numerous opportunities that arise in collecting, selecting, and analyzing data where researchers have some degree of flexibility in the choices they make. This flexibility may not be at all apparent when reading the final published report.

Joseph Simmons, Leif Nelson, and Uri Simonsohn discuss a number of ways in which researchers may exploit such flexibility in order to obtain results that appear to be statistically significant at the magical $p < .05$ level that will make it more likely that the study will not only be deemed to be worth writing up for submission to a journal but also more likely to be accepted for publication if it is submitted.[11]

One example of a QRP is known as *optional stopping*. This refers to the seemingly innocent practice of analyzing one's data as one goes along. It is not surprising that researchers would engage in this practice without feeling that they were doing anything wrong. After all, if one is two-thirds of the way through collecting data in an experiment with two conditions, it is only natural that one would be curious to take a quick peek at how the pattern of results is shaping up, isn't it? The problem is that if one discovers that one already has a statistically significant result in the desired direction, it is then very tempting to simply stop collecting any more data. After all, why waste time and resources when one already has the result one was hoping for? But what if the results appear to be going in the desired direction but are not quite statistically significant? Surely there is no harm in then collecting more data as originally intended? Surprisingly, as Simmons and colleagues demonstrate using computer simulations, such seemingly trivial departures from best practice can boost the false positive rate by about 50 percent.

Simmons and colleagues describe a number of other QRPs in their paper, each of which, taken in isolation, appears to be only a slight departure from best practice. As they point out,

> In the course of collecting and analyzing data, researchers have many decisions to make: Should more data be collected? Should some observations be excluded? Which conditions should be combined and which ones compared? Which control variables should be considered? Should specific measures be combined or transformed or both?

Given these various degrees of freedom, researchers would often analyze their data in many different ways in the hope that they could produce a statistically significant effect of the type desired. This practice was sometimes jokingly referred to as "torturing the data until it confesses." In fairness, it is only recently that researchers have become fully aware of the dangers of such practices in terms of increasing the risk of producing false positive findings. Using computer simulations, Simmons and colleagues demonstrated that by combining a number of QRPs, one could quite quickly reach a point where it was more likely than not that such a spuriously significant result could be found.

Demonstrating the dangers of QRPs using computer simulations was not enough for Simmons and colleagues. They really hammered their message home by reporting the results of two actual studies involving real data. In both cases, the methodology, results, and analyses were truthfully reported. How could it be, then, that the first of their studies reported results that appeared to support a very unlikely hypothesis? This hypothesis was that listening to a children's song would make people feel older. The results of their second study went one step further, supporting an *impossible* hypothesis: that listening to a particular song could actually reduce the age of participants!

Although everything Simmons and colleagues reported in this section of their paper was truthful, their account was far from being the whole truth. For example, in their initial account they failed to mention a long list of additional variables that they collected data on, thus allowing them to perform analyses on many different subsets of variables until they eventually managed to produce the spuriously significant effects reported. The only variables referred to in their original account were, of course, the ones involved in their contrived analyses.

Fortunately, many researchers have taken the lessons of this replicability crisis to heart, taking steps to assess the magnitude of the problem and implementing methods to address it. For example, with respect to the former, Brian Nosek of the University of Virginia organized an ambitious attempt to assess the level of replicability of 100 studies published in three highly regarded journals in 2008. The results of this project, which involved some 270 collaborators, were published in 2015.[12] Only 36.1 percent of the replication studies, carried out using exactly the same methodology as the original studies, replicated the published findings. Even when they did, the effects reported were often smaller than in the original studies.

In terms of actually addressing the problem, many researchers now argue in favor of preregistration of studies. Preregistration requires researchers to state in advance of data collection and analysis the exact methodology and analyses to be employed, thus drastically reducing the problem of undisclosed flexibility. Suggestions have also been made to encourage the publication of replication attempts, and many journals, including the *Journal of Personality and Social Psychology*, have explicitly changed their policies on this issue.[13] In fact, the *JPSP* published the results of a series of seven online investigations of retroactive facilitation of recall by Jeff Galak and colleagues, involving a total of over 3,000 participants. Once again, the effect was not replicated.[14]

Of course, we may never know quite how Bem was able to produce that string of statistically significant results in the first place, but there is fairly strong circumstantial evidence that they are probably the result of QRPs. Bem's previous advice on writing empirical journal articles certainly appears to support this possibility. Eric-Jan Wagenmakers and colleagues provide a couple of telling quotations that point in this direction, arguing that Bem appears to blur the important distinction between exploratory and confirmatory studies.[15]

In Bem's own words:

> The conventional view of the research process is that we first derive a set of hypotheses from a theory, design and conduct a study to test these hypotheses, analyze the data to see if they were confirmed or disconfirmed, and then chronicle this sequence of events in the journal article. . . . But this is not how our enterprise actually proceeds. Psychology is more exciting than that.[16]

Bem goes on to offer the following advice to senior researchers, who are often not directly involved in collecting data. Instead, the data are collected by more junior colleagues, and the senior researcher then goes on to write up and submit the report:

> To compensate for this remoteness from our participants, let us at least become intimately familiar with the record of their behavior: the data. Examine them from every angle. Analyze the sexes separately. Make up new composite indexes. If a datum suggests a new hypothesis, try to find further evidence for it elsewhere in the data. If you see dim traces of interesting patterns, try to reorganize the data to bring them into bolder relief. If there are participants you don't like, or trials, observers, or interviewers who gave you anomalous results, place them aside temporarily and see if any coherent patterns emerge. Go on a fishing expedition for something—anything—interesting.

There is nothing inherently wrong with exploring data in this way. However, it is vitally important that such fishing expeditions are clearly presented as such. It is all too easy to write up a report in such a way that significant effects that were actually discovered in this way are presented as being based on hypotheses that were stated in advance of data collection. Wagenmakers and colleagues present a strong case that Bem is sometimes guilty of blurring this distinction between exploratory and confirmatory analyses.

There is also other suggestive evidence of methodological sloppiness on Bem's part. For example, if the effect sizes obtained in Bem's series are plotted against the number of participants in each study, a statistically significant negative correlation is found. In other words, the smaller the sample, the larger the effect size. It is of note that Bem explicitly states that a minimum of 100 participants would be used in each study and yet his Experiment 9, the one with the biggest effect size, involved only fifty participants. This is consistent with the notion that Bem may have been guilty of optional stopping, as discussed earlier.

We can be absolutely certain that Bem collected data on far more variables than were mentioned in his final report. We know this because we had access to the software he used to run his experiments. In addition to the variables that were reported, Bem also collected data on a range of other variables, such as how much the experimenter liked the participant, how anxious the participant was, how enthusiastic the participant appeared to be, and whether they engaged in biofeedback, meditation, and so on.

It was also apparent that Bem originally divided the words used in his Experiment 9 into common and uncommon words, presumably in the hope that different patterns of results might emerge as a consequence of familiarity, but no mention is made of this variable in the final report. As described by Simmons and colleagues, the collection of data on a range of variables allows the researcher to attempt multiple analyses until a "significant" effect is found.

It is now generally accepted within psychology that many of the effects that have been described in textbooks for decades may in fact be nothing more than unreplicable false positives. Furthermore, replication problems have also been highlighted in many other fields of science.[17] Given this, is it fair to single out parapsychology for particular criticism? I would argue that it is. For one thing, the replicability issue appears to be worse in parapsychology than in any other field of science. In psychology, for example, there are literally hundreds of effects that are extremely robust and reliable, including, to name but a few, the effects of rehearsal on memory, the Stroop effect, and dozens of visual and auditory illusions.[18] In contrast, there is not a single psi effect that is reliable enough to have a reasonable chance of being demonstrated in an undergraduate lab class.

Arguably of even greater importance is the fact that the implications of accepting the results of parapsychological studies that appear to support the existence of psi are much more far-reaching than accepting findings in other areas of science, regardless of whether the effects reported elsewhere were real or spurious.[19] With the benefit of hindsight, I can be fairly certain that some of the effects reported in my publications in mainstream psychology journals over the years were probably spurious, the result of me innocently engaging in the types of QRPs described. At the time that we carried out those studies, analyzed the results, and wrote up our findings, the cumulative impact of numerous minor departures from best practice was not as well appreciated as it is today. Most experimental psychologists of my generation simply did not know any better. However, accepting as real any false positive effects that I may have reported does not require that anyone rejects our currently accepted scientific understanding of the universe.

Accepting the results of apparently positive significant findings supporting the existence of psi would, in the opinion of most mainstream scientists, require precisely such a rejection of our current scientific worldview. It is not absolutely impossible that advocates of the psi hypothesis are correct in their view that our current scientific models should indeed be rejected, or at least drastically revised, in light of their findings, but such a drastic step should not be taken lightly. In Carl Sagan's words, "Extraordinary claims require extraordinary evidence."[20] The quality of evidence produced by parapsychologists to date simply does not reach this high standard.

All sciences are aimed at detecting signals (that is, true effects) against a background of noise. As already stated, two types of error are possible. Type I errors are those where a scientist wrongly concludes that a true effect has been found when, in fact, there isn't one, perhaps as a consequence of QRPs. Type II errors refer to the situation where a scientist concludes that there is no true effect to be found when, in fact, there is, perhaps as a consequence of a poorly designed or underpowered study. Sophisticated techniques have been developed and refined to help scientists to try to avoid both types of error. However, it is in the nature of science that they cannot be eliminated completely. The question arises: What would a science look like if the data being collected contained only noise and no true signals at all? Is it possible it would look something like parapsychology?

The astute reader may well have realized that the preceding discussion raises another question: How many of the effects reported in this book, including those reported by members of the APRU, may themselves be unreplicable? The honest answer is, I simply do not know. I can be fairly confident that many, perhaps most, of them are replicable, either because we have replicated them ourselves or else they have been replicated by other independent researchers. Some of them, such as the induction of false memories using the DRM technique, our poor ability to estimate probabilities, demonstrations of inattentional blindness and population stereotypes, and memory for Roman numerals on clocks and watches (again, to name but a few) are so robust that I routinely use them in public talks, knowing full well that I can rely on them working with my audience. But, in all honesty, I could not guarantee that absolutely all of the results I have discussed would replicate.

If the lessons of the replication crisis are taken to heart by both psychologists and parapsychologists, not to mention the wider scientific community, there is every reason to be hopeful that the rate of false positives in future studies will be reduced. However, it will not be eliminated. For this reason, one of the most important lessons to learn is that one should never take the results of a single study as being absolute proof that a claimed effect is real. Science is never about certainty. It is always a matter of basing one's opinion on the best evidence available at the time but being willing to revise one's opinion if good-quality new evidence contradicting the claim is subsequently produced.

Science is not an established body of facts; it is a method for attempting to approach the truth. On some issues, such as the idea that human activity is causing drastic damage to our climate, we can have a very high degree of confidence that the available evidence overwhelmingly supports the claim. When it comes to claims based on, say, astrology, homeopathy, and other forms of alternative medicine, we can be equally confident in rejecting them given the complete lack of good-quality evidence in their support. The evidence with respect to many claims, however, will be somewhere between these two extremes.

It is not uncommon for scientific hypotheses that were once rejected by the wider scientific community to subsequently be accepted in light of new evidence and vice versa. One of the greatest strengths of the scientific approach is this very willingness to be open to revision in light of new empirical evidence. Whereas faith and uncritical acceptance of authority are at the very heart of religious and (some) political belief systems, they are the absolute opposite of what is required in science. In science, skepticism is valued, and all assertions, no matter who makes them, may be questioned.

There is a famous quotation about democracy that is often attributed to Sir Winston Churchill: "Democracy is the worst form of government, except for all the others." The pedant in me feels obliged to point out that Churchill was, as he acknowledged, quoting an unknown source when he uttered these words. However, in the same spirit, I think it would be fair to say that science is the worst way to investigate how the universe works, except for all the others. Because scientists are only human, the scientific method may not always be applied as well as it should be, and we should be aware of the problems that may arise as a result. But, for my money, the scientific approach is far superior to all others when it comes to the tricky task of assessing controversial claims about both the world around us and the workings of the human mind.

RATIONALITY ISN'T EVERYTHING

Our ability to assess evidence and logical arguments is crucial not only when assessing controversial scientific claims, making consumer decisions, and considering health issues but also when making political judgments, career choices, and even, at times, decisions about personal relationships.[21] Indeed, there is no important area of our lives where such skills should not be applied. But for all that, it is vital that we do not fall into the trap of thinking that rationality is the only important factor in such situations.

We are not robots, and life without emotion would simply not be worth living. It is the joy we feel when good things happen that drives our lives, along with the need to do what we can to minimize the pain that is also an inevitable part of the human condition. We can recognize and embrace these basic truths and set our priorities in life accordingly. Science and rationality cannot directly inform us what our values should be, but once we know what they are, evidence and logic can maximize our chances of living our lives in accordance with those values.

Returning to the main theme of this book, it is probably a mistake to see paranormal and religious beliefs as some kind of deviation from a rational norm. The truth is that the norm for humans is not one of rationality, as I hope this book has shown, and there are many beliefs that may provide psychological benefits even though they are fundamentally wrong.[22] It is simply not the case that a more accurate perception of reality is always an inherently good thing. A huge amount of evidence shows that psychological health tends to be associated with what psychologists call *unrealistic optimism* (also known as the *optimism bias*).[23] If you give a questionnaire to psychologically healthy people asking them what they think the chances are of really good things happening to them, such as winning a major prize, and of really bad things happening to them, like becoming terminally ill, they will tend to overestimate the chances of the good things happening and underestimate the chances of the bad things. Give the same questionnaires to people with clinical depression, and their estimates tend to be much more accurate. Life really is that bad. But the trick is, of course, to accept that reality on an intellectual level but knowingly agree to a spot of self-deception whereby you live your life *as if* it were not the case.

Accepting that we only have one life will inevitably influence the way we live that life. We can either refuse to face that basic fact and put our trust in various deities and the hope of some form of postmortem survival, along with all the other superficially reassuring beliefs that go along with such a worldview, or we can do whatever we can to make the one life we have worth living. That requires us to be able to make difficult decisions when required, using all of the evidence and critical thinking skills at our disposal. But it also requires that we are able to accept and embrace the fundamentally irrational and emotional sides of ourselves.

EPILOGUE: THE LIMITS OF SKEPTICISM

I taught a module on anomalistic psychology to students at Goldsmiths for many years. I would always inform students at the outset that I could, if I so chose, give them a twenty-hour lecture course that would convince them that psi really existed. I was familiar enough with the strongest evidence in support of paranormal claims to be confident that I could do that. Presenting that evidence alone, with no critical analysis, would probably have been enough to convince any reasonable person that at least some phenomena are indeed genuinely paranormal. Instead of doing that, however, I would present my reasons for not being persuaded by such evidence and spend the majority of my time in those lectures exploring possible nonparanormal explanations for such phenomena.

I would also point out that even book-length reviews of parapsychology written by well-informed commentators could sometimes come to diametrically opposed conclusions. For example, here are the words of Dean Radin, a vocal proponent of paranormal claims:

> the effects observed in a thousand psi experiments are not due to chance, selective reporting, variations in experimental quality, or design flaws. They've been independently replicated by competent, conventionally trained scientists at well-known academic, industrial, and government-supported laboratories worldwide for more than a century.[1]

In contrast, psychologist David Marks, an equally vocal critic of parapsychology concluded, after reviewing seven areas of parapsychological research,

"My own beliefs are as they are—toward the extreme of disbelief—because the evidence as I see it warrants nothing more."[2]

My own position was and is somewhere in between these two extremes, although definitely more on the skeptical side. I pretty much agree with the conclusion drawn by parapsychologists Harvey J. Irwin and Caroline Watt at the end of their book-length review:

> as far as spontaneous cases are concerned it seems likely that there are numerous instances of self-deception, delusion, and even fraud. Some of the empirical literature likewise might be attributable to shoddy experimental procedures and to fraudulent manipulation of data. Be this as it may, there is sound phenomenological evidence of parapsychological experiences and experimental evidence of anomalous events too, and to this extent behavioral scientists ethically are obliged to encourage the investigation of these phenomena rather than dismissing them out of hand. If all of the phenomena do prove to be explicable within conventional principles of mainstream psychology surely that is something worth knowing . . . ; and if just one of the phenomena should be found to demand a revision or an expansion of contemporary psychological principles, how enriched behavioral science would be.[3]

THE BORDERLANDS OF SCIENCE

In chapter 1, I argued that the oft-stated maxim that you "cannot prove a negative" is simply wrong as a generalization but true with respect to some specific statements, including "psi does not exist." Logically, it is always possible that convincing proof of the existence of psi will be produced at some future date; for that reason, I have always supported the efforts of experimental parapsychologists engaged in efforts to produce high-quality evidence in support of the psi hypothesis.

Researchers make decisions regarding which topics to investigate based on a host of factors, including their own interests and expertise, the availability of resources to support their research, the interest of the wider scientific community and the general public in their chosen topics, and potential practical benefits produced by their efforts. I have long felt that investigating possible nonparanormal explanations for ostensibly paranormal events was more likely to be fruitful than directly attempting to test the psi hypothesis. I have allocated my research efforts accordingly. However, for me an important part of proper skepticism is to always be open to the possibility that you may be wrong, and therefore I have always devoted some time to directly testing paranormal claims, as described in previous chapters.

At the time of writing, I am engaged in an interesting investigation aimed at testing the hypothesis that lucid dreams may sometimes be precognitive. A lucid dream is one in which the dreamer becomes aware that they are dreaming during the dream itself. In some cases, the dreamer may even be able to exert a degree of control over what happens in the dream. Some people who have never had a lucid dream find it hard to believe that lucid dreaming is a genuine phenomenon, but there is no doubt that it is, as shown by numerous well-controlled studies from sleep laboratories.[4]

This particular investigation came about when I was contacted by an artist named Dave Green. Dave's unique selling point is that he creates his pictures while in the lucid state and then recreates them once he wakes up. Dave's website includes a short video explaining the process. In his words:

> The dream begins with me separating from my body. Now this is a lucid dream so I know what I need to do. I go over to my desk. I grab a dream pen and a dream piece of paper and I attempt to create a drawing. Now this is nothing like drawing in waking life. The image behaves really strangely. Usually I draw just one or two lines, but then the rest of the image gets filled in by the dream. It's like a live interaction between my conscious and my subconscious mind just playing itself out on the page. When I feel like it's done, I memorize the image, then I wake myself up and then, back in my physical body, I go back to my desk and recreate the drawing in waking life.[5]

As if that was not fascinating enough, Dave then let me know that he was collaborating with experimental psychologist Julia Mossbridge and asked if I would like to collaborate on their project. I readily agreed. Julia is convinced that dreams can be precognitive and has carried out experiments the results of which offer support for the reality of precognition. Dave himself is undecided but open to that possibility. Dave and Julia had already carried out a small pilot study—and obtained results that were just about statistically significant. We agreed that it would be a good idea to carry out a second small pilot study primarily with the aim of deciding on the best methodology for a planned larger study.

This is how we carried out that second small pilot study. We decided in advance that we would collect data from five consecutive occasions, each separated by at least a few nights, upon which Dave had a lucid dream. If he did so, he would, while in the lucid state, recall that he was taking part in the precognition study and attempt to create a picture that he hoped would correspond in some way to a target that would not actually be randomly selected until the following day. Upon awakening, he would email me the details, consisting of a brief written description of the dream along with copies of any drawings produced while in the lucid state and reproduced in the waking state.

I would store the dream record digitally on my password-locked computer and only then randomly select a target from a large target pool. This target pool was the "Target of the Week" database, a free online resource produced by Lyn Buchanan for use in parapsychological investigations.[6] It includes hundreds of potential targets, each consisting of interesting stories and news events collected from publicly available sources. Each entry might consist of text describing the target, along with photographs and short video clips. I would then send the link to the target to Dave, who would only then be able to access it. Finally, both the dream record and the target link would then be sent to Julia to record in a spreadsheet on a password-locked computer.

The next phase involved judging the degree of match between each dream record and all of the randomly chosen targets. This phase only began once all data had been logged from the first phase. There are different methods available for rating the degree of match, and it is not necessary to go into such details here. The important point is that such ratings can be analyzed statistically—and the degree of match between each dream record and the corresponding target, based on ratings from independent judges, was found once again to be just about statistically significant. In other words, the results of two small pilot studies do offer support for the idea that lucid dreams may sometimes be precognitive.

We are about to begin a larger investigation using essentially the same methodology but this time collecting data across ten lucid dreaming sessions. Having obtained marginally significant results in two small pilot studies, I am obviously intrigued as to what we will find. Will the results of our next investigation also be statistically significant? If so, how will I react? Although I would be surprised by such a result, would it be enough to shake my overall skepticism regarding the paranormal, given my rather long history of failed attempts to find evidence to support the psi hypothesis? I honestly cannot say. At the very least, I would want to carry out further investigations of the phenomenon. If it proved to be reasonably replicable, I would eventually have to conclude that psi is indeed real and that my current skepticism is misguided.

If, on the other hand, we do not obtain statistically significant results, that would probably be enough for me to conclude that the two marginally significant results in the small pilot studies were probably just statistical blips, perfectly explicable as due to nothing more than chance.

Regardless of the outcome of this or any other parapsychological investigation, it is obvious that most people do not base their level of belief in the paranormal on an assessment of the studies published in parapsychology journals. A much more important factor is that of personal experience. This raises an interesting question for me. I have certainly had experiences that a believer would be inclined to interpret as paranormal in nature but that I, an informed skeptic, have explained in nonparanormal terms. But suppose I had one or more such experiences that I felt could simply not be explained away in nonparanormal terms. Might that be enough to convince even me that psi is actually real?

WHAT IF PSI ONLY OCCURS SPONTANEOUSLY?

Some proponents of the psi hypothesis have suggested that it may only operate in spontaneous and unpredictable ways that can never be reliably captured under controlled laboratory conditions. I found myself reflecting at length on this possibility when I received an email out of the blue with the intriguing title "Not a ghost story." After I had learned more about this intriguing non-ghost-story, via both email and a Zoom call, I asked the sender if I could include an anonymized version of her story in my book, and she kindly agreed. I will tell much of her story by reproducing, with her permission, some extracts from our email exchanges.

That first email began as follows: "I'm sure most of your correspondents protest their sanity at some point, so I'll get that out of the way right up front! I'm a happy woman with loving family and friends. I have two dogs; I own a house; I'm Googleable in a good way." As I had a surname as part of her email address, I naturally immediately Googled the name and figured out the identity of my mystery correspondent, subsequently confirmed by the lady in question. She was indeed a reputable person with a solid reputation in her chosen field. I will refer to her as Belle, although that is not her real name.

In that first email, Belle assured me that she had always been a "healthy skeptic" with respect to "all things connected with the afterlife," but that had changed following the death of her partner, Simon. Simon had died some months earlier while out cycling with Belle. In her words, "Within hours, what are commonly called 'signs' began. These were not generic white feathers and random robins, but signs that were so specific to Simon that I could not ignore them." Belle believed that these signs transcended any explanation in terms of mere coincidence. Intrigued, I asked for more details. Here is a slightly edited version of the email she then sent to me:

> This is going to be long—even though I'll make an effort to be succinct—just because context is important. I'm not going to mention the half-dozen or so incidents of a particular song on the radio or similar things that I feel it's easy to dismiss as coincidence, even if those things happened at what felt like important moments and the coincidence was notable.
>
> My boyfriend Simon was adopted as a baby and was never interested in finding his biological parents. His mum and dad had lost their first son, Kevin, at 2 years old. I learned this early in our relationship, but it was not something he or his family really talked about. After they adopted Simon, his mum and dad went on to have another boy and a girl, and Simon simply considered himself the big brother.
>
> Simon and I were together for nearly 14 years. I had always been very cheerfully single and found him a bit over-enthusiastic at first, but he was so funny, kind and clever that he won me over. He had a teenage son, Ben, and they lived 20 minutes away, which was perfect.

In 2012 Simon got mouth cancer and underwent multiple radical surgeries, chemo, and extensive radiotherapy that left him unable to speak and eat properly. He went through years of treatment, setbacks, and aftermath with the most incredible bravery and good humour. As part of his recovery he rediscovered cycling and got me into it too, and we would ride miles and miles every weekend.

When we saw each other we'd watch TV or go to the movies or stand-up comedy shows. Our favourite TV show was Sky's *Portrait Artist of the Year*, which I wasn't allowed to watch without him. Encouraged into overreaching myself, I once even painted his profile while we watched the show [figure E.1]. It was so bad I didn't finish it and we both agreed I should hang on to the day job.

I love horse racing. Simon always teased me because even though I bet on horse racing every day, I only spend 50p here and £1 there.

We played poker every Sunday night with friends. Simon's poker name was Sweetie because I called him that once and it stuck. The other men would call him that sometimes. Always got a giggle.

Figure E.1
Belle's unfinished portrait of Simon. Image courtesy of Belle.

I bought a trombone with my poker winnings once, but I was so embarrassingly bad that I rarely played it. Simon encouraged me all the time. He bought me a stand and a course of online lessons for one birthday, and later a DVD (which I did not open) called *Play Trombone Today!* He was always asking if I'd practised. I never had.

We never argued. Just never. Never took each other for granted. Every year our relationship grew stronger, kinder, and funnier. There were only two things we disagreed on—heavy metal and the paranormal. Simon was all about the heavy metal and science. His bad taste was something I could do nothing about, but I—even though I was pretty skeptical myself—used to wind him up by insisting that *not* being open to things like ghosts, reincarnation, and telepathy is actually unscientific. It was an argument he couldn't refute. Luckily we were both rabid atheists, so we could always find our way back to common ground.

At the start of the first [COVID-19] lockdown I was diagnosed with cancer myself. I had surgery, radiotherapy, and chemo, but in January last year my cancer had spread anyway, and it looked like curtains. I made my will, bought a plot in a natural burial ground, and insisted on taking my family there as part of being able to talk freely about my death. When Simon visited the burial ground, he asked if he could be buried next to me.

At the eleventh hour I was put onto immunotherapy, which saved my life. Slowly I started to feel better, and around August last year I even got back on my bike—although it was very slow going at first.

Simon would come over on Saturday night, and we would go for a bike ride on Sunday morning.

The night before he died, we watched a movie called *Jeff Who Lives at Home*. It's a small indie film about a slacker who is obsessed with "signs." Throughout the film Jeff follows the sign of "Kevin." Following Kevin takes him through an adventurous day and ends with him saving a man's life. Spoiler alert: the guy's name is Kevin 😊 We both loved the film and chatted about it that night and the next morning as we got ready to go out.

Simon felt cold and a bit weak the whole ride, and it *was* cold and overcast, but it wasn't like him to voice any concerns about himself, so after about seven miles I suggested that we cut the ride short and turn around. As we set off for home, with him setting a nice slow pace, the sun suddenly burst through the clouds, and he said "Oh, sunshine!"

About 30 seconds later I heard a car behind us. Simon was riding about 2 metres from the kerb so I said "car." But he didn't move back in, so I repeated, "Simon. Car." But he veered into the middle of the lane and died, literally on his bike, then fell into the road.

I'll call it a heart attack, although it was something harder to spell.

OK, so this is where things started to be a bit weird . . .

The police drove me home and I gave them his sister's number so she could tell his parents, who live a couple of hours away.

A few hours later his sister called me and I asked how his parents were doing. She said her dad was okay but that her mother was in a terrible state. Then, without any prompting she said: "Remember, Simon was adopted to replace Kevin."

I felt it was an odd thing for her to say at that moment, and the coincidence of the movie we'd watched the night before felt big, but not paranormally so.

I then had to go to tell Simon's son what had happened. Trying to reassure him that Simon had not suffered and that all that could have been done for him had been (three doctors—one an oncologist—were giving him CPR within 30 seconds), I told Ben his last words and it was only when he burst into tears that I remembered that "Sunshine" was what Simon always called him.

Again—a little moment, and a comfort, but not something I'd be writing to you about on its own.

A few days later Ben and I went to see Simon's parents. They're old-school stoical people, and although they were upset, they were still all about making tea and offering biscuits. Out of nowhere his dad asked if it was December 5th that Simon had died, and when I confirmed that it was, he said, "That's the date we lost Kevin."

That weekend I took my mother shopping. While I waited for her in the car park, a man walked towards my car. I noticed him because he was holding something very awkwardly under his arm. As he passed me, I saw it was a bag-for-life, folded so that only one side of it was showing, and held awkwardly, so that his arm was not across it. It was red, and had the words "HELLO SWEETIE" on it.

After he passed my car, he shifted his position to a more natural one, and covered the bag with his arm.

It was at this point that I started to think these things may be signs from Simon.

Figure E.2
Similar profile in the road to Belle's unfinished portrait. Image courtesy of Belle.

A few nights later I had a dream that my sister Chloe and I were at a sort of street parade through a town centre where she lives. She's got a lot of hippie-type friends and she was introducing me to all sorts of jugglers and ballerinas and there were horses everywhere. In the middle of it all, I spotted Simon through the window of a coffee shop. I rushed off the street and hugged him but he was very sad and knew he was about to die. Then the dream switched to him dying in my arms. Regardless, I was happy I had hugged him again because it felt so real, and texted my sister about the dream.

She called and told me that the same time I was having this dream, she and her family were at the Mondol parade in Penzance, where she had just moved, and that there were dancers and jugglers, also many people wearing horse skulls. I'd never heard of the parade but it's a thing!

I was so sad after Simon died that for a couple of weeks I just sat and did nothing. Finally, I decided to go for a walk. There's a three-mile loop outside my house so I set off. I was pretty emotional all the way around and then about a mile from home I saw Simon's profile in the road [figure E.2]. I have walked this loop before and since and never seen anything like this in this spot, whatever the weather conditions, but I immediately thought of the painting I had done of Simon. I didn't bother finishing it around the glasses and this profile also looks unfinished in the same place.

Figure E.3
Photo of Simon with a sparkle in his eye. Image courtesy of Belle.

My mantel clock had stopped on the day that Simon died, and I hadn't had the energy to wind it up, so a few days later I went to do that but the hands wouldn't move. This is a clock I've had for years and it's always kept lovely time, but now it is stuck at 10.23. Simon's time of death (called on the road by the doctor who attended with the ambulance) is listed on his death certificate as 11.04. I estimated at the time that the doctors and I gave Simon CPR and mouth to mouth for about 10–15 minutes before the ambulance arrived. They worked on him for about 30 minutes before he was finally pronounced dead. 10.23 would have been almost exactly the time when he died.

I became seized by the need to paint Simon's portrait. I became so restless I couldn't do anything until I had, so I went and bought proper paints and canvases and an easel. I chose to copy a particular photo of Simon because of the sparkle in his eye [figure E.3]. Apart from that dreadful profile, I hadn't painted for maybe 15 years, and was never very good! Again, as I painted I became overwhelmed with emotion, particularly with self-doubt. I started to sob as I painted, but was suddenly aware of the feeling of something guiding me. Every time I hesitated to make a stroke, I could feel something telling me that it was right, and I would make the stroke and it *was* right. It was the most bizarre experience. I was painting and sobbing, sobbing and painting, and I just kept thinking, "As long as I get the sparkle in his eye, it will be OK."

Figure E.4
Belle's finished portrait of Simon. Image courtesy of Belle.

And it was! Here are the photo and painting [figure E.4], which I'm sure you'll agree is an awful lot better than the painter of that profile had any right to expect—and *does* capture the sparkle in his eye.

When I finished the painting, I realised I'd missed all that day's racing! Only the evening meeting at Wolverhampton was still available, so I opened up my betting app and looked at the first race.

There was a horse running called Sparkle in His Eye.

Came in at 8–1.

With my £1 weighing it down, of course!

A few days later I dreamed that Simon showed me Elvis as a young man in uniform, but with angel wings behind him. In the dream I said to Simon, "Oh, you won your wings!" I was happy for him but sad for me, because I felt that it meant Simon would be moving on now and not send me any more signs.

When I looked at the racing that day there was a horse called Win My Wings, which won at 20–1.

In March I found a childhood toy had fallen off my bookshelf.

He's a soft penguin I got as a toddler, and has sat on the bookshelf in the same place for years.

I picked him up and put him back on the shelf.

The next morning, he had fallen off again.

I put him back.

Third morning, same thing.

On the fourth day, Pengie had flung himself across the room. Please consider:

1. I live alone
2. The penguin is a flightless bird. 😊

So this really made me stop and wonder if it was a sign. I couldn't work out why Simon would be chucking my penguin across the room. As I went to put him back in place, I looked at whatever he had been hiding: it was a DVD . . . *Play Trombone Today!*

I did and I do, and Pengie has stayed put ever since.

So . . . Simon is buried in my grave, and I've bought the plot next to his. I haven't had a sign from him for a little while, but tbh I don't feel the need for any more—I'm so grateful for the ones I've had and they have changed my life. Signs from the arch-atheist have led me to belief—first in the afterlife and finally in God. To *not* believe in the spirit world after what I've experienced would be ridiculous; to not believe in God finally just felt churlish. Since I gave in on that point, I've had a couple of funny religious "signs" but this is about Simon, so I'll leave it at that.

I think it's almost impossible to convince others of the paranormal, but once it happens to you, there's simply no denying it.

Most of the individual "signs" that Belle described can no doubt be explained away as just coincidences, and the resemblance of her first attempt to paint a portrait of Simon to the profile she spotted (and photographed) on the road may be nothing more than a nice example of pareidolia. The flying penguin toy may be harder to explain, but even that may have been open to a nonparanormal, if obscure, explanation had it been properly investigated at the time. But what is undeniably striking about Belle's story is the occurrence of so many odd events in such a relatively short period of time following Simon's death.

What I want you, as a reader, to do is to imagine yourself experiencing a similar run of odd events following the death of a loved one. I am sure many readers of this book would describe themselves as skeptics, but would your skepticism be shaken if you had similar experiences to those described by Belle? Merely reading about Belle's experiences was certainly not enough to shake my own skepticism, but that is because they happened to her. Would I still be so skeptical if they had happened to me? I honestly cannot say.

Although I have no doubt that Belle really was a skeptic before these events occurred, there may be readers who doubt that. After all, we are all familiar with those paranormal anecdotes that begin, "Before this happened to me, I was the most skeptical person you could imagine . . ." and then go on to describe the amazing events that forced them to acknowledge that the paranormal is real.[7] I have yet to come across anyone who began such an anecdote with the words, "Me, I'm the most gullible person you could imagine!" Would a series of amazing coincidences ever be enough to convince a "proper skeptic" to seriously entertain the idea that the paranormal is real? The answer is yes.

FROM SKEPTIC TO ZETETIC

As stated at the beginning of this epilogue, I used to illustrate the opposite extremes of the paranormal belief continuum to my students with quotations from Dean Radin and David Marks. There is absolutely no denying that for much of his career, Marks was a complete skeptic regarding the paranormal. He not only wrote one of the most influential critiques of the paranormal ever, he was also one of the co-founders of New Zealand Skeptics.[8] However, in his recent book, *Psychology and the Paranormal*, he has softened his position considerably, preferring to describe himself as a *zetetic* in contrast to what he describes as his previous "entrenched skeptical stance."[9] He defines a zetetic as "a person who suspends judgment and explores scientific questions by using discussion or dialogue to enquire into a topic."[10]

In *Psychology and the Paranormal*, he reviews evidence from several areas of experimental parapsychology, including telepathy, precognition, and psychokinesis. His conclusion is as damning as it ever was: "In spite of the continuing claims of advocates, the facts show that reliable, replicable laboratory evidence of psi has not been demonstrated." Marks concisely summarizes his new position: "If psi exists, it happens in a spontaneous, unpredictable and uncontrollable manner—e.g., as a remarkable coincidence or other anomalous experience."[11]

Marks describes in detail one such coincidence that he experienced and the profound impact it had on him. He refers to it as "The Chiswick Coincidence."[12] One day in August 2018, Marks found himself with time to kill. He could not go home, as an estate agent from the company Chestertons had arranged a viewing of his apartment for a potential tenant. He decided to go for lunch in a pub next to the Thames called the City Barge. He then opened his Kindle and decided to continue reading a novel he had started a few days earlier. The novel was *The Man Who Was Thursday: A Nightmare* by G. K. Chesterton. The ancestral family of this author founded the estate agents of the same name. As he began to read, Marks realized that a pub described at that point in the novel was the very pub that he had moments ago decided to visit.

Marks describes in total what he refers to as "seven layers of synchronicity" associated with this coincidence. They include the fact that Martin Gardner, who had written the foreword to Marks's own classic, *The Psychology of the Psychic*, had also written the annotations for a special edition of Chesterton's *The Man Who Was Thursday*. Marks had been unaware of this prior to carrying out further research into his Chiswick Coincidence.

Marks estimates that the overall combined probability of his seven synchronicities is somewhere in the region of one in a quintillion (that is, a million million million). He concludes, "These odds are so astronomical that one must consider the possibility of a paranormal explanation. Not to do so would be irrational and contrary to open inquiry."[13] This from an erstwhile vocal skeptic who has written extensively about the psychology of coincidence in his previous publications.

CONCLUDING COMMENT

For me personally, robust, replicable evidence in support of the paranormal from well-controlled scientific studies would always be much more convincing than any number of anecdotal accounts reported by others, no matter how compelling those accounts were to the people reporting them.

Therefore, if psi really does exist but it is in the very nature of psi to elude such controlled experimental demonstrations, it may be that the truth about psi will always be beyond my reach. Or perhaps, like Belle and David Marks, I will one day experience one or more events that I deem to be so remarkable that I decide that psi is real even if it cannot be proved to exist scientifically.

The late Robert Morris pointed out that one of the founding fathers of modern psychology (as well as of the American Society for Psychical Research) had considered this issue of where one draws the line with respect to deciding one's beliefs. I will finish with his wise words:

> William James (1907/1968) makes the distinction between two kinds of people, the tender-minded and the tough-minded (I prefer to call these people hard-headed). Tough-minded people are those whose love for the truth is so great that they want to be sure that truth is all they have, and so they exclude all information, all procedures which do not meet very strict criteria by which they can know they have the truth. On the other hand, the tender-minded are individuals whose love for the truth is so great that they are willing to risk error in taking a chance to grasp the truth. One approach is conservative, the other is liberal. One approach is restrictive, the other approach is expansive. As James points out, whether or not a person is one type or the other is very much a product of his temperament.[14]

Author's Note and Acknowledgments

Originally, I had intended to title this book *Why Weird Stuff Matters*. I still think that this is a pretty good title. In particular, it highlights the fact that, for most people, weird phenomena have a certain fascination and yet are not seen as being particularly important. Media coverage of such topics is usually not treated with the same seriousness as, say, current affairs and politics. As I have argued, I think this is a mistake.

My argument essentially boils down to this: If any alleged paranormal phenomena turn out to be beyond explanation in terms of our current conventional scientific worldview, that would have profound implications. It would mean that the current scientific worldview is either fundamentally flawed or at least in need of major revision. If, on the other hand, claims relating to paranormal phenomena are best explained in terms of psychological factors as opposed to parapsychological factors, then we can learn a lot about the human mind by taking their investigation seriously. Either way, it is worth doing.

So why did I opt for the somewhat naughtier title of *The Science of Weird Shit*? When people asked me the title of my next book, I'd tell them "*Why Weird Stuff Matters*" and outline my reasons for doing so. I would sometimes add, purely as a joke, "or I might just call it *The Science of Weird Shit*." Almost without exception, people would respond, "Oh, I'd buy that!" With future sales in mind, that explains the title of the book you now hold in your hands.

My aim was to produce an accessible and, I hope, entertaining introduction to the field of anomalistic psychology. This book is an attempt to explain why I hold skeptical beliefs about paranormal and related phenomena, having once been a believer in those same phenomena. It is much more personal than most of my previous writings on these topics.

I should also say a few words about what this book is *not*. It is not an attempt to comprehensively review any of the topics covered. I am aware that this will annoy readers who are more sympathetic toward paranormal claims. They may feel that my presentation of material sometimes does not address what they feel to be the strongest evidence in favor of the paranormal. As stated, that was not my aim in these pages. Instead, I am simply telling the reader what I believe about each topic and why. I usually presented more academic and comprehensive discussions of most topics elsewhere, and I provide references to these publications for the interested reader.

Over the years, numerous people have helped to shape my thinking, to a greater or lesser extent, on the topics covered in this book. Any list of such people is bound to be incomplete (and I apologize in advance to anyone I should have included but didn't), but I must extend my thanks to at least the following: James Alcock, "Belle," Barry Beyerstein, Susan Blackmore, Jason Braithwaite, Rob Brotherton, David Clarke, Andrew Colman, Duncan Colvin, Susan Crawley, Neil Dagnall, Geoffrey Dean, Dan Denis, Edzard Ernst, Hilary Evans, Kendrick Frazier, Fiona Gabbert, David Robert Grimes, Wendy Grossman, Erlendur Haraldsson, Karen Hatton, Michael Heap, Terence Hines, Kate Holden, Al Hopwood, Jim Houran, Deborah Hyde, Ray Hyman, Harvey Irwin, Stanley Krippner, Gustav Kuhn, Paul Kurtz, Stephen Law, Scott Lilienfeld, Elizabeth Loftus, Carla MacKinnon, David Marks, Michael Marshall, Richard McNally, Bob Morris, Marie-Catherine Mousseau, Joe Nickell, James Ost, Henry Otgaar, Lawrence Patihis, Massimo Polidoro, James Randi, Stuart Ritchie, Paul Rogers, Barbara Rowlands, Julia Santomauro, Louis Savva, Brian Sharpless, Simon Singh, Anna Stone, Vic Tandy, Michael Thalbourne, Christopher Thresher-Andrews, Matt Tompkins, Stuart Vyse, Rosie Waterhouse, Caroline Watt, Mark Williams, Krissy Wilson, and Richard Wiseman.

Special thanks go to Kat French-Richards, Alice Gregory, Deborah Hyde, Anne Richards, and Lucy Richards, each of whom read one or more draft chapters and provided me with valuable feedback. Any remaining errors in the text are entirely their fault. Also, I would like to thank Mark Gottlieb of the Trident Media Group for his support and encouragement and Matthew Browne of MIT Press for the same plus seemingly inexhaustible patience.

All images supplied by Mary Evans Picture Library unless otherwise stated.

Finally, sincere thanks for emotional support throughout the writing process to my constant writing companions: Ted (the dog) and Tom (the cat).

Chris French
Greenwich, London

Notes

PREFACE

1. Erich von Däniken's books include *Chariots of the Gods* (London: Souvenir Press, 1969), *Return to the Stars* (London: Souvenir Press, 1970), and *The Gold of the Gods* (London: Souvenir Press, 1973).
2. Ronald Story, *The Space-Gods Revealed: A Close Look at the Theories of Erich von Däniken* (New York: Harper & Row, 1976).
3. Up to the time of writing, I have only actually met Uri Geller in the flesh on one occasion. Some years ago, I appeared with him on BBC Radio 2's *Jeremy Vine Show* discussing the paranormal. Unlike most discussions of such topics on daytime radio or TV, we had around half an hour to argue as opposed to just a few minutes, so we covered quite a lot of ground. During our conversation I referred to some relevant research by my good friend and fellow skeptic, Professor Richard Wiseman. After the show, as we walked down the stairs to leave the building, Uri turned to me and said, "I'll tell you what I can't stand about that Richard Wiseman." Intrigued, I replied, "What's that, Uri?" Pausing to find the right word, Uri answered, without any trace of irony, "He's such a publicity-seeker!"
4. John G. Taylor, *Science and the Supernatural: An Investigation of Paranormal Phenomena Including Psychic Healing, Clairvoyance, Telepathy, and Precognition by a Distinguished Physicist and Mathematician* (New York: Dutton, 1980).
5. For the most comprehensive critiques of Uri Geller's claims, see James Randi's *The Truth About Uri Geller* (Buffalo, NY: Prometheus Books, 1982) and David Marks's *The Psychology of the Psychic*, 2nd ed. (Amherst, NY: Prometheus Books, 2000).
6. James E. Alcock, *Parapsychology: Science or Magic?* (Oxford: Pergamon Press, 1981).
7. Sadly, James Randi passed away during the writing of this book. I described what he meant to me personally in this piece: Chris French, "What James Randi Meant to Me: Chris French Reflects on the Passing of the Amazing Randi," *The Skeptic*, October 22, 2020, https://www.skeptic.org.uk/2020/10/what-james-randi-meant-to-me-chris-french-reflects-on-the-passing-of-the-amazing-randi.

8. This reminds me of an account given in Alcock's book (*Parapsychology*, 59) of what happened at one of Randi's performances:

> The Amazing Randi was giving a performance in which he did all the well-known Geller feats explaining that what the audience saw was done by trickery but that Geller, when doing the same things, tells his audience that paranormal powers are at work. In the midst of Randi's extremely skilful presentation, a spectator (who as it turns out was a university professor) jumped angrily to his feet and loudly denounced Randi as a fraud. To this attack, Randi replied that he was indeed a fraud, that everything he had done, as he had several times stated, was done by trickery. But his accuser was not easily quieted; he proclaimed that Randi was a fraud because he really was using psychic powers, but was keeping this fact from the audience!

9. Sometimes when giving public talks on anomalistic psychology, I will include one of my favorite jokes about Mr. Geller: "Uri Geller has been in the news again. Apparently, he scratched his neck and his head fell off." I then point out that this is the clean version of the joke.

10. You can read an account of this experience in Sue's colorful autobiography, *In Search of the Light: The Adventures of a Parapsychologist* (Amherst, NY: Prometheus Books, 1996).

11. From blurb on the back of Richard Wiseman's *Psychology: Why It Matters* (Cambridge: Polity Press, 2022).

12. Kendrick Frazier, "It's CSI Now, Not CSICOP," *Skeptical Inquirer*, December 4, 2006, https://skepticalinquirer.org/exclusive/its_csi_now_not_csicop.

13. Psychometrics should not be confused with psychometry. *Psychometrics* is the term given to the measurement of psychological variables such as intelligence, personality, aptitudes, attitudes, and so on. *Psychometry* is a form of clairvoyance, specifically the alleged ability to be able to ascertain the past history of an object merely by handling it. Psychometrics is a valuable tool in psychological investigations. In contrast, psychometry is bullshit.

14. Christopher C. French, "Factors Underlying Belief in the Paranormal: Do Sheep and Goats Think Differently?" *The Psychologist* 5, no. 7 (July 1992): 295–299.

15. Christopher C. French, "Population Stereotypes and Belief in the Paranormal: Is There a Relationship?" *Australian Psychologist* 27, no. 1 (March 1992): 57–58.

16. More on what exactly anomalistic psychology is in the next chapter.

17. E.g., Christopher C. French, "Dying to Know the Truth: Visions of a Dying Brain, or Just False Memories?" *Lancet* 358, no. 9298 (December 15, 2001): 2010–2011; Michael A. Thalbourne and Christopher C. French, "Paranormal Belief, Manic-Depressiveness, and Magical Ideation: A Replication," *Personality and Individual Differences* 18, no. 2 (February 1995): 291–292; Michael A. Thalbourne and Christopher C. French, "The Sheep-Goat Variable and Belief in Non-Paranormal Anomalous Phenomena," *Journal of the Society for Psychical Research* 62, no. 1 (January 1997): 41–45.

18. On cognition and emotion, see, e.g., Anne Richards, Christopher C. French, and Fiona Randall, "Anxiety and the Use of Strategies in the Processing of an Emotional Sentence-Picture Verification Task," *Journal of Abnormal Psychology* 105, no. 1 (February 1996): 132–136. On functions of the cerebral hemispheres, see, e.g., Anne Richards, Christopher C. French, and Rose Dowd, "Hemisphere Asymmetry and the Processing of Emotional Words in Anxiety," *Neuropsychologia* 33, no. 7 (July 1995): 835–841.

19. Kevin Dutton, *Black and White Thinking: The Burden of a Binary Brain in a Complex World* (London: Transworld, 2020).

20. Oscar Wilde, *The Importance of Being Earnest*, act 1 (New York: Methuen, 1909).

21. For a review of relevant literature, see Christopher C. French and Anna Stone, *Anomalistic Psychology: Exploring Paranormal Belief and Experience* (Basingstoke: Palgrave Macmillan, 2014), 53–68.

22. Stuart Ritchie, *Science Fictions: Exposing Fraud, Negligence and Hype in Science* (London: The Bodley Head, 2020).

23. Research in my own unit, the Anomalistic Psychology Research Unit (APRU) at Goldsmiths, was based on a similar approach. The only difference was one of emphasis. At the KPU, parapsychological research was the primary focus, and anomalistic research was secondary. The opposite is true of the APRU.

24. Elsewhere, I have referred to the more simplistic, black-and-white version of skepticism as Type I and the more nuanced version as Type II: Christopher C. French, "Reflections on Pseudoscience and Parapsychology: From Here to There and (Slightly) Back Again," in *Pseudoscience: The Conspiracy Against Science*, ed. Allison B. Kaufman and James C. Kaufman (Cambridge, MA: MIT Press, 2018), 375–391.

25. One of my favorite things about that special issue was the front cover. It featured a photograph of me and my merry little band of APRU-nauts standing on the playing field at the back of Goldsmiths on a misty autumn morning. Steve Yesson, the technician who took the photograph, had the inspired idea of Photoshopping in a ghostly image of Elvis in the background.

26. In chronological order, they were Kate Holden, Julia Nunn, Victoria Hamilton, and Lindsay Kallis.

27. Also available in the pictorial style of *The Simpsons*.

28. *The Skeptic*, https://www.skeptic.org.uk.

CHAPTER 1

1. Leonard Zusne and Warren H. Jones, *Anomalistic Psychology: A Study of Extraordinary Phenomena of Behavior and Experience* (Hillsdale, NJ: Lawrence Erlbaum Associates, 1982). A second edition was published seven years later: Leonard Zusne and Warren H. Jones, *Anomalistic Psychology: A Study of Magical Thinking* (Hillsdale, NJ: Lawrence Erlbaum Associates, 1989).

2. Wikipedia, s.v. "anomalistic psychology," last modified September 23, 2022, 1:56, accessed May 27, 2023, https://en.wikipedia.org/wiki/Anomalistic_psychology.

3. French and Stone, *Anomalistic Psychology*.

4. *Cambridge Dictionary*, s.v. "paranormal," accessed May 27, 2023, https://dictionary.cambridge.org/dictionary/english/paranormal.

5. Strictly speaking, *clairvoyance* refers to picking up visual information, in contrast to, say, *clairaudience*, which would refer to obtaining auditory information. In practice, the term *clairvoyance* is used to cover the alleged ability to remotely receive all types sensory information.

6. When I first became a skeptic, I believed that parapsychology was not a real science but a pseudoscience. Over the years, I gradually changed my mind, and I now believe that parapsychology at its best is just as scientific as, say, psychology (at its best). Science is first and foremost a method for obtaining knowledge, not an established body of "facts." My view on this is very much a minority opinion among skeptics, however.

7. Note that many scientists who are skeptical regarding the claim that aliens have physically visited our planet are comfortable with the idea that intelligent life may well have evolved elsewhere in our galaxy and beyond.

8. See, e.g., Martin Kottmeyer, "Fairies," in *The Encyclopedia of the Paranormal*, ed. Gordon Stein (Amherst, NY: Prometheus Books, 1996), 265–271; Robert Sheaffer, "Do Fairies Exist?" *The Zetetic* 2, no. 1 (Fall/Winter 1977): 45–52.

9. It is ironic in the extreme that the creator of the world's most rational fictional detective should himself be one of the most gullible people who ever lived.

10. For more on the psychology of superstitions, see, e.g., Stuart Vyse, *Believing in Magic: The Psychology of Superstition*, rev. ed. (Oxford: Oxford University Press; 2014); Stuart Vyse, *Superstition: A Very Short Introduction* (Oxford: Oxford University Press, 2019).

11. In Germany, some believe that a black cat crossing your path from left to right will bring good luck, but a black cat crossing your path from right to left is a bad omen. Perhaps the European Union should standardize across the continent?

12. At the risk of totally undermining any credibility I may have in the eyes of my fellow skeptics, a personal anecdote illustrates the power of such magical thinking. The very first time I ever appeared as a sceptic on daytime TV many decades ago, I was naturally a bit nervous. I had heard that Jeremy Paxman, a well-known presenter on British television, always wore red socks. "Paxo" was renowned for his tough interviewing style. No one intimidated him. Now, although I was not going to be interviewed by the great man himself, I still needed a bit of a confidence boost. With my tongue firmly in my cheek, I decided, "If red socks are good enough for Paxo, they're good enough for me!" That first appearance went well. However, the next time I was due to appear on TV, I found myself hunting around for the same pair of red socks as I got ready. I had to make a conscious decision that red socks were not actually necessary for a good TV performance and choose another pair!

13. I am reminded here of a famous story (of dubious authenticity) about the Nobel Prize–winning physicist Niels Bohr. There are several versions of the story but all with the same punchline. It is said that a colleague once noticed a horseshoe hanging up in Bohr's office. "Surely you don't believe that old wives' tale about horseshoes bringing good luck, do you?" snorted the colleague dismissively. "I am told," replied Bohr, "that they do so whether you believe in them or not."

14. Practitioners of complementary medicine believe that their therapies should be used alongside conventional forms of treatment. Practitioners of alternative medicine argue for the more dangerous position of using their forms of treatment *instead of* conventional forms of treatment.

15. I say "most of these claims" rather than "all" because there are some isolated examples of well-documented cases of genuine conspiracies within the medical community. Arguably the most notorious example was the Tuskegee syphilis experiment, which ran from 1932 to 1972. This totally unethical study by the United States Public Health Service investigated the symptoms of untreated syphilis in a large sample of Black American males. Even when effective treatment for syphilis became available in 1947, it was not offered to the participants in the study. By the time the investigation was closed, following a leak to the press, many of the participants had died of syphilis, forty of their wives had been infected, and nineteen of their children had been born with congenital syphilis.

16. See, e.g., Robert Brotherton and Christopher C. French, "Conspiracy Theories," in *Parapsychology: The Science of Unusual Experience*, 2nd ed., ed. David Groome and Ron Roberts (London: Psychology Press, 2017), 158–176.

17. *Cambridge Dictionary*, s.v. "parapsychology," accessed May 27, 2023, https://dictionary.cambridge.org/dictionary/english/parapsychology.

18. Christopher C. French, "Why I Study Anomalistic Psychology," *The Psychologist* 14, no. 7 (July 2001): 356–357.

19. In case you've forgotten the definition given in the preface, *psi* is the term parapsychologists use to refer to anything paranormal.

20. Ciaran McGlone, "BMG Halloween Poll: A Third of Brits Believe in Ghosts, Spirits or Other Types of Paranormal Activity," BMG Research, October 30, 2017, https://www.bmgresearch.co.uk/bmg-halloween-poll-third-brits-believe-ghosts-spirits-types-paranormal-activity.

21. YouGov, "Halloween Paranormal," 2019, https://bit.ly/2OjGVJh.

22. Franz Höllinger and Timothy B. Smith, "Religion and Esotericism Among Students: A Cross-Cultural Comparative Study," *Journal of Contemporary Religion* 17, no. 2 (2002): 229–249.

23. E.g., Susan J. Blackmore, "A Postal Survey of OBEs and Other Experiences," *Journal of the Society for Psychical Research* 52 (1984): 225–244; David Clarke, "Experience and Other Reasons Given for Belief and Disbelief in Paranormal and Religious Phenomena," *Journal of the Society for Psychical Research* 60 (1995): 371–384; John Palmer, "A Community Mail Survey

of Psychic Experiences," *Journal of the American Society for Psychical Research* 73 (1979): 221–251.

24. Geoffrey Dean, Arthur Mather, David Nias, and Rudolf Smit, *Understanding Astrology: A Critical Review of a Thousand Empirical Studies 1900–2020* (Amsterdam: AinO Publications, 2022).

25. What's the Harm?, http://whatstheharm.net.

26. Farley's number of deaths include an estimated 365,000 premature deaths caused by the South African President's AIDS denialist policies between 2000 and 2005; the number of injuries includes over 200 chiropractic patients suffering serious harm, including strokes and paraplegia, between 2001 and 2007; and the economic damages include people losing millions of dollars to Nigerian email scams.

27. "Psychic Services Industry in the US," IBISWorld, April 9, 2021, https://www.ibisworld.com/united-states/market-research-reports/psychic-services-industry/.

28. Richard L. Nahin, Patricia M. Barnes, and Barbara J. Stussman, "Expenditures on Complementary Health Approaches: United States, 2012," *National Health Statistics Reports* (Hyattsville, MD: National Center for Health Statistics, 2016).

29. "Wellness Industry Statistics & Facts," Global Wellness Institute, accessed May 27, 2023, https://globalwellnessinstitute.org/press-room/statistics-and-facts.

30. YouGov, "Halloween Paranormal," 2019, https://bit.ly/2OjGVJh.

31. For a review of relevant literature, see French and Stone, *Anomalistic Psychology*.

32. For the benefit of readers with little familiarity with statistics, *standard deviation* is a measure of the variability of scores around a mean score, thus giving an index of the degree to which scores cluster around the mean. The higher the standard deviation, the less tightly the scores are clustered.

33. See, e.g., Steve E. Hartman, "Another View of the Paranormal Belief Scale," *Journal of Parapsychology* 63, no. 2 (June 1999): 131–141; Tony R. Lawrence, "How Many Factors of Paranormal Belief Are There? A Critique of the Paranormal Belief Scale," *Journal of Parapsychology* 59, no. 1 (March 1995): 3–25.

34. I have chosen to write about developmental psychology at somewhat greater length in this chapter compared to the other subdisciplines of psychology simply because, despite dealing with some fascinating issues, it gets little coverage in later chapters.

35. See, e.g., Jacqueline D. Woolley, "Thinking About Fantasy: Are Children Fundamentally Different Thinkers and Believers from Adults?" *Child Development* 68, no. 6 (December 1997): 991–1011; Jacqueline D. Woolley and Maliki E. Ghossainy, "Revisiting the Fantasy-Reality Distinction: Children as Naïve Skeptics," *Child Development* 84, no. 5 (September/October 2013): 1496–1510.

36. Why do imaginary playmates often have such weird names? My niece, Chrysanthi, had one called "Mitty Petty" and my nephew, Kahan, had one called "Emergency."

37. Christine H. Legare, E. Margaret Evans, Karl S. Rosengren, and Paul L. Harris, "The Coexistence of Natural and Supernatural Explanations Across Cultures and Development," *Child Development* 83, no. 3 (May/June 2012): 779–793.

CHAPTER 2

1. "*Buffy the Vampire Slayer*: The Gentlemen," YouTube video, 3:03, https://www.youtube.com/watch?v=KKfNuMWO128.
2. Chris French, "The Waking Nightmare of Sleep Paralysis," *Guardian*, October 2, 2009, https://www.theguardian.com/science/2009/oct/02/sleep-paralysis.
3. David J. Hufford, *The Terror That Comes in the Night: An Experience-Centered Study of Supernatural Assault Traditions* (Philadelphia: University of Pennsylvania Press, 1982).
4. Shelley R. Adler, *Sleep Paralysis: Night-Mares, Nocebos, and the Mind-Body Connection* (New Brunswick, NJ: Rutgers University Press, 2011); Brian A. Sharpless and Karl Doghramji, *Sleep Paralysis: Historical, Psychological, and Medical Perspectives* (Oxford: Oxford University Press, 2015).
5. J. Allan Cheyne, Steve D. Rueffer, and Ian R. Newby-Clark, "Hypnagogic and Hypnopompic Hallucinations during Sleep Paralysis: Neurological and Cultural Construction of the Night-Mare," *Consciousness and Cognition* 8, no. 3 (September 1999): 319–337.
6. Carla's sleep paralysis experiences are also discussed by "clinical neuropsychologist-turned-writer" Paul Broks in his wonderful book, *The Darker the Night, the Brighter the Stars* (New York: Crown, 2018).
7. *Devil in the Room* is available on Carla's website: https://carla-mackinnon-d8hy.squarespace.com/sleepparalysis/. (Not to be confused with *The Devil in the Room*, a US horror movie released in 2020.)
8. Alice Gregory, *Nodding Off: The Science of Sleep* (London: Bloomsbury, 2020).
9. Gregory, *Nodding Off*, 91.
10. Brian A. Sharpless, "Exploding Head Syndrome," *Sleep Medicine Reviews* 18, no. 6 (December 2014): 489–493; Brian A. Sharpless, Dan Denis, Rotem Perach, Christopher C. French, and Alice M. Gregory, "Exploding Head Syndrome: Clinical Features, Theories About Etiology, and Prevention Strategies in a Large International Sample," *Sleep Medicine* 75 (November 2020): 251–255.
11. Brian A. Sharpless, Kevin S. McCarthy, Dianne L. Chambless, Barbara L. Milrod, Shabad-Ratan Khalsa, and Jacques P. Barber, "Isolated Sleep Paralysis and Fearful Isolated Sleep Paralysis in Out-Patients with Panic Attacks," *Journal of Clinical Psychology* 66, no. 12 (December 2010): 1292–1306; Sharpless and Doghramji, *Sleep Paralysis*.
12. American Academy of Sleep Medicine, *International Classification of Sleep Disorders: Diagnostic and Coding Manual*, 3rd ed. (Darien, IL: American Academy of Sleep Medicine, 2014).

13. American Psychiatric Association, *Diagnostic and Statistical Manual of Mental Disorders: DSM-V*, 5th ed. (Arlington, VA: American Psychiatric Association, 2013).
14. Christopher C. French and Julia Santomauro, "Something Wicked This Way Comes: Causes and Interpretations of Sleep Paralysis," in *Tall Tales About the Mind and Brain: Separating Fact from Fiction*, ed. Sergio Della Sala (Oxford: Oxford University Press, 2007), 380–398.
15. G. B. Goode, "Sleep Paralysis," *Archives of Neurology* 6 (March 1962): 228–234.
16. Robert C. Ness, "The Old Hag Phenomenon as Sleep Paralysis: A Biocultural Interpretation," *Culture, Medicine and Psychiatry* 2, no. 1 (March 1978):15–39.
17. Brian A. Sharpless and Jacques P. Barber, "Lifetime Prevalence Rates of Sleep Paralysis: A Systematic Review," *Sleep Medicine Review* 15, no. 5 (October 2011): 311–315.
18. Kazuhiko Fukuda, "One Explanatory Basis for the Discrepancy of the Reported Prevalences of Sleep Paralysis among Healthy Respondents," *Perceptual and Motor Skills* 77, no. 3, pt. 1 (December 1993): 803–807.
19. Up until fairly recently, textbooks referred to four sleep stages, but modern sleep researchers have combined the very similar stages 3 and 4 into a single stage.
20. Eye muscles, unlike most muscle systems in the body, are still operative during REM sleep.
21. Yasuo Hishikawa and Ziro Kaneko, "Electroencephalographic Study on Narcolepsy," *Electroencephalography and Clinical Neurophysiology* 18, no. 3 (February 1965): 249–259.
22. Michele Terzaghi, Pietro Luca Ratti, Francesco Manni, and Raffaele Manni, "Sleep Paralysis in Narcolepsy: More Than Just a Motor Dissociative Phenomenon?," *Neurological Science* 33, no. 1 (February 2012): 169–172.
23. Cheyne, Rueffer, and Newby-Clark, "Hypnagogic and Hypnopompic Hallucinations during Sleep Paralysis."
24. Pierre Maquette, Jean-Marie Péters, Joël Aerts, Guy Delfiore, Christian Degueldre, André Luxen, and Georges Franck, "Functional Neuroanatomy of Human Rapid-Eye-Movement Sleep and Dreaming," *Nature* 383, no. 6596 (September 1996): 163–166; J. Allan Hobson, Robert Stickgold, and Edward F. Pace-Schott, "The Neurophysiology of REM Sleep Dreaming," *NeuroReport* 9, no. 3 (February 1998): R1–R14.
25. J. Allan Cheyne, "The Ominous Numinous: Sensed Presence and 'Other' Hallucinations," *Journal of Consciousness Studies* 8, no. 5–7 (May 2001): 133–150.
26. Cheyne, Rueffer, and Newby-Clark, "Hypnagogic and Hypnopompic Hallucinations during Sleep Paralysis," 331.
27. Dan Denis, Christopher C. French, and Alice M. Gregory, "A Systematic Review of Variables Associated with Sleep Paralysis," *Sleep Medicine Reviews* 38 (April 2018): 141–157.
28. Dan Denis, Christopher C. French, Melanie N. Schneider, and Alice M. Gregory, "Subjective Sleep-Related Variables in Those Who Have and Have Not Experienced Sleep Paralysis," *Journal of Sleep Research* 27, no. 5 (October 2018): 1–10.

29. Dan Denis, Christopher C. French, Richard Rowe, Helena M. S. Zavos, Patrick M. Nolan, Michael J. Parsons, and Alice M. Gregory, "A Twin and Molecular Genetics Study of Sleep Paralysis and Associated Factors," *Journal of Sleep Research* 24, no. 4 (August 2015): 438–446.

30. Samad E. J. Golzari, Kazem Khodadoust, Farid Alakbarli, Kamyar Ghabili, Ziba Islambulchilar, Mohammadali M. Shoja, Majid Khalili, Feridoon Abbasnejad, Niloufar Sheikholeslamzadeh, Nasrollah Moghaddam Shahabi, Seyed Fazel Hosseini, and Khalil Ansarin, "Sleep Paralysis in Medieval Persia—the *Hidayat* of Akhawayni (?–938 AD)," *Neuropsychiatric Disease and Treatment* 8 (June 2012): 231.

31. Sharpless and Doghramji, *Sleep Paralysis*, Appendix A, 217–228.

32. Devon E. Hinton, David J. Hufford, and Laurence J. Kirmayer, "Culture and Sleep Paralysis," *Transcultural Psychiatry* 42, no. 1 (March 2005): 9.

33. Owen Davies, "The Nightmare Experience, Sleep Paralysis, and Witchcraft Accusations," *Folklore* 114, no. 2 (August 2003): 181–203.

34. Hiroko Arikawa, Donald I. Templer, Ric Brown, W. Gary Cannon, and Shan Thomas-Dodson, "The Structure and Correlates of Kanashibari," *Journal of Psychology* 133, no. 4 (1999): 369–375.

35. Anna Schegoleva, "Sleepless in Japan: The *Kanashibari* Phenomenon" (presentation, Postgraduate Research Seminar in Japanese Studies, Oxford Brookes University Research Centre, July 28, 2001).

36. Yun-Kwok Wing, Sharon Therese Lee, and Char-Nie Chen, "Sleep Paralysis in Chinese: Ghost Oppression Phenomenon in Hong Kong," *Sleep* 17, no. 7 (October 1994): 609–613.

37. Cheyne, Rueffer, and Newby-Clark, "Hypnagogic and Hypnopompic Hallucinations during Sleep Paralysis."

38. Alejandro Jiminez-Genchi, Victor M. Avila-Rodriguez, Frida Sanchez-Rojas, Blanca E. Terrez, and Alejandro Nenclares-Portocarrero, "Sleep Paralysis in Adolescents: The 'A Dead Body Climbed on Top of Me' Phenomenon in Mexico," *Psychiatry and Clinical Neurosciences* 63, no. 4 (August 2009): 546–549.

39. Ryan Hurd, *Sleep Paralysis: A Guide to Hypnagogic Visions and Visitors of the Night* (Los Altos, CA: Hyena Press, 2011).

40. Jack Lockwood, *The Maverick Ghost Hunter* (Capulin, CO: Xlibris, 2010).

41. Shelley R. Adler, "Sudden Unexpected Nocturnal Death Syndrome among Hmong Immigrants: Examining the Role of the 'Nightmare,'" *Journal of American Folklore* 104, no. 411 (Winter 1991): 54–71; Shelley R. Adler, "Ethnomedical Pathogenesis and Hmong Immigrants' Sudden Nocturnal Deaths," *Culture, Medicine, and Psychiatry* 18, no. 1 (March 1994): 23–59.

42. Nicolas Bruno, https://nicolasbrunophotography.com.

43. Corinne Purtill, "Why Everyone Around the World Is Having the Same Nightmare," *Quartz*, accessed May 28, 2023, https://qz.com/quartzy/1444843/what-is-sleep-paralysis. An

informative short film about sleep paralysis accompanies this article. It includes an interview with yours truly suggesting that the hat man imagery in sleep paralysis episodes may be based on Freddy K. The highlights for me, however, are the sequences where, at the cameraman's request, I am filmed walking around near my home at night wearing a hat, accompanied by my trusty labrador, Ted, who was keen to get in on the action.

44. Matthew Tompkins, "The Strange Case of the Phantom Pokemon," *BBC Future*, March 23, 2017, https://www.bbc.com/future/article/20170323-the-strange-case-of-the-phantom-pokemon.

45. Quotation extracted from Sharpless and Doghramji, *Sleep Paralysis*, 48.

46. Brian Andrew Sharpless and Jessica Lynn Grom, "Isolated Sleep Paralysis: Fear, Prevention, and Disruption," *Behavioral Sleep Medicine* 14, no. 2 (2016): 134–139.

47. Eric Suni, "Sleep Hygiene," SleepFoundation.org, accessed May 28, 2023, https://www.sleepfoundation.org/articles/sleep-hygiene.

48. Cheryl M. Paradis, Steven Friedman, and Marjorie Hatch, "Isolated Sleep Paralysis in African Americans with Panic Disorder," *Cultural Diversity and Mental Health* 3, no 1 (1997): 69–76.

49. See also Baland Jalal, "How to Make the Ghosts in My Bedroom Disappear? Focused-Attention Meditation Combined with Muscle relaxation (MR Therapy)—A Direct Treatment Intervention for Sleep Paralysis," *Frontiers in Psychology* 7 (January 2016): 28; Brian A. Sharpless and Karl Doghramji, "Commentary: How to Make the Ghosts in My Bedroom Disappear? Focused-Attention Meditation Combined with Muscle Relaxation (MR Therapy)—A Direct Treatment Intervention for Sleep Paralysis," *Frontiers in Psychology* 8 (April 2017): 506.

CHAPTER 3

1. For an excellent and accessible introduction to the topic of consciousness, see Susan Blackmore and Emily T. Troscianko, *Consciousness: An Introduction*, 3rd ed. (Abingdon: Routledge, 2018).

2. William Grey, "Philosophy and the Paranormal. Part 2: Skepticism, Miracles, and Knowledge," *Skeptical Inquirer* 18, no. 3 (Spring 1994): 288–294.

3. David Hume, *Enquiries Concerning Human Understanding and Concerning the Principles of Morals*, 3rd ed. (Oxford: Clarendon Press, 1777/1975), 115–116.

4. R. C. Finucane, *Ghosts: Appearances of the Dead and Cultural Transformation* (Amherst, NY: Prometheus Books, 1996), 4.

5. Some readers, a minority I hope, may of course have thought of a floating sheet with two eyes in it.

6. For further examples, see Robert A. Baker and Joe Nickell, *Missing Pieces: How to Investigate Ghosts, UFOs, Psychics, & Other Mysteries* (Buffalo, NY: Prometheus Books, 1992); Milbourne Christopher, *Seers, Psychics and ESP* (London: Cassell, 1970); Joe Nickell, *Entities: Angels,*

Spirits, Demons, and Other Alien Beings (Amherst, NY: Prometheus Books, 1995); Benjamin Radford, *Investigating Ghosts: The Scientific Search for Spirits* (Corrales, NM: Rhombus, 2017).

7. James Randi, "The Columbus Poltergeist Case: Part I. Flying Phones, Photos, and Fakery," *Skeptical Inquirer* 9, no. 3 (Spring 1985): 221–235.

8. He died in prison on March 12, 2021.

9. Robert L. Morris, "Review of *The Amityville Horror*, by Jay Anson. Prentice-Hall, Englewood Cliffs, N.J., 1977. 201 pages, $7.95," *Skeptical Inquirer* 2, no. 2 (Spring/Summer 1978): 95–96.

10. Wikipedia, s.v. "*The Amityville Horror*," accessed May 29, 2023, https://en.wikipedia.org/wiki/The_Amityville_Horror.

11. For further discussion of such factors, see, for example: Neil Dagnall, Kenneth G. Drinkwater, Ciarán O'Keeffe, Annalisa Ventola, Brian Laythe, Michael A. Jawer, Brandon Massullo, Giovanni B. Caputo, and James Houran, "Things That Go Bump in the Literature: An Environmental Appraisal of 'Haunted Houses,'" *Frontiers in Psychology* 11 (June 2020): 1328; James Houran and Rense Lange, eds., *Hauntings and Poltergeists: Multidisciplinary Perspectives* (Jefferson, NC: McFarland, 2001); Brian Laythe, James Houran, Neil Dagnall, Kenneth G. Drinkwater, and Ciarán O'Keeffe, *Ghosted! Exploring the Haunting Reality of Paranormal Encounters* (Jefferson, NC: McFarland, 2022); Peter A. McCue, "Theories of Haunting: A Critical Overview," *Journal of the Society for Psychical Research* 66, no. 866 (January 2002): 1–21; Radford, *Investigating Ghosts*.

12. Vic Tandy and Tony R. Lawrence, "The Ghost in the Machine," *Journal of the Society for Psychical Research* 62, no. 851 (April 1998): 360–364.

13. See, for example, Kate Buck, "Vigilante Shed Cleaner Revealed to Be House Proud Mouse," *Metro*, March 18, 2019, https://metro.co.uk/2019/03/18/vigilante-shed-cleaner-revealed-to-be-house-proud-mouse-8926281. Online article includes link to cute house-proud mouse in action.

14. Rense Lange and James Houran, "Context-Induced Paranormal Experiences: Support for Houran and Lange's Model of Haunting Phenomena," *Perceptual and Motor Skills* 84, no. 3, Pt. 2 (June 1997): 1455–1458.

15. Richard Wiseman, Caroline Watt, Emma Greening, Paul Stevens, and Ciarán O'Keeffe, "An Investigation into the Alleged Haunting of Hampton Court Palace: Psychological Variables and Magnetic Fields," *Journal of Parapsychology* 66, no. 4 (December 2002): 388–408.

16. Louise C. Johns, "Hallucinations in the General Population," *Current Psychiatry Reports* 7, no. 3 (May 2005): 162–167.

17. Charles Heriot-Maitland, Matthew Knight, and Emmanuelle Peters, "A Qualitative Comparison of Psychotic-Like Phenomena in Clinical and Non-Clinical Populations," *British Journal of Clinical Psychology* 51, no. 1 (March 2012): 37–53.

18. Robert Todd Carroll, *The Skeptic's Dictionary: A Collection of Strange Beliefs, Amusing Deceptions & Dangerous Delusions* (Hoboken, NJ: Wiley, 2003), 275.

19. Cited in Peter Brugger, Marianne Regard, Theodor Landis, Norman Cook, Denise Krebs, and Joseph Niederberger, "'Meaningful' Patterns in Visual Noise: Effects of Lateral Stimulation and the Observer's Belief in ESP," *Psychopathology* 26, no. 5–6 (January 1993): 261–265.

20. Functional magnetic resonance imaging monitors brain activity by measuring the amount of blood flowing to different regions, whereas magnetoencephalography does so by measuring the magnetic fields naturally produced by the brain's electrical activity.

21. Susan G. Wardle, Jessica Taubert, Lina Teichmann, and Chris I. Baker, "Rapid and Dynamic Processing of Face Pareidolia in the Human Brain," *Nature Communications* 11, no. 4518 (September 2020): https://doi.org/10.1038/s41467-020-18325-8.

22. For the stories behind some of the most famous examples of pareidolia, see Buzz Poole, *Madonna of the Toast* (New York: Mark Batty, 2007).

23. Wikipedia, s.v. "Perceptions of Religious Imagery in Natural Phenomena," accessed May 29, 2023, https://en.wikipedia.org/wiki/Perceptions_of_religious_imagery_in_natural_phenomena.

24. Brugger et al., "'Meaningful' Patterns in Visual Noise."

25. Tapani Riekki, Marjaana Lindeman, Marja Aleneff, Anni Halme, and Antti Nuortimo, "Paranormal and Religious Believers Are More Prone to Illusory Face Perception Than Skeptics and Non-Believers," *Applied Cognitive Psychology* 27, no. 2 (October 2012): 150–155.

26. This study has one of the best titles for a paper to be found in the literature. Daniel J. Simons and Christopher F. Chabris, "Gorillas in Our Midst: Sustained Inattentional Blindness for Dynamic Events," *Perception* 28, no. 9 (September 1999): 1059–1074.

27. Anne Richards, Moa Gunnarsson Hellgren, and Christopher C. French, "Inattentional Blindness, Absorption, Working Memory Capacity, and Paranormal Belief," *Psychology of Consciousness: Theory, Research, and Practice* 1, no. 1 (2014): 60–69.

28. The first appearance was in Arien Mack and Irvin Rock, *Inattentional Blindness* (Cambridge, MA: MIT Press, 1998). Tony Cornell's work: A. D. Cornell, "An Experiment in Apparitional Observation and Findings," *Journal of the Society for Psychical Research* 40, no. 701 (1959): 120–124; A. D. Cornell, "Further Experiments in Apparitional Observation," *Journal of the Society for Psychical Research* 40, no. 706 (1960): 409–418.

29. Matthew Tompkins, "The Strange Tale of an X-rated Haunting," *BBC Future*, October 24, 2016, https://www.bbc.com/future/article/20161024-the-strange-tale-of-an-x-rated-haunting.

30. Elizabeth F. Loftus, *Eyewitness Testimony* (London: Harvard University Press, 1979); Mark L. Howe, Lauren M. Knott, and Martin A. Conway, *Memory and Miscarriages of Justice* (London: Routledge, 2018).

31. Christopher C. French, "Fantastic Memories: The Relevance of Research into Eyewitness Testimony and False Memories for Reports of Anomalous Experiences," *Journal of Consciousness Studies* 10, nos. 6–7 (2003): 153–174.

32. Christopher C. French and James Ost, "Beliefs about Memory, Childhood Abuse, and Hypnosis Amongst Clinicians, Legal Professionals and the General Public," in *Wrongful Allegations of Sexual and Child Abuse*, ed. Ros Burnett (Oxford: Oxford University Press, 2016), 143–154; James Ost and Christopher C. French, "How Misconceptions About Memory May Undermine Witness Testimony," in *Witness Testimony in Sexual Cases: Evidential, Investigative and Scientific Perspectives*, eds. Pamela Radcliffe, Gisli H. Gudjonsson, Anthony Heaton-Armstrong, and David Wolchover (Oxford: Oxford University Press, 2016), 361–373.

33. Daniel J. Simons and Christopher F. Chabris, "What People Believe About How Memory Works: A Representative Survey of the U.S. Population," *PLoS ONE* 6, no. 8: e22757; Daniel J. Simons and Christopher F. Chabris, "Common (Mis)Beliefs About Memory: A Replication and Comparison of Telephone and Mechanical Turk Survey Methods," *PLOS ONE* 7, no. 12: e51876.

34. When I first realized that the four was represented in this unexpected way on clocks and watches with Roman numerals, I spent a considerable amount of time trying to ascertain why this was. It turns out that there several theories but in fact no one really knows. The most plausible explanation is that it is simply more aesthetically pleasing to balance the "VIII" on the left of the face with "IIII" on the right rather than the less chunky "IV." Also, I am aware that there are a few exceptions to this general rule. You do not need to email me to tell me about the clock on the tower housing Big Ben.

35. Christopher C. French and Anne Richards, "Clock* This! An Everyday Example of a Schema-Driven Error in Memory," *British Journal of Psychology* 84, no. 2 (May 1993): 249–253.

36. Just to be clear, my tongue was firmly in my cheek as I wrote that sentence. In reality, we have no way of knowing who had that conversation with Lucy—or indeed if anyone did. My point here is that even professional psychologists with a good knowledge of the unreliability of memory are just as prone in everyday life to insist that their memory is 100 percent accurate as anyone else.

37. Ruth Brandon, *The Spiritualists: The Passion for the Occult in the Nineteenth and Twentieth Centuries* (London: Weidenfeld and Nicolson, 1983); Ray Hyman, "A Critical Historical Overview of Parapsychology," in *A Skeptic's Handbook of Parapsychology*, ed. Paul Kurtz (Buffalo, NY: Prometheus Books, 1985), 3–96.

38. S. J. Davey, "The Possibilities of Mal-Observation and Lapse of Memory from a Practical Point of View: Experimental Investigation," *Proceedings of the Society for Psychical Research* 4 (1887): 405–495.

39. Hyman, "A Critical Historical Overview of Parapsychology," 27.

40. Theodore Besterman, "The Psychology of Testimony in Relation to Paraphysical Phenomena: Report of an Experiment," *Proceedings of the Society for Psychical Research* 124 (1932): 363–387.

41. Richard Wiseman, Emma Greening, and Matthew Smith, "Belief in the Paranormal and Suggestion in the Seance Room," *British Journal of Psychology* 94, no. 3 (August 2003): 285–297.

42. Wiseman, Greening, and Smith, "Belief in the Paranormal and Suggestion in the Seance Room," 295.

43. Richard Wiseman and Robert L. Morris, "Recalling Pseudo-Psychic Demonstrations," *British Journal of Psychology* 86, no. 1 (February 1995): 113–125.

44. Magicians often have several methods at their disposal to achieve what appears to be the same effect. This comes in handy if, say, they perform the same trick several times in front of the same people. Viewers may gradually guess at the technique being used if they are able to view the trick multiple times. By switching to an alternative technique that appears to rule out the first hypothesized technique, viewers can be thrown off the scent. There are, for example, several techniques that a skilled conjurer might employ to appear to psychokinetically bend metal.

45. Readers may care to note that these days, if you know where to buy them, it is possible to purchase self-bending spoons for use in your magic shows, but these were not available in the early days of Geller's career.

46. Richard Wiseman and Emma Greening, "'It's Still Bending': Verbal Suggestion and Alleged Psychokinetic Ability," *British Journal of Psychology* 96, no. 1 (February 2005): 115–127.

47. Just for the record, the key definitely does not continue to bend.

48. Elizabeth F. Loftus, David G. Miller, and Helen J. Burns, "Semantic Integration of Verbal Information into Visual Memory," *Journal of Experimental Psychology: Human Learning and Memory* 4, no. 1 (January 1978): 19–31.

49. For example, Fiona Gabbert, Amina Memon, Kevin Allan, and Daniel B. Wright, "Say It to My Face: Examining the Effects of Socially Encountered Misinformation," *Legal and Criminological Psychology* 9, no. 2 (September 2004): 215–227.

50. Fiona Gabbert, Amina Memon, and Kevin Allan, "Memory Conformity: Can Eyewitnesses Influence Each Other's Memories for an Event?," *Applied Cognitive Psychology* 17, no. 5 (April 2003): 533–543.

51. Krissy Wilson and Christopher C. French, "Magic and Memory: Using Conjuring to Explore the Effects of Suggestion, Social Influence and Paranormal Belief on Eyewitness Testimony for an Ostensibly Paranormal Event," *Frontiers in Psychology* 5 (November 2014): 1289.

52. Richard Wiseman, Caroline Watt, Paul Stevens, Emma Greening, and Ciarán O'Keeffe, "An Investigation into Alleged 'Hauntings,'" *British Journal of Psychology* 94, no. 2 (May 2003): 195–211.

53. The strength of the background electromagnetic field in any unshielded location on Earth depends on a range of factors, including the planet's own constantly fluctuating field, the field around electronic devices, and even the underlying geology of the area.

54. See, e.g., Michael A. Persinger, "Geophysical Variables and Behavior: LV. Predicting the Details of Visitor Experiences and the Personality of Experients: The Temporal Lobe Factor," *Perceptual and Motor Skills* 68, no.1 (February 1989): 55–65.

55. Jason J. Braithwaite, "Neuromagnetic Effects on Anomalous Cognitive Experiences: A Critical Appraisal of the Evidence for Induced Sensed-Presence and Haunt-Type Experiences," *NeuroQuantology* 8, no. 1 (December 2010): 517–530; Jason J. Braithwaite, "Magnetic Fields, Hallucinations and Anomalous Experiences: A Skeptical Critique of the Current Evidence," *The Skeptic* 22, no. 4 (Summer 2011): 38–45.

56. Craig Aaen-Stockdale, "Neuroscience for the Soul," *The Psychologist* 25, no. 7 (July 2012): 520–523.

57. Katherine Makarec and Michael A. Persinger, "Electroencephalographic Validation of a Temporal Lobe Signs Inventory," *Journal of Research in Personality* 24, no. 3 (September 1990): 323–337.

58. Braithwaite, "Neuromagnetic Effects"; Braithwaite, "Magnetic Fields."

59. C. M. Cook and M. A. Persinger, "Experimental Induction of the 'Sensed Presence' in Normal Subjects and an Exceptional Subject," *Perceptual and Motor Skills* 85, no. 2 (November 1997): 683–693; C. M. Cook and M. A. Persinger, "Geophysical Variables and Behavior: XCII. Experimental Elicitation of the Experience of a Sentient Being by Right Hemispheric, Weak Magnetic Fields: Interaction with Temporal Lobe Sensitivity," *Perceptual and Motor Skills* 92, no. 2 (April 2001): 447–448; Michael A. Persinger, "The Sensed Presence Within Experimental Settings: Implications for Male and Female Concept of Self," *Journal of Psychology* 137, no. 1 (January 2003): 5–16.

60. M. A. Persinger, S. G. Tiller, and S. A. Koren, "Experimental Simulation of a Haunt Experience and Elicitation of Paroxysmal Electroencephalographic Activity by Transcerebral Complex Magnetic Fields: Induction of a Synthetic 'Ghost'?," *Perceptual and Motor Skills* 90 no. 2 (April 2000): 659–674.

61. Susan Blackmore, "Alien Abduction," *New Scientist* 144 (November 19, 1994): 19–21.

62. Aaen-Stockdale, "Neuroscience for the Soul," 522.

63. Pehr Granqvist, Mats Fredrikson, Patrik Unge, Andrea Hagenfeldt, Sven Valind, Dan Larhammar, and Marcus Larsson, "Sensed Presence and Mystical Experiences Are Predicted by Suggestibility, Not by the Application of Transcranial Weak Complex Magnetic Fields," *Neuroscience Letters* 379, no. 1 (April 2005): 1–6.

64. M. A. Persinger and S. A. Koren, "A Response to Granqvist et al. 'Sensed Presence and Mystical Experiences Are Predicted by Suggestibility, Not by the Application of Transcranial Weak Complex Magnetic Fields,'" *Neuroscience Letters* 380, no. 3 (June 2005): 346–347; see also Marcus Larsson, Dan Larhammar, Mats Fredrikson, and Pehr Granqvist, "Reply to M. A. Persinger and S. A. Koren's Response to Granqvist et al. 'Sensed Presence and Mystical Experiences Are Predicted by Suggestibility, Not by the Application of Transcranial Weak Complex Magnetic Fields,'" *Neuroscience Letters* 380, no. 3 (June 2005): 348–350.

65. Braithwaite, "Neuromagnetic Effects," 527.

66. I first met Vic when I worked at Coventry Polytechnic (now Coventry University) between 1981 and 1982 as a temporary lecturer. At that time, he was a technician in the same

department. We got on very well, and I often spent my coffee breaks chatting with him. I had no idea that many years later he would become well known in the world of the paranormal for coming up with a new theory of hauntings. Vic Tandy and Tony R. Lawrence, "The Ghost in the Machine," *Journal of the Society for Psychical Research* 62, no. 851 (April 1998): 360–364.

67. Vic Tandy "Something in the Cellar," *Journal of the Society for Psychical Research* 64 (2000): 129–140.

68. Jason J. Braithwaite and Maurice Townsend, "Good Vibrations: The Case for a Specific Effect of Infrasound in Instances of Anomalous Experience Has Yet to Be Empirically Demonstrated," *Journal of the Society for Psychical Research* 70 (October 2006): 211–224.

69. Christopher C. French, Usman Haque, Rosie Bunton-Stasyshyn, and Rob Davis, "The 'Haunt' Project: An Attempt to Build a 'Haunted' Room by Manipulating Complex Electromagnetic Fields and Infrasound," *Cortex* 45, no. 5 (May 2009): 619–629.

70. French et al., "The 'Haunt' Project," 624–625.

71. M. A. Persinger and C. F. De Sano, "Temporal Lobe Signs: Positive Correlations with Imaginings and Hypnosis Induction Profiles," *Psychological Reports* 58, no. 2 (April 1986): 347–350.

CHAPTER 4

1. Sybo A. Schouten, "An Overview of Quantitatively Evaluated Studies with Mediums and Psychics," *Journal of the American Society for Psychical Research* 88 (July 1994): 221–254.

2. Schouten, "An Overview," 221.

3. Marco Aurélio Vinhosa Bastos Jr., Paulo Roberto Haidamus de Oliveira Bastos, Lidia Maria Gonçalves, Igraíne Helena Scholz Osório, and Giancarlo Lucchetti, "Mediumship: Review of Quantitatives [sic] Studies Published in the 21st Century," *Archives of Clinical Psychiatry* 42, no. 5 (October 2015): 129–138.

4. The two studies that reported higher accuracy were Julie Beischel and Gary E. Schwartz, "Anomalous Information Reception by Research Mediums Demonstrated Using a Novel Triple-Blind Protocol," *Explore (NY)* 3, no. 1 (January–February 2007): 23–27; Julie Beischel, Mark Boccuzzi, Michael Biuso, and Adam J. Rock, "Anomalous Information Reception by Research Mediums under Blinded Conditions. II: Replication and Extension," *Explore (NY)* 11, no. 2 (March–April 2015): 36–42. The three that did not were Emily Williams Kelly and Dianne Arcangel, "An Investigation of Mediums Who Claim to Give Information About Deceased Persons," *Journal of Nervous and Mental Disease* 199, no. 1 (January 2011): 11–17; Ciarán O'Keeffe and Richard Wiseman, "Testing Alleged Mediumship: Methods and Results," *British Journal of Psychology* 96, no. 2 (May 2005): 165–179; Christian Jensen, Etzel Cardeña, and Devin Terhune, "A Controlled Long-Distance Test of a Professional Medium," *European Journal of Parapsychology* 24, no. 1 (2009): 53–67.

5. Ray Hyman, "'Cold Reading': How to Convince Strangers That You Know All About Them," *The Zetetic* 1 (Spring/Summer 1977): 18–37; reprinted in Ray Hyman, *The Elusive Quarry: A Scientific Appraisal of Psychical Research* (Buffalo, NY: Prometheus Books, 1989), 402–419.

6. Ian Rowland, *The Full Facts Book of Cold Reading*, 7th ed. (London: Ian Rowland., 2019). This book is available from various online booksellers, but Ian would prefer it if you ordered it directly from the following website: https://www.coldreadingsuccess.com.

7. Bertram R. Forer, "The Fallacy of Personal Validation: A Classroom Demonstration of Gullibility," *Journal of Abnormal and Social Psychology* 44, no. 1 (January 1949): 118–123.

8. C. R. Snyder, Randee Jae Shenkel, and Carol R. Lowery, "Acceptance of Personality Interpretations: The "Barnum Effect" and Beyond," *Journal of Consulting and Clinical Psychology* 45, no. 1 (1977): 104–114; D. H. Dickson and I. W. Kelly, "The 'Barnum Effect' in Personality Assessment: A Review of the Literature," *Psychological Reports* 57, no. 2 (October 1985): 367–382; Adrian Furnham and Sandra Schofield, "Accepting Personality Test Feedback: A Review of the Barnum Effect," *Current Psychological Research & Reviews* 6, no. 2 (Summer 1987): 162–178.

9. Susan J. Blackmore, "Probability Misjudgment and Belief in the Paranormal: A Newspaper Survey," *British Journal of Psychology* 88, no. 4 (November 1997): 683–689.

10. Krissy Wilson and Christopher C. French, "Misinformation Effects for Psychic Readings and Belief in the Paranormal," *Imagination, Cognition and Personality* 28, no. 2 (2008–2009): 155–171.

11. Francesca Cookney, "Top Five Brilliant Examples of Flawless Sherlock Holmes Logic," *Mirror*, January 2, 2014, https://www.mirror.co.uk/lifestyle/staying-in/top-five-brilliant-examples-flawless-2980770.

12. Adam J. Powell and Peter Moseley, "When Spirits Speak: Absorption, Attribution, and Identity among Spiritualists Who Report 'Clairaudient' Voice Experiences," *Mental Health, Religion & Culture* 23, no. 10 (July 2020): 841-856.

13. Matt Roper, "The Spooky Truth: Most Haunted Exposed As a Fake by Star That Claims Derek Acorah Show Features 'Showmanship and Dramatics,'" *Mirror*, May 15, 2012, https://www.mirror.co.uk/tv/tv-news/most-haunted-exposed-fake-star-833433.

14. Roper, "The Spooky Truth."

15. Proper Ouija boards can be quite beautiful things. They consist of a polished wooden board marked with letters and numbers, as well as the words "yes" and "no." Sitters each rest a finger on a heart-shaped piece of wood known as a planchette that moves over the board pointing at letters in response to the sitters' questions.

16. In the interests of total transparency, I cannot honestly claim that I never got spooked by a Ouija session. For an account of one such occasion, see my contribution to the edited collection by Karen Stollznow, *Would You Believe It? Mysterious Tales from People You'd Least Expect* (Denver, CO: Karen Stollznow, 2017): 77–81.

17. Carroll, *The Skeptic's Dictionary*, 172.
18. As demonstrated in season 5 of the National Geographic series, *Brain Games*: https://www.youtube.com/watch?v=PRo8TytvIDw.
19. Konstantin Raudive, *Breakthrough* (New York: Taplinger, 1971).
20. Michael A. Nees and Charlotte Phillips, "Auditory Pareidolia: Effects of Contextual Priming on Perceptions of Purportedly Paranormal and Ambiguous Auditory Stimuli," *Applied Cognitive Psychology* 29, no. 1 (January/February 2015): 129–134.
21. E. L. Smith, "The Raudive Voices—Objective or Subjective? A Discussion," *Journal of the Society for Psychical Research* 46 (1972): 192–200; D. J. Ellis, "Listening to the 'Raudive Voices,'" *Journal of the Society for Psychical Research* 48 (1975): 31–42; Joe Banks, "Rorschach Audio: Ghost Voices and Perceptual Creativity," *Leonardo Music Journal* 11, no. 2 (December 2001): 77–83; Christopher C. French, *Paranormal Perception? A Critical Evaluation* (London: Institute for Cultural Research, 2001).

CHAPTER 5

1. E.g., David Basey, "The Size of Things," *British Astronomical Association*, November 16, 2018, https://britastro.org/node/13975.
2. Douglas Adams, *The Hitchhiker's Guide to the Galaxy* (London: Pan Books, 1979).
3. Jim Al-Khalili, ed., *Aliens: Science Asks: Is There Anyone Out There?* (London: Profile, 2016).
4. Lydia Saad, "Americans Skeptical of UFOs, But Say Government Knows More," *Gallup*, September 6, 2019, https://news.gallup.com/poll/266441/americans-skeptical-ufos-say-government-knows.aspx.
5. Chris Jackson, "A Quarter of Americans Believe That Crashed Ufo Spacecrafts Are Held at Area 51 in Southern Nevada," *Ipsos*, October 3, 2019, https://www.ipsos.com/en-us/news-polls/americans-believe-crashed-ufo-spacecrafts-held-at-area-51.
6. Will Dahlgreen, "You Are Not Alone: Most People Believe That Aliens Exist," *YouGov*, September 24, 2015, https://yougov.co.uk/topics/lifestyle/articles-reports/2015/09/24/you-are-not-alone-most-people-believe-aliens-exist.
7. Charles Berlitz and William M. Moore, *The Roswell Incident* (New York: Grosset & Dunlap, 1980); Kevin D. Randle and Donald R. Schmitt, *UFO Crash at Roswell* (New York: Avon, 1991); Stanton T. Friedman and Don Berliner, *Crash at Corona: The U.S. Military Retrieval and Cover-Up of a UFO* (New York: Paragon House, 1992); Kevin Randle and Donald Schmitt, *The Truth About the UFO Crash at Roswell* (New York: M. Evans, 1994).
8. Philip J. Klass, *The Real Roswell Crashed-Saucer Coverup* (Amherst, NY: Prometheus Books, 1997); Kal K. Korff, *The Roswell UFO Crash: What They Don't Want You to Know* (Amherst, NY: Prometheus Books, 1997).

9. Occam's (or Ockham's) razor is a principle rightly beloved of skeptics. It is attributed to the English monk William of Ockham (c. 1287–1347) and states that "entities should not be multiplied without necessity," or, more simply put, whenever more than one explanation is available for a phenomenon, the simplest is to be preferred.

10. David Clarke, *The UFO Files: The Inside Story of Real-Life Sightings,* 2nd ed. (London: Bloomsbury, 2012), 107.

11. Ian Ridpath, "The Rendlesham Forest UFO Case," accessed May 30, 2023, http://www.ianridpath.com/ufo/rendlesham.html. See also Jenny Randles, Andy Roberts, and David Clarke, *The UFOs That Never Were* (London: London House, 2000), 164–222.

12. *BBC News*, "Meteor Captured on Doorbell Cameras in England," March 2, 2021, https://www.bbc.co.uk/news/uk-56241705.

13. Peter Brookesmith, *UFO: The Complete Sightings catalogue* (London: Blandford, 1995).

14. Robert Durant, "Public Opinion Polls and UFOs," in *UFO 1947–1997: Fifty Years of Flying Saucers*, ed. Hilary Evans and Dennis Stacy (London: John Brown, 1997), 230–239.

15. As reported in Robert Sheaffer, *UFO Sightings: The Evidence* (Amherst, NY: Prometheus Books, 1998), 139.

16. Sheaffer, *UFO Sightings*; Philip J. Klass, *UFOs: The Public Deceived* (Buffalo, NY: Prometheus Books, 1983).

17. Joe Nickell, *Camera Clues: A Handbook for Photographic Investigation* (Lexington: University Press of Kentucky, 1994).

18. For a great example of the use of such techniques, see Kal K. Korff, *The Billy Meier Story: Spaceships of the Pleiades* (Amherst, NY: Prometheus Books, 1995).

19. A nice example of such a misinterpretation was reported in the UK as I was writing this chapter: Michael Moran, "Google Maps User Spots 'Giant UFO Mothership'—Can You Guess What It Really Is," *Daily Star*, February 28, 2021, https://www.dailystar.co.uk/news/weird-news/google-maps-user-spots-giant-23580039.

20. Medical examination and experimentation has become one of the most common themes reported as part of alien abduction narratives. It has been commented that the aliens' curiosity regarding human reproductive organs could be satisfied more easily by them obtaining a good anatomy textbook. *Gray's Anatomy*, first published in 1858 and now in its 40th edition, would be the obvious choice.

21. John G. Fuller, *The Interrupted Journey* (New York: Dell, 1966).

22. Sheaffer, *UFO Sightings*; Philip J. Klass, *UFO Abductions: A Dangerous Game,* updated edition (Buffalo, NY: Prometheus Books, 1989).

23. Whitley Strieber, *Communion: A True Story* (New York: Morrow, 1987).

24. Klass, *UFO Abductions*.

25. Strieber, *Communion*, 172–173.

26. Budd Hopkins, *Intruders: The Incredible Visitations at Copley Woods* (New York: Random House, 1987); Budd Hopkins, *Missing Time: A Documented Study of UFO Abductions* (New York: Richard Marek, 1981).

27. In a study by Peter Hough and Paul Rogers, almost half of the female abductees in their sample claimed to have been impregnated by the aliens, although it was not clear whether this was alleged to be the result of artificial procedures similar to in vitro fertilization or forced sexual contact with the aliens. See Peter Hough and Paul Rogers, "Individuals Who Report Being Abducted by Aliens: Investigating the Differences in Fantasy Proneness, Emotional Intelligence and the Big Five Personality Factors," *Imagination, Cognition, and Personality* 27, no. 2 (October 2007), 139–161.

28. John E. Mack, *Abduction: Human Encounters with Aliens* (New York: Charles Scribner's Sons, 1994).

29. Examples of proven or probable hoaxes in ufology include the alleged abduction of Travis Walton (see Klass, *UFO Abductions*, 25–37); the "evidence" produced by Billy Meier to support his claims of alien contact (Korff, *The Billy Meier Story*); the Gulf Breeze UFO photographs (Sheaffer, *UFO Sightings*, 100–102); and the Roswell "alien autopsy" hoax (see articles by Joe Nickell, C. Eugene Emery, Trey Stokes, and Joseph A. Bauser reprinted in Kendrick Frazier, Barry Karr, and Joe Nickell, eds., *The UFO Invasion: The Roswell Incident, Alien Abductions, and Government Coverups* (Amherst, NY: Prometheus Books, 1997), 135–157).

30. Susan Blackmore, *The Meme Machine* (Oxford: Oxford University Press, 1999), 175.

31. Kevin D. Randle, Russ Estes, and William P. Cone, *The Abduction Enigma: The Truth Behind the Mass Alien Abductions of the Late Twentieth Century* (New York: Tom Doherty Associates, 1999), 322–327.

32. Richard J. McNally, Natasha B. Lasko, Susan A. Clancy, Michael L. Macklin, Roger K. Pitman, and Scott P. Orr, "Psychophysiological Responding during Script-Driven Imagery in People Reporting Abduction by Space Aliens," *Psychological Science* 15, no. 7 (July 2004): 493–497.

33. Robert E. Bartholomew, Keith Basterfield, and George S. Howard, "UFO Abductees and Contactees: Psychopathology or Fantasy Proneness?," *Professional Psychology: Research and Practice* 22, no. 3 (June 1991): 215–222; Ted Bloecher, Aphrodite Clamar, and Budd Hopkins, *Summary Report on the Psychological Testing of Nine Individuals Reporting UFO Abduction Experiences* (Mount Rainier, MD: Fund for UFO Research, 1985); Mack, *Abduction*; June O. Parnell and R. Leo Sprinkle, "Personality Characteristics of Persons Who Claim UFO Experiences," *Journal of UFO Studies* 2, no. 1 (1990): 45–58; M. Rodeghier, J. Goodpaster, and S. Blatterbauer, "Psychosocial Characteristics of Abductees: Results from the CUFOS Abduction Project," *Journal of UFO Studies* 3 (1991): 59–90; Nicholas P. Spanos, Patricia A. Cross, Kirby Dickson, and Susan C. DuBreuil, "Close Encounters: An Examination of UFO Experiences," *Journal of Abnormal Psychology* 102, no. 4 (November 1993): 624–632.

34. Jatinder Takhar and Sandra Fisman, "Alien Abduction in PTSD," *Journal of the American Academy of Child and Adolescent Psychiatry* 34, no. 8 (August 1995): 974–975; Susan Marie Powers, "Alien Abduction Narratives," in *Broken Images, Broken Selves: Dissociative Narratives in Clinical Practice*, ed. Stanley Krippner and Susan Marie Powers (Washington, DC: Bruner/Mazel, 1997), 199–215; Rodeghier et al., "Psychosocial Characteristics."

35. J. Stone-Carmen, "A Descriptive Study of People Reporting Abduction by Unidentified Flying Objects (UFOs)," in *Alien Discussions: Proceedings of the Abduction Study Conference Held at MIT*, ed. Andrea Pritchard, David E. Pritchard, John E. Mack, Pam Kasey, and Claudia Yapp (Cambridge, MA: North Cambridge Press, 1994), 309–315.

36. Parnell and Sprinkle, "Personality Characteristics," 45.

37. Christopher C. French, Julia Santomauro, Victoria Hamilton, Rachel Fox, and Michael A. Thalbourne, "Psychological Aspects of the Alien Contact Experience," *Cortex* 44, no. 10 (November/December 2008): 1387–1395.

38. Joseph Glicksohn and Terry R. Barrett, "Absorption and Hallucinatory Experience," *Applied Cognitive Psychology* 17, no. 7 (November 2003): 833–849; French, "Fantastic Memories"; Christopher C. French and Krissy Wilson, "Incredible Memories: How Accurate Are Reports of Anomalous Events?," *European Journal of Parapsychology* 21 (2006): 166–181.

39. Elizabeth Loftus and Katherine Ketcham, *The Myth of Repressed Memory: False Memories and Allegations of Sexual Abuse* (New York: St. Martin's Press, 1994); Paul R. McHugh, *Try to Remember: Psychiatry's Clash over Meaning, Memory, and Mind* (New York: Dana Press, 2008); Richard Ofshe and Ethan Watters, *Making Monsters: False Memories, Psychotherapy, and Sexual Hysteria* (London: Andre Deutsch, 1995); Mark Pendergrast, *The Repressed Memory Epidemic: How It Happened and What We Need to Learn from It* (Cham, Switzerland: Springer, 2017);

40. Richard J. McNally, *Remembering Trauma* (Cambridge, MA: Harvard University Press, 2003).

41. C. J. Brainerd and V. F. Reyna, *The Science of False Memory* (New York: Oxford University Press, 2005).

42. James Deese, "On the Prediction of Occurrence of Particular Verbal Intrusions in Immediate Recall," *Journal of Experimental Psychology* 58, no. 1 (July 1959): 17–22; Henry L. Roediger III and Kathleen B. McDermott, "Creating False Memories: Remembering Words Not Presented on Lists," *Journal of Experimental Psychology: Learning, Memory, and Cognition* 21, no. 4 (July 1995). 803–814.

43. Elizabeth F. Loftus and Jacqueline E. Pickrell, "The Formation of False Memories," *Psychiatric Annals* 25, no. 12 (December 1995): 720–725.

44. Elizabeth F. Loftus, "Imagining the Past," *The Psychologist* 14, no. 11 (November 2001): 584–587.

45. Kimberley A. Wade, Maryanne Garry, J. Don Read, and D. Stephen Lindsay, "A Picture Is Worth a Thousand Lies: Using False Photographs to Create False Childhood Memories," *Psychonomic Bulletin & Review* 9, no. 3 (September 2002): 597–603.

46. Hans F. M. Crombag, Willem A. Wagenaar, and Peter J. van Koppen, "Crashing Memories and the Problem of 'Source Monitoring,'" *Applied Cognitive Psychology* 10, no. 2 (April 1996): 95–104.

47. James Ost, Aldert Vrij, Alan Costall, and Ray Bull, "Crashing Memories and Reality Monitoring: Distinguishing between Perceptions, Imaginations and 'False Memories,'" *Applied Cognitive Psychology* 16, no. 2 (February 2002): 125–134; Pär A. Granhag, Leif A. Strömwall, and James F. Billings, "'I'll Never Forget the Sinking Ferry': How Social Influence Makes False Memories Surface," in *Much Ado About Crime: Chapters on Psychology and Law*, ed. Miet Vanderhallen, Geert Vervaeke, Peter Jan Van Koppen, and Johan Goethals (Brussels: Uitgeverij Politeia, 2003), 129–140.

48. French, "Fantastic Memories"; French and Wilson, "Incredible Memories."

49. Auke Tellegen and Gilbert Atkinson, "Openness to Absorbing and Self-Altering Experiences ('Absorption'), a Trait Related to Hypnotic Susceptibility," *Journal of Abnormal Psychology* 83, no. 3 (June 1974): 268.

50. Glicksohn and Barrett, "Absorption and Hallucinatory Experience"; Robert Nadon and John F. Kihlstrom, "Hypnosis, Psi, and the Psychology of Paranormal Experience," *Behavioral and Brain Sciences* 10, no. 4 (December 1987): 597–599; Richards, Hellgren, and French, "Inattentional Blindness"; Glicksohn and Barrett, "Absorption and Hallucinatory Experience"; John Palmer and Ivo van der Velden, "ESP and 'Hypnotic Imagination': A Group Free-Response Study," *European Journal of Parapsychology* 4, no. 4 (May 1983): 413–434; Nicholas P. Spanos and Patricia Moretti, "Correlates of Mystical and Diabolical Experiences in a Sample of Female University Students," *Journal for the Scientific Study of Religion* 27, no. 1 (March 1988): 105–116; Susan A. Clancy, Richard J. McNally, Daniel L. Schacter, Mark F. Lenzenweger, and Roger K. Pitman, "Memory Distortion in People Reporting Abduction by Aliens," *Journal of Abnormal Psychology* 111, no. 3 (August 2002): 455–461; French et al., "Psychological Aspects."

51. On dissociativity and paranormal belief, see Harvey J. Irwin, "Paranormal Beliefs and Proneness to Dissociation," *Psychological Reports* 75 (1994): 1344–1346; Toni Makasovski and Harvey J. Irwin, "Paranormal Belief, Dissociative Tendencies and Parental Encouragement of Imagination in Childhood," *Journal of the American Society for Psychical Research* 93, no. 3 (July 1999): 233–247; Ronald J. Pekala, V. K. Kumar, and Geddes Marcano, "Anomalous/Paranormal Experiences, Hypnotic Susceptibility, and Dissociation," *Journal of the American Society for Psychical Research* 89, no. 4 (1995): 313–332; Shelley L. Rattet and Krisanne Bursik, "Investigating the Personality Correlates of Paranormal Belief and Precognitive Experience," *Personality and Individual Differences* 31, no. 3 (August 2001): 433–444; Uwe Wolfradt, "Dissociative Experiences, Trait Anxiety and Paranormal Beliefs," *Personality and Individual Differences* 23, no. 1 (July 1997): 15–19. On paranormal experiences, see Pekala et al., "Anomalous/Paranormal Experiences"; Douglas G. Richards, "A Study of the Correlations between Subjective Psychic Experiences and Dissociative Experiences," *Dissociation* 4, no. 2 (June 1991): 83–91; Colin A. Ross and Shaun Joshi, "Paranormal Experiences in the General Population," *Journal of Nervous and Mental Disease* 180, no. 6 (June 1992): 357–361; Colin

A. Ross, Lynne Ryan, Harrison Voigt, and Lyle Eide, "High and Low Dissociators in a College Student Population," *Dissociation* 4, no. 3 (September 1991): 147–151. On alien contact, see Susan Marie Powers, "Dissociation in Alleged Extraterrestrial Abductees," *Dissociation* 7, no. 1 (March 1994): 44–50; French et al., "Psychological Aspects."

52. Stephen Jay Lynn, Judith Pintar, and Judith W. Rhue, "Fantasy Proneness, Dissociation, and Narrative Construction," in *Broken Images, Broken Selves: Dissociative Narratives in Clinical Practice*, ed. Stanley Krippner and Susan Marie Powers (Washington, DC: Bruner/Mazel, 1997), 274–302.

53. Krissy Wilson and Christopher C. French, "The Relationship between Susceptibility to False Memories, Dissociativity, and Paranormal Belief and Experience," *Personality and Individual Differences* 41, no. 8 (December 2006): 1493–1502.

54. Neil Dagnall, Andrew Parker, and Gary Munley, "News Events, False Memory and Paranormal Belief," *European Journal of Parapsychology* 22 (September 2008): 173–188.

55. Sheryl C. Wilson and Theodore X. Barber, "The Fantasy-Prone Personality: Implications for Understanding Imagery, Hypnosis, and Parapsychological Phenomena," in *Imagery: Current Theory, Research and Application*, ed. Anees A. Sheikh (New York: Wiley, 1983), 340–387.

56. Steven Jay Lynn and Judith W. Rhue, "Fantasy Proneness: Hypnosis, Developmental Antecedents, and Psychopathology," *American Psychologist* 43, no. 1 (January 1988): 35–44.

57. On paranormal belief, see Heather R. Auton, Jacqueline Pope, and Gus Seeger, "It Isn't That Strange: Paranormal Belief and Personality Traits," *Social Behavior and Personality* 31, no. 7 (November 2003): 711–720; Kathryn Gow, Adam Lane, and David Chant, "Personality Characteristics, Beliefs and the Near-Death Experience," *Australian Journal of Clinical and Experimental Hypnosis* 31, no. 2 (2003): 128–152; Kathryn Gow, Tracey Lang, and David Chant, "Fantasy Proneness, Paranormal Beliefs and Personality Features in Out-of-Body Experiences," *Contemporary Hypnosis* 21, no. 3 (September 2004): 107–125; Harvey J. Irwin, "Fantasy-Proneness and Paranormal Beliefs," *Psychological Reports* 66, no. 2 (April 1990): 655–658; Harvey J. Irwin, "A Study of Paranormal Belief, Psychological Adjustment and Fantasy-Proneness," *Journal of the American Society for Psychical Research* 85 (1991): 317–331; Tony Lawrence, Claire Edwards, Nicholas Barraclough, Sarah Church, and Francesca Hetherington, "Belief and Experience: Childhood Trauma and Childhood Fantasy," *Personality and Individual Differences* 19, no. 2 (August 1995): 209–215; Paul Rogers, Pamela Qualter, and Gemma Phelps, "The Mediating and Moderating Effects of Loneliness and Attachment Style on Belief in the Paranormal," *European Journal of Parapsychology* 22, no. 2 (2007): 138–165. On susceptibility to paranormal experiences, see Gow, Lane, and Chant, "Personality Characteristics"; Gow, Lang, and Chant, "Fantasy Proneness"; S. A. Myers, H. R. Austrin, J. Grisso, and R. Nickeson, "Personality Characteristics as Related to Out-of-Body Experiences," *Journal of Parapsychology* 47 (1983): 131–144.

58. Bartholomew, Basterfield, and Howard, "UFO Abductees and Contactees." See also Robert E. Bartholomew and George S. Howard, *UFOs and Alien Contact: Two Centuries of Mystery* (Amherst, NY: Prometheus Books, 1998).

59. Joe Nickell, "A Study of Fantasy Proneness in the Thirteen Cases of Alleged Encounters in John Mack's *Abduction*," *Skeptical Inquirer* 20, no. 3 (May/June 1996): 18–20.

60. S. A. Myers, "The Wilson-Barber Inventory of Childhood Memories and Imaginings: Children's Form and Norms for 1337 Children and Adolescents," *Journal of Mental Imagery* 7, no. 1 (1983): 83–94; Rodeghier et al., "Psychosocial Characteristics"; Spanos et al., "Close Encounters."

61. French et al., "Psychological Aspects."

62. Hough and Rogers, "Individuals who Report Being Abducted by Aliens"; Harald Merckelbach, Robert Horselenberg, and Peter Muris, "The Creative Experiences Questionnaire: A Brief Self-Report Measure of Fantasy Proneness," *Personality and Individual Differences* 31, no. 6 (October 2001): 987–995.

63. Kenneth Ring and Christopher J. Rosing, "The Omega Project: A Psychological Survey of Persons Reporting Abductions and Other UFO Encounters," *Journal of UFO Studies* 2 (1990): 59–98.

64. Mack, *Abduction*; Ring and Rosing, "The Omega Project."

65. Keith Basterfield, "Paranormal Aspects of the UFO Phenomenon: 1975–1999," *Australian Journal of Parapsychology* 1 (2001): 30–55; Thomas E. Bullard, *UFO Abductions: The Measure of a Mystery* (Mount Rainier, MD: Fund for UFO Research, 1987); Ann Druffel and D. Scott Rogo, *The Tujunga Canyon Contacts* (Eaglewood Cliffs, NJ: Prentice Hall, 1980); Hilary Evans, *The Evidence for UFOs* (Wellingborough: Aquarian, 1983); Hilary Evans, *From Other Worlds: The Truth About Aliens, Abductions, UFOs and the Paranormal* (London: Carlton, 1998); Mack, *Abduction*; Jenny Randles, *Abduction* (London: Robert Hale, 1988); Berthold Eric Schwarz, *UFO Dynamics: Psychiatric and Psychic Aspects of the UFO Syndrome* (Moore Haven, FL: Rainbow Books, 1983); John Spencer, *Gifts of the Gods? UFOs, Alien Visitors or Psychic Phenomena* (London: Virgin, 1994); Jacques Vallee, *UFOs: The Psychic Solution* (St. Albans: Panther, 1977); Keith Basterfield and Michael A. Thalbourne, "Belief in, and Alleged Experience of, the Paranormal in Ostensible UFO Abductees," *Australian Journal of Parapsychology* 2 (2002): 2–18; French et al., "Psychological Aspects."

66. Clancy et al., "Memory Distortion."

67. Thomas E. Bullard, "Hypnosis and UFO Abductions: A Troubled Relationship," *Journal of UFO Studies* 1 (1989): 3–40.

68. Simons and Chabris, "What People Believe"; see also Simons and Chabris, "Common (Mis)beliefs."

69. Mark R. Kebbell and Graham F. Wagstaff, "Hypnotic Interviewing: The Best Way to Interview Eyewitnesses?," *Behavioral Sciences and the Law* 16, no. 1 (January 1998): 115–129; Graham F. Wagstaff, "Forensic Aspects of Hypnosis," in *Hypnosis: The Cognitive-Behavioral Perspective*, ed. Nicholas P. Spanos and John F. Chaves (Buffalo, NY: Prometheus Books, 1989): 340–357.

70. Loftus and Ketcham, *The Myth of Repressed Memory*; McNally, *Remembering Trauma*.

71. Andrew M. Colman, *Facts, Fallacies and Frauds in Psychology* (London: Hutchinson, 1987).

72. Michael D. Yapko, "Suggestibility and Repressed Memories of Abuse: A Survey of Psychotherapists' Beliefs," 36, no. 3 (January 1994): 163–171; see also Michael D. Yapko, *Suggestions of Abuse: True and False Memories of Childhood Sexual Trauma* (New York: Simon & Schuster, 1994).

73. Yapko, "Suggestibility and Repressed Memories of Abuse," 163.

74. James Ost, Simon Easton, Lorraine Hope, Christopher C. French, and Daniel B. Wright, "Latent Variables Underlying the Memory Beliefs of Chartered Clinical Psychologists, Hypnotherapists and Undergraduate Students," *Memory* 25, no. 1 (January 2017): 57–68.

75. As described by Peter Brookesmith, *Alien Abductions* (London: Barnes and Noble, 1998), 96–97.

76. Alvin H. Lawson, "Perinatal Imagery in UFO Abduction Reports," *Journal of Psychohistory* 12, no. 2 (Fall 1984): 211–239.

77. Henry Otgaar, Ingrid Candel, Harald Merckelbach, and Kimberley A. Wade, "Abducted by a UFO: Prevalence Information Affects Young Children's False Memories for an Implausible Event," *Applied Cognitive Psychology* 23, no. 1 (January 2009): 115–125.

78. Clancy et al., "Memory Distortion."

79. Budd Hopkins, David M. Jacobs, and Ron Westrum, *Unusual Personal Experiences: An Analysis of the Data From Three National Surveys Conducted by the Roper Organization* (Las Vegas: Bigelow Holding Company, 1992).

80. Budd Hopkins, David M. Jacobs, and Ron Westrum, *Unusual Personal Experiences*, 56–57.

81. Goodreads, "Arthur C. Clarke," accessed June 3, 2023, https://www.goodreads.com/quotes/157737-i-m-sure-the-universe-is-full-of-intelligent-life-it-s.

82. Goodreads, "Arthur C. Clarke," accessed June 3, 2023, https://www.goodreads.com/quotes/41383-two-possibilities-exist-either-we-are-alone-in-the-universe.

83. Goodreads, "Ellen DeGeneres," accessed June 3, 2023, https://www.goodreads.com/quotes/3086-the-only-thing-that-scares-me-more-than-space-aliens.

CHAPTER 6

1. For a more detailed account, see Christopher C. French, "Reincarnation Claims," in *Parapsychology: The Science of Unusual Experience*, 2nd ed., ed. David Groome and Ron Roberts (London: Routledge, 2017), 82–95.

2. I find the concept of reincarnation so fascinating that I decided to check out the Reincarnation Society. The membership fee was no less than $1,200! I dithered for a while but then decided that I would take the plunge and join—after all, you only live once! (NB: This is a corny dad joke shamelessly stolen from science writer Simon Singh. It is not a fact.)

3. Michael A. Thalbourne, *A Glossary of Terms Used in Parapsychology* (Charlottesville, VA: Puente Publications, 2003), 107.

4. Erlendur Haraldsson, "Popular Psychology, Belief in Life After Death and Reincarnation in the Nordic Countries, Western and Eastern Europe," *Nordic Psychology* 58, no. 2 (July 2006): 171–180.

5. Data collected 1999–2000 except for Norway (around 1990).

6. Data collected 1999–2002 except for United Kingdom and Switzerland (1990–1993).

7. Data collected 1999–2002.

8. See, for example, Brian Butterworth, "What Makes a Prodigy?," *Nature Neuroscience* 4, no. 1 (January 2001): 11–12; Dean Keith Simonton, "Creative Geniuses, Polymaths, Child Prodigies, and Autistic Savants: The Ambivalent Function of Interests and Obsessions," in *The Science of Interest*, ed. Paul A. O'Keefe and Judith M. Harackiewicz (Cham, Switzerland: Springer, 2017): 175–185; Alan S. Brown, *The Déjà Vu Experience* (New York: Psychology Press, 2004).

9. For example, Brian L. Weiss, *Many Lives, Many Masters: The True Story of a Prominent Psychiatrist, His Young Patient, and the Past-Life Therapy That Changed Both Their Lives* (London: Piatkus Books, 1988); Brian L. Weiss, *Through Time into Healing: How Past Life Regression Therapy Can Heal Mind, Body, and Soul* (London: Piatkus Books, 1992).

10. John C. Norcross, Gerald P. Koocher, and Ariele Garofalo, "Discredited Psychological Treatments and Tests," *Professional Psychology: Research and Practice* 37, no. 5 (October 2006): 515–522.

11. Robert A. Baker, *They Call It Hypnosis* (Buffalo, NY: Prometheus Books, 1990); Robert A. Baker, *Hidden Memories: Voices and Visions from Within* (Buffalo, NY: Prometheus Books, 1992); French, "Fantastic Memories"; Nicholas P. Spanos, *Multiple Identities and False Memories: A Sociocognitive Perspective* (Washington, DC: American Psychological Association, 1996).

12. Morey Bernstein, *The Search for Bridey Murphy* (Garden City, NY: Doubleday, 1956).

13. Martin Gardner, *Fads and Fallacies in the Name of Science* (New York: Dover, 1957); Melvin Harris, *Sorry, You've Been Duped! The Truth Behind Classic Mysteries of the Paranormal* (London: Weidenfeld & Nicolson, 1986).

14. Jeffrey Iverson, *More Lives Than One?* (London: Pan Books, 1977).

15. Harris, *Sorry, You've Been Duped!*

16. Joan Evans, *Life in Medieval France* (London: Phaidon, 1957).

17. Spanos, *Multiple Identities and False Memories*; N. P. Spanos, C. A. Burgess, and M. F. Burgess, "Past-Life Identities, UFO Abductions, and Satanic Ritual Abuse: The Social Construction of Memories," *International Journal of Clinical and Experimental Hypnosis* 42, no. 4 (October 1994): 433–446; Nicholas P. Spanos, Evelyn Menary, Natalie J. Gabora, Susan C. DuBreuil,

and Bridget Dewhirst, "Secondary Identity Enactments During Hypnotic Past-Life Regression: A Sociocognitive Perspective," *Journal of Personality and Social Psychology* 61, no. 2 (August 1991): 308–320.

18. Cynthia A. Meyersburg, Ryan Bogdan, David A. Gallo, and Richard J. McNally, "False Memory Propensity in People Reporting Recovered Memories of Past Lives," *Journal of Abnormal Psychology* 118, no. 2 (May 2009): 399–404.

19. Cynthia A. Meyersburg and Richard J. McNally, "Reduced Death Distress and Greater Meaning in Life among Individuals Reporting Past Life Memory," *Personality and Individual Differences* 50, no. 8 (2011): 1218–1221.

20. Ian Stevenson, "A Case of the Psychotherapist's Fallacy: Hypnotic Regression to 'Previous Lives,'" *American Journal of Clinical Hypnosis* 36, no. 3 (January 1994): 188–193.

21. For an alternative account that is more sympathetic to the possibility that reincarnation genuinely occurs, see Roy Stemmen, "Lebanese Research among the Druse," *Reincarnation International* (June 1998): 11–21.

22. For further critical discussion of the concept of reincarnation, see Paul Edwards, *Reincarnation: A Critical Examination* (Amherst, NY: Prometheus Books, 1996).

23. For more on Druse history and culture, see Nejla M. Abu-Izzeddin, *The Druzes: A New Study of Their History, Faith and Society* (Leiden: Brill, 1984); Robert Brenton Betts, *The Druze*, rev. ed. (New Haven, CT: Yale University Press, 2009); Kais M. Firro, *A History of the Druze* (Leiden: Brill, 1992); Sami Nasib Makarem, *The Druze Faith* (Delmar, NY: Caravan Books, 1974).

24. Cf. Marwan Dwairy, "The Psychosocial Function of Reincarnation among Druze in Israel," *Culture, Medicine and Psychiatry* 30, no. 1 (May 2006): 29–53.

25. Anne Bennett, "Reincarnation, Sect Unity, and Identity among the Druze," *Ethnology* 45, no. 2 (Spring 2006): 87–104; Dwairy, "The Psychosocial Function"; Roland Littlewood, "Social Institutions and Psychological Explanations: Druze Reincarnation as a Therapeutic Resource," *British Journal of Medical Psychology* 74, no. 2 (June 2001): 213–222; Eli Somer, Carmit Klein-Sela, and Keren Or-Chen, "Beliefs in Reincarnation and the Power of Fate and Their Association with Emotional Outcomes among Bereaved Parents of Fallen Soldiers," *Journal of Loss and Trauma* 16, no. 5 (September 2011): 459–475; Ian Wilson, *The After Death Experience* (London: Corgi, 1987).

CHAPTER 7

1. From Harvey J. Irwin and Caroline A. Watt, *An Introduction to Parapsychology*, 5th ed. (Jefferson, NC: McFarland, 2007), 157–158.

2. French and Stone, *Anomalistic Psychology*, 277–278.

3. Raymond A. Moody, *Life After Life* (Covington, GA: Mockingbird, 1975).

4. For example, Raymond A. Moody, *Reflections on Life After Life* (Covington, GA: Mockingbird, 1977).

5. The interested reader may like to consult the following publications for more detailed discussion, from a variety of perspectives, regarding the true nature of NDEs: Lee W. Bailey and Jenny Yates, eds., *The Near-Death Experience: A Reader* (New York: Routledge, 1996); Susan Blackmore, *Dying to Live: Science and the Near-Death Experience* (London: Grafton, 1993); Susan J. Blackmore, *Seeing Myself: The New Science of Out-of-Body Experiences* (London: Robinson, 2017); Ornella Corazza, *Near-Death Experiences: Exploring the Mind-Body Connection* (London: Routledge, 2008); Peter Fenwick and Elizabeth Fenwick, *The Truth in the Light: An Investigation of Over 300 Near-Death Experiences* (London: Headline, 1995); Mark Fox, *Religion, Spirituality and the Near-Death Experience* (London: Routledge, 2003); Christopher C. French, "Near-Death Experiences in Cardiac Arrest Survivors," *Progress in Brain Research* 150 (January 2005): 351–367; Bruce Greyson, "Near-Death Experiences," in *Varieties of Anomalous Experience: Examining the Scientific Evidence*, 2nd ed., ed. Etzel Cardeña, Steven Jay Lynn, and Stanley Krippner (Washington, DC: American Psychological Association: 2014), 333–367; Bruce Greyson, *After: A Doctor Explores What Near-Death Experiences Reveal About Life and Beyond* (London: Bantam, 2021); Harvey J. Irwin and Caroline A. Watt, *An Introduction to Parapsychology*; Craig D. Murray, ed., *Psychological Scientific Perspectives on Out-of-Body and Near-Death Experiences* (New York: Nova Science Inc., 2009); Melvyn Morse, *Closer to the Light* (London: Souvenir, 1990); Melvyn Morse, *Transformed by the Light: The Powerful Effect of Near-Death Experiences on People's Lives* (New York: Ballantine, 1992); Sam Parnia, *What Happens When We Die: A Ground-Breaking Study into the Nature of Life and Death*, 2nd ed. (London: Hay House, 2008); Michael B. Sabom, *Recollections of Death: A Medical Investigation* (London: Corgi, 1982); Michael B. Sabom, *Light and Death: One Doctor's Fascinating Account of Near-Death Experiences* (Grand Rapids, MI: Zondervan, 1998); G. M. Woerlee, *Mortal Minds: A Biology of the Soul and the Dying Experience* (Utrecht: de Tijdstroom, 2003).

6. These elements are ineffability; hearing oneself pronounced dead; feelings of peace and quiet; hearing unusual noises; seeing a dark tunnel; being "out of the body"; meeting "spiritual beings"; experiencing a bright light as a "being of light"; panoramic life review; experiencing a realm in which all knowledge exists; experiencing cities of light; experiencing a realm of bewildered spirits; experiencing a "supernatural rescue"; sensing a border or limit; coming back into the body.

7. Kenneth Ring, *Life at Death: A Scientific Investigation of the Near-Death Experience* (New York: Coward, McCann, and Geoghegan, 1980).

8. Bruce Greyson and Nancy Evans Bush, "Distressing Near-Death Experiences," *Psychiatry: Interpersonal and Biological Processes* 55, no. 1 (March 1992): 95–110.

9. Bruce Greyson, "The Near-Death Experience Scale: Construction, Reliability, and Validity," *Journal of Nervous and Mental Disease* 171, no. 6 (June 1983): 369–375.

10. George Gallup Jr., with William Proctor, *Adventures in Immortality: A Look Beyond the Threshold of Death* (New York: McGraw-Hill, 1982).

11. Pim van Lommel, Ruud van Wees, Vincent Meyers, and Ingrid Elfferich, "Near-Death Experience in Survivors of Cardiac Arrest: A Prospective Study in the Netherlands," *The Lancet* 358, no. 9298 (December 15, 2001): 2039–2045; Sam Parnia, D. G. Waller, R. Yeates, and Peter Fenwick, "A Qualitative and Quantitative Study of the Incidence, Features and Aetiology of Near Death Experiences in Cardiac Arrest Survivors," *Resuscitation* 48, no. 2 (February 2001): 149–156; Janet Schwaninger, Paul R. Eisenberg, Kenneth B. Schectman, and Alan N. Weiss, "A Prospective Analysis of Near-Death Experiences in Cardiac Arrest Patients," *Journal of Near-Death Studies* 20, no. 4 (June 2002): 215–232; Bruce Greyson, "Incidence and Correlates of Near-Death Experiences in a Cardiac Care Unit," *General Hospital Psychiatry* 25, no. 4 (July–August 2003): 269–276.

12. Penny Sartori, "The Incidence and Phenomenology of Near-Death Experiences," *Network Review (Scientific and Medical Network)* 90 (2006): 23–25; Zalika Klemenc-Ketis, Janko Kersnik, and Stefek Grmec, "The Effect of Carbon Dioxide on Near-Death Experiences in Out-of-Hospital Cardiac Arrest Survivors: A Prospective Observational Study," *Critical Care* 14, no. 2 (April 8, 2010): R56; Sam Parnia, Ken Spearpoint, Gabriele de Vos, Peter Fenwick, Diana Goldberg, Jie Yang, et al. "AWARE—AWAreness during REsuscitation—A Prospective Study," *Resuscitation* 85, no. 12 (2014): 1799–1805.

13. Christopher C. French, "Dying to Know the Truth: Visions of a Dying Brain, or False Memories?," *The Lancet* 358, no. 9298 (December 15, 2001): 2010–2011.

14. Bruce Greyson, "Consistency of Near-Death Experience Accounts over Two Decades: Are Reports Embellished over Time?," *Resuscitation* 73, no. 3 (June 2007): 407–411; Charlotte Martial, Vanessa Charland-Verville, Héléna Cassol, Vincent Didone, Martial Van Der Linden, and Steven Laureys, "Intensity and Memory Characteristics of Near-Death Experiences," *Consciousness and Cognition* 56 (November 2017): 120–127; Lauren E. Moore and Bruce Greyson, "Characteristics of Memories for Near-Death Experiences," *Consciousness and Cognition* 51 (May 2017): 116–124; Marie Thonnard, Vanessa Charland-Verville, Serge Brédart, Hedwige Dehon, Didier Ledoux, Steven Laureys, and Audray Vanhaudenhuyse, "Characteristics of Near-Death Experiences Memories as Compared to Real and Imagined Events Memories," *PLOS ONE* 8, no. 3 (March 27, 2013): e57620.

15. Chris Roe, "Near-Death Experiences," in *Parapsychology: The Science of Unusual Experiences*, 2nd ed., ed. David Groome and Ron Roberts (London: Routledge, 2017), 65–81.

16. Stanislav Grof and Joan Halifax, *The Human Encounter with Death* (New York: Dutton, 1977); Carl Sagan, *Broca's Brain: Reflections on the Romance of Science* (New York: Random House, 1979).

17. Susan J. Blackmore, "Birth and the OBE: An Unhelpful Analogy," *Journal of the American Society for Psychical Research* 77, no. 3 (1983): 229–238.

18. Russell Noyes Jr. and Roy Kletti, "Depersonalization in Response to Life-Threatening Danger," *Comprehensive Psychiatry* 18, no. 4 (July–August 1977): 375–384.

19. I. R. Judson and E. Wiltshaw, "A Near-Death Experience," *The Lancet* 322, no. 8349 (1983): 561–562.

20. Blackmore, *Seeing Myself*, 250.

21. Judson and Wiltshaw, "A Near-Death Experience," 561.

22. Melvin L. Morse, David Venecia, and Jerrold Milstein, "Near-Death Experiences: A Neurophysiologic Explanatory Model," *Journal of Near-Death Studies* 8, no. 1 (January 1989): 45–53.

23. Karl L. R. Jansen, "Near-Death Experience and the NMDA Receptor," *British Medical Journal* 298 (1989): 1708–1709; Karl L. R. Jansen, "The Ketamine Model of the Near-Death Experience: A Central Role for the N-Methyl-D-Aspartate Receptor," *Journal of Near-Death Studies* 16, no. 1 (January 1997): 79–95; Karl L. R. Jansen, *Ketamine: Dreams Realities* (Sarasota, FL: Multidisciplinary Association for Psychedelic Studies, 2001).

24. R. Strassman, "Endogenous Ketamine-Like Compounds and the NDE: If So, So What?," *Journal of Near-Death Studies* 16, no. 1 (January 1997): 27–41; Peter Fenwick, "Is the Near Death Experience Only N-Methyl-D-Aspartate Blocking?," *Journal of Near-Death Studies* 16, no. 1 (January 1997): 43–53.

25. James E. Whinnery, "Psychophysiologic Correlates of Unconsciousness and Near-Death Experiences," *Journal of Near-Death Studies* 15, no. 4 (Summer 1997): 231–258.

26. Ladislas Joseph Meduna, *Carbon Dioxide Therapy: A Neurophysiological Treatment of Nervous Disorders* (Springfield, IL: Charles C. Thomas, 1950).

27. Blackmore, *Dying to Live*; Blackmore, *Seeing Myself*.

28. Keith Augustine, "Psychophysiological and Cultural Correlates Undermining a Survivalist Interpretation of Near-Death Experiences," *Journal of Near-Death Studies* 26, no. 2 (Winter 2007): 89–125.

29. Orrin Devinsky, Edward Feldmann, Kelly Burrowes, and Edward Bromfield, "Autoscopic Phenomena with Seizures," *Archives of Neurology* 46, no. 10 (October 1989): 1080–1088; P. Vuilleumier, P. A. Despland, G. Assal, and F. Regli, "Voyages astraux et hors du corps: héautoscopie, extase et hallucinations expérientelles d'origine épileptique [Astral and Out-of-Body Voyages: Heautoscopy, Ecstasis and Experimental Hallucinations of Epileptic Origin]," *Revue Neurologique* 153 no. 2 (March 1997): 115–119; Blackmore, *Dying to Live*, 206.

30. Aaen-Stockdale, "Neuroscience for the Soul."

31. On questioning the link between temporal lobe epilepsy and mystical experiences, see Aaen-Stockdale, "Neuroscience for the Soul"; Bruce Greyson, Donna K. Broshek, Lori L. Derr, and Nathan B. Fountain, "Mystical Experiences Associated with Seizures," *Religion, Brain & Behavior* 5, no. 3 (2015): 182–196.

32. Susan J. Blackmore, "Out-of-Body Experiences," in *The Encyclopedia of the Paranormal*, ed. Gordon Stein (Amherst, NY: Prometheus Books, 1996), 471–483.

33. For a more detailed critical assessment of such cases, see Keith Augustine, "Does Paranormal Perception Occur in Near-Death Experiences?," *Journal of Near-Death Studies* 25 (2007): 203–236.

34. Kimberley Clark, "Clinical Interventions with Near-Death Experiencers," in *The Near-Death Experience: Problems, Prospects, Perspectives*, ed. Bruce Greyson and Charles P. Flynn (Springfield, IL: Charles C. Thomas), 242–255.

35. Hayden Ebbern, Sean Mulligan, and Barry L. Beyerstein, "Maria's Near Death Experience: Waiting for the Other Shoe to Drop," *Skeptical Inquirer* 20, no. 4 (July–August 1996): 27–33.

36. Sabom, *Light and Death*.

37. G. M. Woerlee, "An Anaesthesiologist Examines the Pam Reynolds Story. Part 1. Background Considerations," *The Skeptic* 18, no. 1 (Spring 2005): 14–17; G. M. Woerlee, "An Anaesthesiologist Examines the Pam Reynolds Story. Part 2. The Experience," *The Skeptic* 18, no. 2 (Summer 2005): 16–20.

38. G. M. Woerlee, "Could Pam Reynolds Hear? A New Investigation into the Possibility of Hearing during This Famous Near-Death Experience," *Journal of Near-Death Studies* 30, no. 1 (Fall 2011): 3–25. (The speakers, fitted into earplugs, are used to present clicking sounds during the operation thus allowing brainstem function to be assessed via vestibular evoked potentials.)

39. Van Lommel et al., "Near-Death Experience in Survivors of Cardiac Arrest," 2044.

40. Blackmore, *Seeing Myself*.

41. Van Lommel et al., "Near-Death Experience in Survivors of Cardiac Arrest," 2041.

42. Rudolf H. Smit, "Corroboration of the Dentures Anecdote Involving Veridical Perception in a Near-Death Experience," *Journal of Near-Death Studies* 27, no. 1 (Fall 2008): 47–61.

43. Van Lommel et al., "Near-Death Experience in Survivors of Cardiac Arrest," 2044.

44. Larry Dossey, *Recovering the Soul: A Scientific and Spiritual Search* (New York: Bantam, 1989), 18.

45. NDE hoaxes, although rare, do occur. For example, in 2015 Alex Malarkey admitted that the story told in the bestselling book *The Boy Who Came Back From Heaven*, published in 2010, was untrue. Tragically, Alex had been left paralyzed as a young child following a car accident. The book, written by his father Kevin but with Alex listed as a coauthor, described his journey to heaven, including descriptions of angels and of meeting Jesus. As Alex subsequently admitted on his blog at the age of sixteen, "I said I went to heaven because I thought it would get me attention. I did not die. I did not go to heaven." For further details, see Michelle Dean, "The Boy Who Didn't Come Back from Heaven: Inside a Bestseller's 'Deception,'" *Guardian*, January 21, 2015, https://www.theguardian.com/books/2015/jan/21/boy-who-came-back-from-heaven-alex-malarkey.

46. Keith Augustine, "Near-Death Experiences with Hallucinatory Features," *Journal of Near-Death Studies* 26, no. 1 (Fall 2007): 3–31.

47. Parnia et al., "AWARE—AWAreness during REsuscitation."

48. Greyson, "Incidence and Correlates of Near-Death Experiences," 275.

49. Van Lommel et al., "Near-Death Experience in Survivors of Cardiac Arrest," 2044.

50. Sam Parnia and Peter Fenwick, "Near Death Experiences in Cardiac Arrest: Visions of a Dying Brain or Visions of a New Science of Consciousness," *Resuscitation* 52, no. 1 (January 2002): 5–11.

51. French, "Dying to Know the Truth"; French, "Near-Death Experiences in Cardiac Arrest Survivors"; Christopher C. French, "Near-Death Experiences and the Brain," in *Psychological Scientific Perspectives on Out-of-Body and Near-Death Experiences*, ed. Craig D. Murray (New York: Nova Science, 2009), 187–203.

52. Lahmir S. Chawla, Seth Akst, Christopher Junker, Barbara Jacobs, and Michael G. Seneff, "Surges of Electroencephalogram Activity at the Time of Death: A Case Series," *Journal of Palliative Medicine* 12, no. 12 (October 2009): 1095–1100.

53. Jimo Borjigin, UnCheol Lee, Tiecheng Liu, Dinesh Pal, Sean Huff, Daniel Klarr et al., "Surge of Neurophysiological Coherence and Connectivity in the Dying Brain," *Proceedings of the National Academy of Sciences* 110, no. 35 (August 27, 2013): 14432–14437.

54. Jason J. Braithwaite, "Towards a Cognitive Neuroscience of the Dying Brain," *The Skeptic* 21, no. 2 (Summer 2008): 8–16.

55. Blackmore, *Seeing Myself*.

56. Olaf Blanke, Theodore Landis, Laurent Spinelli, and Margitta Seeck, "Out-of-Body Experience and Autoscopy of Neurological Origin," *Brain* 127, Part 2 (February 2004): 243–258; Silvio Ionta, Lukas Heydrich, Bigna Lenggenhager, Michael Mouthon, Eleonora Fornari, Dominique Chapuis et al., "Multisensory Mechanisms in Temporo-Parietal Cortex Support Self-Location and First-Person Perspective," *Neuron* 70, no. 2 (April 28, 2011): 363–374.

57. Wilder Penfield, "The Role of the Temporal Cortex in Certain Psychical Phenomena," *Journal of Mental Science* 101, no. 424 (July 1955): 451–465; F. Tong, "Out-of-Body Experiences: From Penfield to Present," *Trends in Cognitive Sciences* 7, no. 3 (March 2003): 104–106.

58. Olaf Blanke, Stéphanie Ortigue, Theodore Landis, and Margitta Seeck, "Stimulating Illusory Own-Body Perceptions," *Nature* 419, no. 6904 (September 19, 2002): 269–270.

59. Dirk De Ridder, Koen Van Laere, Patrick Dupont, Tomas Menovsky, and Paul Van de Heyning, "Visualizing Out-of-Body Experience in the Brain," *New England Journal of Medicine* 357, no. 18 (November 2007): 1829–1833.

60. Lukas Heydrich, Christophe Lopez, Margitta Seeck, and Olaf Blanke, "Partial and Full Own-Body Illusions of Epileptic Origin in a Child with Right Temporoparietal Epilepsy," *Epilepsy & Behavior* 20, no. 3 (February 2011): 583–586.

61. Matthew Botvinick and Jonathan Cohen, "Rubber Hands 'Feel' Touch That Eyes See," *Nature* 391, no. 6669 (February 19, 1998): 756.

62. Bigna Lenggenhager, Tej Tadi, Thomas Metzinger, and Olaf Blanke, "Video Ergo Sum: Manipulating Bodily Self-Consciousness," *Science* 317, no. 5841 (August 24, 2007): 1096–1099.

63. H. Henrik Ehrsson, "The Experimental Induction of Out-of-Body Experiences," *Science* 317, no. 5841 (August 24, 2007): 1048.

64. Bigna Lenggenhager, Michael Mouthon, and Olaf Blanke, "Spatial Aspects of Bodily Self-Consciousness," *Consciousness and Cognition* 18, no. 1 (March 2009): 110–117.

65. Valeria I. Petkova and H. Henrik Ehrsson, "If I Were You: Perceptual Illusion of Body Swapping," *PLOS ONE* 3, no. 12 (December 3, 2008): e3832.

CHAPTER 8

1. Christopher C. French, "Factors Underlying Belief in the Paranormal: Do Sheep and Goats Think Differently?," *The Psychologist* 5, no. 7 (July 1992): 296, 298.

2. Christopher C. French and Krissy Wilson, "Cognitive Factors Underlying Paranormal Beliefs and Experiences," in *Tall Tales About the Mind and Brain: Separating Fact from Fiction*, ed. Sergio Della Sala (Oxford: Oxford University Press, 2007), 3–22.

3. I would refer interested readers who would like to delve deeper into this area to the following publications: James E. Alcock, *Belief: What It Means to Believe and Why Our Convictions Are So Compelling* (Amherst, NY: Prometheus Books, 2018); David Groome, Michael Eysenck, and Robin Law, *The Psychology of the Paranormal* (London: Routledge, 2019); Christopher C. French, "The Psychology of Belief and Disbelief in the Paranormal," in *Extrasensory Perception: Support, Skepticism, and Science, Vol. 1: History, Controversy, and Research*, ed. Edwin C. May and Sonali Bhatt Marwaha (Santa Barbara, CA: Praeger, 2015), 129–151; French and Stone, *Anomalistic Psychology*; Terence Hines, *Pseudoscience and the Paranormal*, 2nd ed. (Amherst, NY: Prometheus Books, 2003); Bruce Hood, *Supersense: Why We Believe in the Unbelievable* (New York: HarperCollins, 2009); Matthew Hutson, *The 7 Laws of Magical Thinking: How Irrationality Makes Us Happy, Healthy, and Sane* (Oxford: Oneworld Publications, 2012); Harvey J. Irwin, *The Psychology of Paranormal Belief: A Researcher's Handbook* (Hatfield: University of Hertfordshire Press, 2009); Marks, *The Psychology of the Psychic*; David F. Marks, *Psychology and the Paranormal* (Los Angeles: Sage, 2020); Michael Shermer, *Why People Believe Weird Things: Pseudoscience, Superstition, and Other Confusions of Our Time*, rev. ed. (New York: Owl Books, 2002); Michael Shermer, *The Believing Brain: From Ghosts and Gods to Politics and Conspiracies—How We Construct Beliefs and Reinforce Them as Truths* (New York: Times Books, 2011); Stuart Vyse, *Believing in Magic: The Psychology of Superstition*, updated ed. (Oxford: Oxford University Press, 2014); Richard Wiseman, *Paranormality: Why We See What Isn't There* (London: Macmillan, 2011).

4. With the exception of my sister-in-law's name, I have changed the names and dates in the interests of confidentiality.

5. David Hand, *The Improbability Principle: Why Coincidences, Miracles and Rare Events Happen Every Day* (London: Bantam, 2014).

6. Simon Hoggart and Mike Hutchinson, *Bizarre Beliefs* (London: Richard Cohen Books, 1995).

7. Ruma Falk, "Judgment of Coincidences: Mine versus Yours," *American Journal of Psychology* 102, no. 4 (Winter 1989): 477–493.

8. Louis Pauwels and Jacques Bergier, *Morning of the Magicians* (Vermont: Destiny Books, 1960).

9. Martin Plimmer and Brian King, *Beyond Coincidence: Stories of Amazing Coincidences and the Mystery and Mathematics That Lie Behind Them* (Cambridge: Icon Books, 2004).

10. David Mikkelson, "Laura Buxton Balloon Coincidence," *Snopes*, April 29, 2013, https://www.snopes.com/fact-check/whether-balloon.

11. Hand, *The Improbability Principle*.

12. Hand, *The Improbability Principle*.

13. For examples of mathematicians, see Persi Diaconis and Frederick Mosteller, "Methods for Studying Coincidences," *Journal of the American Statistical Association* 84, no. 408 (December 1989): 853–861; Hand, *The Improbability Principle*; John Allen Paulos, *Innumeracy: Mathematical Illiteracy and Its Consequences* (London: Penguin, 1988). For examples of psychologists, Mark K. Johansen and Magda Osman, "Coincidences: A Fundamental Consequence of Rational Cognition," *New Ideas in Psychology* 39 (October 2015): 34–44; Mark K. Johansen and Magda Osman, "Coincidence Judgment in Causal Reasoning: How Coincidental Is This?," *Cognitive Psychology* 120 (August 2020): 101290; Michiel van Elk, Karl Friston, and Harold Bekkering, "The Experience of Coincidence: An Integrated Psychological and Neurocognitive Perspective," in *The Challenge of Chance: A Multidisciplinary Approach from Science and the Humanities*, eds. Klaas Landsman and Ellen van Wolde (Cham, Switzerland: Springer, 2016): 171–185; Caroline A. Watt, "Psychology and Coincidences," *European Journal of Parapsychology* 8 (1990–91): 66–84.

14. Daniel Kahneman, Paul Slovic, and Amos Tversky, eds., *Judgment Under Uncertainty: Heuristics and Biases* (Cambridge: Cambridge University Press, 1982).

15. A practical tip: Don't ever organize a party where the guests are randomly selected. No one will know each other, they will probably have very little in common, and it will be a rubbish party. Also, your guests will think you are weird for doing so.

16. 23—yes, that's right, only 23!

17. Hand, *The Improbability Principle*, 115–116.

18. Alcock, *Parapsychology*.

19. Norris McWhirter and Ross McWhirter, *Dunlop Illustrated Encyclopedia of Facts* (New York: Bantam, 1969), 492.

20. I stole this from Simon Singh, too, along with the corny reincarnation joke in note 2 in chapter 6.

21. John Allen Paulos, *Innumeracy: Mathematical Illiteracy and Its Consequences* (London: Penguin, 1988).

22. French, "Population Stereotypes and Belief in the Paranormal"; Marks, *The Psychology of the Psychic*.

23. David Paradine Productions, *Beyond Belief*. Available on YouTube: https://www.youtube.com/watch?v=EIfP4FyIkx8, https://www.youtube.com/watch?v=2bC_1wj9XgM, and https://www.youtube.com/watch?v=OnhRkRWKluw, all accessed June 4, 2023.

24. Frederick H. Lund, "Extra-Sensory-Perception Another Name for Free Association?," *Journal of General Psychology* 20, no. 1 (January 1939): 235–238.

25. Maya Bar-Hillel, "The Base-Rate Fallacy in Probability Judgments," *Acta Psychologica* 44, no. 3 (May 1980): 211–233; Daniel Kahneman and Amos Tversky, "On the Psychology of Prediction," *Psychological Review* 80, no. 4 (July 1973): 237–251.

26. Maya Bar-Hillel, "The Base-Rate Fallacy in Probability Judgments," 211.

27. From Toby Prike, Michelle M. Arnold, and Paul Williamson, "The Relationship between Anomalistic Belief, Misperception of Chance and the Base Rate Fallacy," *Thinking and Reasoning* 26, no. 3 (August 2020): 447–477.

28. David V. Budescu, "A Markov Model for Generation of Random Binary Sequences," *Journal of Experimental Psychology: Human Perception and Performance* 13, no. 1 (February 1987): 25–39; W. A. Wagenaar, "Generation of Random Sequences by Human Subjects: A Critical Survey of the Literature," *Psychological Bulletin* 77, no. 1 (January 1972): 65–72.

29. Interested readers are referred to Paul Rogers, "Paranormal Believers' Proneness to Probability Reasoning Biases: A Review of the Empirical Literature," in *Aberrant Beliefs and Thinking: Current Issues in Thinking and Reasoning*, ed. Niall Galbraith (Hove, UK: Psychology Press): 114–131; see also Prike, Arnold, and Williamson, "The Relationship between Anomalistic Belief, Misperception of Chance and the Base Rate Fallacy."

30. For example, Peter Brugger, Theodor Landis, and Marianne Regard, "A 'Sheep-Goat Effect' in Repetition Avoidance: Extra-Sensory Perception as an Effect of Subjective Probability?," *British Journal of Psychology* 81, no. 4 (November 1990): 455–468; Peter Brugger, Marianne Regard, and Theodor Landis, "Belief in Extrasensory Perception and Illusory Control: A Replication," *Journal of Psychology* 125, no. 4 (1991): 501–502; Neil Dagnall, Kenneth Drinkwater, Andrew Denovan, Andrew Parker, and Kevin Rowley, "Misperception of Chance, Conjunction, Framing Effects and Belief in the Paranormal: A Further Evaluation," *Applied Cognitive Psychology* 30, no. 3 (March 2016): 409–419; Andrew Denovan, Neil Dagnall, Kenneth Drinkwater, and Andrew Parker, "Latent Profile Analysis of Schizotypy and Paranormal Belief: Associations with Probabilistic Reasoning Performance," *Frontiers in Psychology* 9, no. 35 (January 2018); Kenneth Drinkwater, Neil Dagnall, Andrew Denovan, Andrew Parker, and Peter Clough, "Predictors and Associates of Problem-Reaction-Solution: Statistical Bias, Emotion-Based Reasoning, and Belief in the Paranormal," *SAGE Open* 8, no. 1 (January–March 2018): 1–11; Carrie A. Leonard and Robert J. Williams, "Fallacious Beliefs: Gambling Specific and Belief in the Paranormal," *Canadian Journal of Behavioural Science* 51, no. 1 (January 2019): 1–11; Toby Prike, Michelle M. Arnold, and Paul Williamson, "Psychics, Aliens,

or Experience? Using the Anomalistic Belief Scale to Examine the Relationship between Type of Belief and Probabilistic Reasoning," *Consciousness and Cognition* 53 (August 2017): 151–164; Prike, Arnold, and Williamson, "The Relationship between Anomalistic Belief, Misperception of Chance and the Base Rate Fallacy."

31. Susan Blackmore and Tom Troscianko, "Belief in the Paranormal: Probability Judgements, Illusory Control, and the 'Chance Baseline Shift,'" *British Journal of Psychology* 76, no. 4 (November 1985): 459–468; Mark Blagrove, Christopher C. French, and Gareth Jones, "Probabilistic Reasoning, Affirmative Bias and Belief in Precognitive Dreams," *Applied Cognitive Psychology* 20, no. 1 (January 2006): 65–83; Paola Bressan, "The Connection between Random Sequences, Everyday Coincidences, and Belief in the Paranormal," *Applied Cognitive Psychology* 16, no. 1 (January 2002): 17–34; Neil Dagnall, Andrew Parker, and Gary Munley, "Paranormal Belief and Reasoning," *Personality and Individual Differences* 43, no. 6 (October 2007): 1406–1415; Neil Dagnall, Kenneth Drinkwater, Andrew Parker, and Kevin Rowley, "Misperception of Chance, Conjunction, Belief in the Paranormal and Reality Testing: A Reappraisal," *Applied Cognitive Psychology* 28, no. 5 (September 2014): 711–719; Jochen Musch and Katja Ehrenberg, "Probability Misjudgment, Cognitive Ability, and Belief in the Paranormal," *British Journal of Psychology* 93, no. 3 (May 2002): 169–177.

32. Maxwell J. Roberts and Paul B. Seager, "Predicting Belief in Paranormal Phenomena: A Comparison of Conditional and Probabilistic Reasoning," *Applied Cognitive Psychology* 13, no. 5 (October 1999): 443–450.

33. Amos Tversky and Daniel Kahneman, "Judgments of and by Representativeness," in *Judgment Under Uncertainty: Heuristics and Biases*, ed. Daniel Kahneman, Paul Slovic, and Amos Tversky (Cambridge: Cambridge University Press, 1982), 84–98; Amos Tversky and Daniel Kahneman, "Extensional versus Intuitive Reasoning: The Conjunction Fallacy in Probability Judgment," in *Heuristics and Biases: The Psychology of Intuitive Judgment*, ed. Thomas Gilovich, Dale W. Griffin, and Daniel Kahneman (Cambridge: Cambridge University Press, 2002), 19–48.

34. Paul Rogers, Tiffany Davis, and John Fisk, "Paranormal Belief and Susceptibility to the Conjunction Fallacy," *Applied Cognitive Psychology* 23, no. 4 (May 2009): 524–542; Paul Rogers, John E. Fisk, and Dawn Wiltshire, "Paranormal Belief and the Conjunction Fallacy: Controlling for Temporal Relatedness and Potential Surprise Differentials in Component Events," *Applied Cognitive Psychology* 25, no. 5 (September/October 2011): 692–702; Paul Rogers, John E. Fisk, and Emma Lowrie, "Paranormal Believers' Susceptibility to Confirmatory versus Disconfirmatory Conjunctions," *Applied Cognitive Psychology* 30, no. 4 (July/August 2016): 628–634; Paul Rogers, John E. Fisk, and Emma Lowrie, "Paranormal Belief, Thinking Style Preference and Susceptibility to Confirmatory Conjunction Errors," *Consciousness and Cognition* 65 (October 2018): 182–196. See also Robert Brotherton and Christopher C. French, "Belief in Conspiracy Theories and Susceptibility to the Conjunction Fallacy," *Applied Cognitive Psychology* 28 (2014): 238–248; Dagnall, Drinkwater, Andrew Parker, and Rowley, "Misperception of Chance, Conjunction, Belief in the Paranormal and

Reality Testing"; Dagnall, Drinkwater, Denovan, Parker, and Rowley, "Misperception of Chance, Conjunction, Framing Effects and Belief in the Paranormal"; Prike, Arnold, and Williamson, "Psychics, Aliens, or Experience?"

35. For example, Dagnall, Parker, and Munley. "Paranormal Belief and Reasoning."

36. Johansen and Osman, "Coincidences"; Johansen and Osman, "Coincidence Judgment in Causal Reasoning."

37. Johansen and Osman, "Coincidences," 36.

CHAPTER 9

1. Russell Targ and Harold E. Puthoff, "Information Transfer under Conditions of Sensory Shielding," *Nature* 251 (1974): 602–607.

2. Some remote viewing studies do not involve senders attempting to telepathically transmit information from a randomly selected location but simply ask the individual to psychically obtain such information on the basis of the location's map coordinates, thus testing for possible clairvoyant abilities.

3. David F. Marks and Richard Kammann, *The Psychology of the Psychic* (Buffalo, NY: Prometheus Books, 1980), 24.

4. See, for example, Jonathan St. B. T. Evans, *Bias in Human Reasoning: Causes and Consequences* (Hove, UK: Psychology Press, 1989); Jonathan St. B. T. Evans, Ruth M. J. Byrne, and Stephen E. Newstead, *Human Reasoning: The Psychology of Deduction* (Hove, UK: Psychology Press, 1993).

5. Michael Wierzbicki, "Reasoning Errors and Belief in the Paranormal," *Journal of Social Psychology* 125, no. 4 (August 1985): 489–494; Harvey J. Irwin, "Reasoning Skills of Paranormal Believers," *Journal of Parapsychology* 55, no. 3 (September 1991): 281–300.

6. Roberts and Seager, "Predicting Belief in Paranormal Phenomena."

7. Caroline Watt and Richard Wiseman, "Experimenter Differences in Cognitive Correlates of Paranormal Belief and in Psi," *Journal of Parapsychology* 66, no. 4 (December 2002): 371–385.

8. Michael A. Thalbourne and Peter Delin, "A Common Thread Underlying Belief in the Paranormal, Mystical Experience and Psychopathology," *Journal of Parapsychology* 58, no. 1 (January 1994): 3–38. See also Michael A. Thalbourne, "Transliminality: A Fundamental Mechanism in Psychology and Parapsychology," *Australian Journal of Parapsychology* 10, no. 1 (January 2010): 70–81.

9. Michael A. Thalbourne and John Maltby, "Transliminality, Thin Boundaries, Unusual Experiences, and Temporal Lobe Lability," *Personality and Individual Differences* 44, no. 7 (May 2008): 1618.

10. For example, James Grier Miller, "The Role of Motivation in Learning without Awareness," *American Journal of Psychology* 53, no. 2 (April 1940); Alcock, *Parapsychology*; Stuart Wilson,

"Psi, Perception without Awareness, and False Recognition," *Journal of Parapsychology* 66, no. 3 (September 2002): 271–289.

11. Alcock, *Parapsychology*.

12. Susan E. Crawley, Christopher C. French, and Steven A. Yesson, "Evidence for Transliminality from a Subliminal Card-Guessing Task," *Perception* 31, no. 7 (July 2002): 887–892.

13. Keith E. Stanovich and Richard F. West, "Individual Differences in Reasoning: Implications for the Rationality Debate," *Behavioral and Brain Sciences* 23, no. 5 (October 2000): 645–665.

14. Daniel Kahneman, *Thinking, Fast and Slow* (London: Penguin, 2011).

15. Shermer, *The Believing Brain*.

16. Klaus Conrad, *Die Beginnende Schizophrenie: Versuch Einer Gestaltanalyse des Wahns* (Stuttgart: Thieme, 1958).

17. Justin L. Barrett, *Why Would Anyone Believe in God?* (Plymouth, UK: AltaMira Press, 2004).

CHAPTER 10

1. Philip Escoffey's website is: http://www.escoffey.com.

2. Ray Hyman, "Dowsing," in *The Encyclopedia of the Paranormal*, ed. Gordon Stein (Amherst, NY: Prometheus Books, 1996), 222.

3. Ray Hyman and Evon Z. Vogt, "Water Witching: Magical Ritual in Contemporary United States," in *The Elusive Quarry: A Scientific Appraisal of Psychical Research*, ed. Ray Hyman (Buffalo, NY: Prometheus Books), 323–338.

4. Simon Hoggart and Mike Hutchinson, *Bizarre Beliefs* (London: Richard Cohen, 1995).

5. Herman H. Spitz, *Nonconscious Movements: From Mystical Messages to Facilitated Communication* (Mahwah, NJ: Lawrence Erlbaum, 1997).

6. For example, William F. Barrett, "On a 'Magnetic Sense,'" *Nature* 29 (1884): 476–477; William F. Barrett, "On the Detection of Hidden Objects by Dowsers," *Journal of the Society for Psychical Research* 14 (1910): 183–193; William F. Barrett and Theodore Besterman, *The Divining Rod: An Experimental and Psychological Investigation* (London: Methuen and Co., 1926).

7. For example, James Randi, *Flim-Flam! Psychics, ESP, Unicorns and Other Delusions* (Buffalo, NY: Prometheus Books, 1982).

8. A clip from the documentary showing our test is available on YouTube: https://www.youtube.com/watch?v=_VAasVXtCOI.

9. Sally Le Page, "In 2017, UK Water Companies Still Rely on 'Magic,'" *Medium*, November 20, 2017, https://sallylepage.medium.com/in-2017-uk-water-companies-still-rely-on-magic-6eb62e036b02.

10. Matthew Weaver, "UK Water Firms Admit Using Divining Rods to Find Leaks and Pipes," *Guardian*, November 21, 2017, https://www.theguardian.com/business/2017/nov/21/uk-water-firms-admit-using-divining-rods-to-find-leaks-and-pipes; "In Defence of Dowsing," *Guardian*, November 27, 2017, https://www.theguardian.com/environment/2017/nov/27/in-defence-of-dowsing-to-detect-water.

11. Yvonne Roberts, "The Man Who Can Read Babies' Minds," *Guardian*, June 19, 2006, https://www.theguardian.com/media/2006/jun/19/familyandrelationships.tvandradio.

12. Derek Ogilvie, *The Baby Mind Reader: Amazing Psychic Stories from the Man Who Can Read Babies' Minds* (London: Harper Element, 2006).

13. James Randi's One Million Dollar Paranormal Challenge ran from 1996 until 2015. The money was on offer to anyone who could demonstrate paranormal powers under properly controlled conditions. Usually, Randi insisted, not unreasonably, that claimants passed a preliminary test of the claimed ability carried out by people whom Randi knew and trusted before being allowed to enter the formal million dollar challenge itself. Our test of dowsers reported in the previous section and our test of Patricia Putt reported in the next section were two such preliminary tests. In exceptional cases, such as with Derek Ogilvie, the preliminary test was waived.

14. Chris French, "Scientists Put Psychic's Paranormal Claims to the Test," *Guardian*, May 12, 2009, https://www.theguardian.com/science/2009/may/12/psychic-claims-james-randi-paranormal.

15. Alison Smith, "Patricia Putt MDC Test: Protocol Failure?," James Randi Education Foundation (blog), May 9, 2009, http://archive.randi.org/site/index.php/swift-blog/549-patricia-putt-mdc-test-protocol-failure.html.

16. Richard Wiseman, "Testing a Medium: Results," accessed June 5, 2023, https://richardwiseman.wordpress.com/2009/05/06/testing-a-medium-results/.

17. Chris French, "Halloween Challenge: Psychics Submit Their Powers to a Scientific Trial," *Guardian*, October 31, 2012, https://www.theguardian.com/science/2012/oct/31/halloween-challenge-psychics-scientific-trial.

18. Logistically, even a simple test such as this one was quite complicated to run. We had to make sure, for example, that all volunteer sitters arrived and left at the appropriate times, that sitters and psychics were always kept apart, and that all stages of the test were appropriately recorded. Thanks for assistance are therefore due to Ursula Blaszko, Deborah Bowden, Rob Brotherton, Duncan Colvin, Dan Denis, Maurice Douglas, Dave Hughes, Benjamin Kuper-Smith, Molly Maclean, and James Munroe.

CHAPTER 11

1. A full description of this study is included in Louie Savva's PhD thesis, "Is Some of the Evidence for Ostensible Precognition Indicative of Darwinian Adaptation to Retrocausal Influences?," Goldsmiths, University of London.

1. Christopher C. French and James Ost, "Beliefs about Memory, Childhood Abuse, and Hypnosis Amongst Clinicians, Legal Professionals and the General Public," in *Wrongful Allegations of Sexual and Child Abuse*, ed. Ros Burnett (Oxford: Oxford University Press, 2016), 143–154; James Ost and Christopher C. French, "How Misconceptions About Memory May Undermine Witness Testimony," in *Witness Testimony in Sexual Cases: Evidential, Investigative and Scientific Perspectives*, ed. Pamela Radcliffe, Gisli H. Gudjonsson, Anthony Heaton-Armstrong, and David Wolchover (Oxford: Oxford University Press, 2016), 361–373; Lawrence Patihis, Lavina Y. Ho, Elizabeth Loftus, and Mario E. Herrera, "Memory Experts' Beliefs About Repressed Memory," *Memory* 29, no. 6 (July 2021): 823–828.

2. Chris French, "The Unseen Force That Drives Ouija Boards and Fake Bomb Detectors," *Guardian*, April 27, 2013, https://www.theguardian.com/science/2013/apr/27/ouija-boards-dowsing-rods-bomb-detectors; French and Stone, *Anomalistic Psychology*.

3. Caroline Hawley and Meirion Jones, "Export Ban for Useless 'Bomb Detector,'" *BBC News*, January 22, 2010, http://news.bbc.co.uk/1/hi/programmes/newsnight/8471187.stm.

4. John W. Jacobson, James A. Mulick, and Allen A. Schwartz, "A History of Facilitated Communication: Science, Pseudoscience, and Antiscience. Science Working Group on Facilitated Communication," *American Psychologist* 50, no. 9 (September 1995): 750–765; Mark P. Mostert, "Facilitated Communication since 1995: A Review of Published Studies," *Journal of Autism and Developmental Disorders* 31, no. 3 (July 2001): 287–313; Mark P. Mostert, "Facilitated Communication and Its Legitimacy—Twenty-First Century Developments," *Exceptionality* 18, no. 1 (January 2010): 31–41; Ralf W. Schlosser, Susan Balandin, Bronwyn Hemsley, Teresa Iacono, Paul Probst, and Stephen Tetzchner, "Facilitated Communication and Authorship: A Systematic Review," *Augmentative and Alternative Communication* 30, no. 4 (November 2014): 359–368; Richard L. Simpson and Brenda Smith Myles, "Facilitated Communication and Children with Disabilities: An Enigma in Search of a Perspective," *Focus on Exceptional Children* 27, no. 9 (May 1995): 1–16; Herman Spitz, *Nonconscious Movements*; Daniel M. Wegner, *The Illusion of Conscious Will* (Cambridge, MA: MIT Press, 2017).

5. Daryl J. Bem, "Feeling the Future: Experimental Evidence for Anomalous Retroactive Influences on Cognition and Affect," *Journal of Personality and Social Psychology* 100, no. 3 (February 2011): 407–425.

6. For example, James Alcock, "Back from the Future: Parapsychology and the Bem Affair," *Skeptical Inquirer*, January 6, 2011, https://skepticalinquirer.org/exclusive/back-from-the-future; Eric-Jan Wagenmakers, Ruud Wetzels, Denny Borsboom, and Han L. J. van der Maas, "Why Psychologists Must Change the Way They Analyze Their Data: The Case of Psi: Comment on Bem (2011)," *Journal of Personality and Social Psychology* 100, no. 3 (February 2011): 426–432.

7. Ben Goldacre, "Backwards Step on Looking into the Future," *Guardian*, April 23, 2011, https://www.theguardian.com/commentisfree/2011/apr/23/ben-goldacre-bad-science.

8. Stuart J. Ritchie, Richard Wiseman, and Christopher C. French, "Failing the Future: Three Unsuccessful Attempts to Replicate Bem's 'Retroactive Facilitation of Recall' Effect," *PLOS ONE* 7, no. 3 (March 2012): e33423.

9. See, for example, Chris French, "Precognition Studies and the Curse of the Failed Replications," *Guardian*, March 15, 2012, https://www.theguardian.com/science/2012/mar/15/precognition-studies-curse-failed-replications; Harold Pashler and Eric-Jan Wagenmakers, "Editors' Introduction to the Special Section on Replicability in Psychological Science: A Crisis of Confidence?," *Perspectives on Psychological Science* 7, no. 6 (November 2012): 528–530; Stuart J. Ritchie, Richard Wiseman, and Christopher C. French, "Replication, Replication, Replication," *The Psychologist* 25, no. 5 (May 2012): 346–348.

10. For a clear and comprehensive discussion of QRPs and other factors that can contribute to bad science, see Stuart Ritchie, *Science Fictions: Exposing Fraud, Bias, Negligence and Hype in Science* (London: Bodley Head, 2020).

11. Joseph P. Simmons, Leif D. Nelson, and Uri Simonsohn, "False-Positive Psychology: Undisclosed Flexibility in Data Collection and Analysis Allows Presenting Anything as Significant," *Psychological Science* 22, no. 11 (October 2011): 1359–1366.

12. Open Science Collaboration, "Estimating the Reproducibility of Psychological Science," *Science* 349, no. 6251 (August 28, 2015): aac4716.

13. For example, Sander L. Koole and Daniël Lakens, "Rewarding Replications: A Sure and Simple Way to Improve Psychological Science," *Perspectives on Psychological Science* 7, no. 6 (November 2012): 608–614; Daniel J. Simons, "Replication: Where Do We Go from Here?," *The Psychologist* 25, no. 5 (May 2012): 5.

14. Jeff Galak, Robyn A. LeBoeuf, Leif D. Nelson, and Joseph P. Simmons, "Correcting the Past: Failures to Replicate Psi," *Journal of Personality and Social Psychology* 103, no. 6 (December 2012): 933–948.

15. Wagenmakers et al., "Why Psychologists Must Change the Way They Analyze Their Data."

16. Daryl J. Bem, "Writing an Empirical Article," in *Guide to Publishing in Psychology Journals*, ed. Robert J. Sternberg (Cambridge: Cambridge University Press), 3–16.

17. See, for example: in medicine, John P. A. Ioannidis, "Contradicted and Initially Stronger Effects in Highly Cited Clinical Research," *JAMA* 294, no. 2 (13 July 2005): 218–228; Tara Haelle, "Dozens of Major Cancer Studies Can't Be Replicated," *Science News*, December 7, 2021, https://www.sciencenews.org/article/cancer-biology-studies-research-replication-reproducibility; in economics, Colin F. Camerer et al., "Evaluating Replicability of Laboratory Experiments in Economics," *Science* 351, no. 6280 (March 3, 2016): 1433–1436; in other fields, Monya Baker, "1,500 Scientists Lift the Lid on Reproducibility," *Nature* 533, no. 7604 (May 25, 2016): 542–454.

18. The Stroop effect is named after John Ridley Stroop, the first person to describe the phenomenon almost a century ago. It refers to the fact that we find it more difficult to identify the

physical color of color words when the physical color of the words is incongruent with the meaning of the words (e.g., *red* written in green ink vs. *red* written in red ink).

19. Christopher C. French, "Reflections of a (Relatively) Moderate Skeptic," in *Debating Psychic Experience: Human Potential or Human Illusion?*, ed. Stanley Krippner and Harris L. Friedman (Santa Barbara, CA: Praeger, 2010), 53–64.

20. Carl Sagan, *Broca's Brain: The Romance of Science* (London: Hodder and Stoughton, 1979), 62.

21. David Robert Grimes, *The Irrational Ape: Why Flawed Logic Puts Us All at Risk, and How Critical Thinking Can Save the World* (London: Simon & Schuster, 2019); Carol Tavris and Elliot Aronson, *Mistakes Were Made (But Not by Me): Why We Justify Foolish Beliefs, Bad Decisions, and Hurtful Acts*, rev. ed. (Boston: Mariner, 2015).

22. Stuart Vyse, *The Uses of Delusion: Why It's Not Always Rational to Be Rational* (New York: Oxford University Press, 2022).

23. Tali Sharot, *The Optimism Bias: A Tour of the Irrationally Positive Brain* (New York: Vintage, 2012).

EPILOGUE

1. Dean Radin, *The Conscious Universe: The Scientific Truth of Psychic Phenomena* (New York: HarperEdge, 1997), 275.

2. Marks, *The Psychology of the Psychic*, 308.

3. Harvey J. Irwin and Caroline A. Watt, *An Introduction to Parapsychology*, 5th ed. (Jefferson, NC: McFarland, 2007), p. 261.

4. See, e.g., Stephen LaBerge, "Lucid Dreaming: Paradoxes of Dreaming Consciousness," in *Varieties of Anomalous Experience: Examining the Scientific Evidence*, 2nd ed., ed. Etzel Cardeña, Steven Jay Lynn, and Stanley Krippner (Washington, DC: American Psychological Association, 2014), 145–173.

5. See https://dave-green.co.uk.

6. See https://crviewer.com/targets/targetindex.php.

7. Robin Wooffitt, *Telling Tales of the Unexpected: The Organization of Factual Discourse* (Hemel Hempstead, UK: Harvester Wheatsheaf, 1992); Peter Lamont, "Paranormal Belief and the Avowal of Prior Scepticism," *Theory & Psychology* 17, no. 5 (October 2007): 681–696; Anna Stone, "An Avowal of Prior Scepticism Enhances the Credibility of an Account of a Paranormal Event," *Journal of Language and Social Psychology* 33, no. 3 (2014): 260–281.

8. Marks and Kammann, *The Psychology of the Psychic*; Marks, *The Psychology of the Psychic*.

9. David F. Marks, *Psychology and the Paranormal* (London: Sage, 2020), 333n4.

10. Marks, *Psychology and the Paranormal*, 336.

11. Marks, *Psychology and the Paranormal*, 311, 307.

12. Marks, *Psychology and the Paranormal*, 66–77.
13. Marks, *Psychology and the Paranormal*, 73.
14. William James, "The Present Dilemma in Philosophy," in *The Writings of William James: A Comprehensive Edition*, ed. John J. McDermott (New York: Modern Library, 1968; original work published 1907). Quoted in Hoyt L. Edge, Robert L. Morris, John Palmer, and Joseph H. Rush, *Foundations of Parapsychology: Exploring the Boundaries of Human Capability* (Boston: Routledge & Kegan Paul, 1986), 320.

Index

Aaen-Stockdale, Craig, 104
Absorption, 26, 84–85, 105, 122, 156, 158, 160, 164
Abu-Dyab, Rabih, 185–187
Acorah, Derek, 112, 123, 264
Adams, Douglas, 134
Adamski, George, 147
ADE-651, 285–286
Adler, Shelley R., 36, 60, 61
Affirmation of the consequent, 238
Alcock, James ("Jim"), xiv, xvi, xix, 222
Alien abduction, 16, 58, 147, 149–156, 175, 178, 238–239, 284–285
Alien implants, 154
Aliens, xxi, 9, 22, 135–170
Alpdrück, 58
Al Younis, Jad, 181
American Society for Psychical Research, 112, 316
Amityville Horror, 76–77
Amygdala, 52
Angels, 9–10
Anomalistic psychology, xiv, xx, xxi, xxiii, xxv, 2, 15–19, 26, 247, 283–284, 285, 301
Anomalistic Psychology Research Unit (APRU), xxv, xxvi, 106, 247, 249, 259, 298
Anoxia, 210
Anson, Jay, 76–77

Apophenia, 244
Area 51, 135
Arnold, Kenneth, 135, 142
Ascher, Rodney, 43, 63
Asimov, Isaac, xix
Astrology, 12–13, 15, 19, 21
Atkinson, Gilbert, 160
Aubrey, Nell, 40
Augustine, Keith, 204
Australian Sheep-Goat Scale (ASGS), 23, 107, 108
Autobiographical memory, 88, 195
Autoscopy, 54, 199
AWARE (AWAreness during REsuscitation) project, 205
Ayahuasca, 211
Azande, 31

Barber, Jacques, 49
Barber, Theodore X., 161–162
Barbieri, Suzanne, 263
Bar-Hillel, Maya, 231
Barnum effect, 115–118
Barrett, Justin, 244
Bartholomew, Robert, 162
Base rate fallacy (a.k.a. base rate neglect), 231
Bastos Jr., Marco Aurélio Vinhosa, 114
Beattie, Elaine, 256
Beaumont, J. Graham, xiv, xx

Beischel, Julie, 114
Belief bias, 237–238
Belle, 306–314, 316
Bem, Daryl, 288–291
Bergier, Jacques, 218
Bernstein, Morey, 175–176
Besterman, Theodore, 94
Bigfoot, 8
Biklen, Douglas, 287
Birthday problem, 221–222
Bishop, Bridget, 56
Black-and-white thinking, xxii
Blackmore, Susan, xvii, xxi, 104, 117, 154, 196, 197, 199, 200, 203–204, 210
Blanke, Olaf, 211–212
Bloxham, Arnall, 176
Boas, Antonio Villas, 149
Bokhari, Akhawayni, 55
Borjigin, Jimo, 209
Bottom-up processing, 88, 240
Bowie, David, 218, 221
Boyer, Christina, 76
Braithwaite, Jason, 102–103, 105, 106, 209
Brazel, William "Mac," 135–136
Britain's Psychic Challenge, 247–251
British False Memory Society, 158
Brown, Derren, xxvii
Browne, Sylvia, 111
Brugger, Peter, 82
Bruno, Nicolas, 62
Buchanan, Lyn, 304
Bullard, Thomas E., 164
Bunton-Stasyshyn, Rosie, 256, 257
Bush, Nancy Evans, 193
Buxton, Laura, 219

Cabs problem, 231
Cannabis, 211
Carpenter, William Benjamin, 125
Carroll, Robert Todd, 81
Categorical syllogism, 237
Chabris, Christopher, 83, 165

Chawla, Lakhmir, 208
Chesterton, G. K., 315
Cheyne, J. Allen, 38, 51, 53–54
Child prodigies, 173
Childhood sexual abuse, 157, 158, 284–285
Childhood trauma, 160–161, 162, 163
Chiswick Coincidence, The, 315
Churchill, Winston, 299
Clairvoyance, 6, 13
Clancy, Susan, 164, 168
Clark, Kimberley, 201
Clarke, Arthur C., 170
Clinical psychology, 27
Close encounters of the fifth kind, 147
Close encounters of the first kind, 142
Close encounters of the fourth kind, 147
Close encounters of the second kind, 142, 144–145
Close encounters of the third kind, 139, 147
Cognition, 27
Cognitive biases, xx, 20, 28, 215–216, 235–245, 283
Cognitive psychology, 26, 27
Coincidence, 216–234, 236, 268, 313
Cold reading, 28, 114–122, 251, 259
Columbus poltergeist, 74–76
Coman, Richard, 56
Committee for Skeptical Inquiry (CSI), xix, xxvi
Committee for the Scientific Investigation of Claims of the Paranormal (CSICOP), xix, xxvi
Communication with the dead, 18, 111–131
Complementary and alternative medicine (CAM), 13–14, 21, 284
Conan Doyle, Sir Arthur, 11
Conditional syllogism, 238–239
Confirmation bias, 132
Conjunction fallacy, 232–233
Conrad, Klaus, 244
Consciousness, 6, 69–70, 88, 131, 191, 192, 195, 200–209, 210, 240–242

Conspiracies, 14
Contactees, 147
Context, 79–80
Cooper, Alice, xvi
Cornell, Tony, 86–87, 91
Cortical disinhibition, 199
Costain, Thomas B., 177
Cottingley fairies, 10–11
Cox, Brian, xxvii
Cox, Sarah Elizabeth, 33–35, 38
Crashing memories technique, 159, 161
Crawley, Susan, 241
Critical thinking, 20, 283
Crombag, Hans, 159
Crookes, William, 113
Crossley, Rosemary, 287
Crowley, Aleister, 218–219
Cryptids, 8, 22
Cryptomnesia, 178
Cryptozoology, 8–9
Crystal balls, 12
Crystals, 13

Dab tsog, 60–61
Dagnall, Neil, 161, 233
Dark matter, 6
Darwin, Charles, 112–113
Davey, S. John, 93
Davies, Neil, xxvii
Davies, Owen, 56
da Vinci, Leonardo, 81
Dawkins, Richard, xxvii, 104, 252, 256, 257
Deductive reasoning, 239
Deese, James, 159
DeFeo Jr., Ronald, 76, 77
DeGeneres, Ellen, 170
Déjà vu, 173–174, 179
Delin, Peter, 240
de Maupassant, Guy, 63–64
Demons, 9, 16, 22, 52–53, 67, 77, 193
Denis, Dan, 35, 54
Depersonalization, 160, 196

Derealization, 160
De Ridder, Dirk, 212
Descartes, René, 69
Developmental Psychology, 26, 28
Dissociativity, 26, 156, 158, 160–161, 163–164, 168
Distance healing, 13
Divination, 12, 15–16
Dobson, Barrie, 177
Doghramji, Karl, 36, 56, 58, 66
Donnelly, Steve, xxvi
Dossey, Larry, 203–204
Double-blind, xxiv, 104, 250, 252, 256, 287
Downer, Robert, 56
Dowsing, 252–258, 285–286
Drake equation, 134
Drake, Frank, 134
DRM method, 158–159, 164, 179, 298
Druse (also spelled "Druze"), 171, 179–189
Dualism, 69, 195
Dying brain hypothesis, 196–214

Easter Bunny, 10, 29
Ectoplasm, 91
Edward, John, 111
Ehrsson, Henrik, 213–214
Electromagnetic fields, 101–105, 106–109
Electronic voice phenomenon (EVP), 7, 127–131
Endorphins, 197, 198
Environmental factors, 100–109
Escoffey, Philip, 248
Evans, Jane, 176–177
Evans, Joan, 177
Evolutionary psychology, 28
Exorcism, 9
Exploding head syndrome (EHS), 47
Extraordinary life forms, 23
Extrasensory perception (ESP), xviii, xxi, 6, 13, 23, 227, 229, 236, 241–242
Extraterrestrial hypothesis, 9
Eyewitness testimony, 87, 93–100

Facilitated communication (a.k.a. progressive kinesthetic feedback or supported typing), 286–287
Fairies, 10
False memories, 16, 59, 100, 156, 157–169, 175–179, 194–195, 211, 284
False pregnancy (pseudocyesis), 155, 161
Fantasy proneness, 26, 156, 158, 160, 161–164, 168, 178
Faraday, Michael, 125, 127
Farley, Tim, 20
Feifer, George, 217–218
Fenwick, Peter, 208
Fermi, Enrico, 134
Fermi's paradox, 134
Finucane, R. C., 70
Fish, Marjorie, 150
Fitzgerald, F. Scott, 63
Fleming, Alexander, 234
Folklore, 10
Forer, Bertram R., 115
Fox sisters, 111
Fox, Kate, 91
Fox, Margaretta, 91, 93
Fraud, xxii, 114–123, 302
Frazier, Kendrick, xix
French's First Law, 77
French's Second Law, 122
Frost, David, 228
Fry, Colin, 112, 264
Fry, Stephen, xxvii
Fukuda, Kazuhiko, 49
Fuller, John G., 151–152
Fuseli, Henry, 61–62

Gabbert, Fiona, 99
Galak, Jeff, 294
Gallup, George, 141
Gambler's fallacy, 232
Gardiner, Ian, 36–37
Gardner, Martin, xix, 315
Geller, Uri, xiv, xvii, 6, 95–96, 97, 228–229

Ghost oppression, 58
Ghosts, xxi, 7, 16, 22, 26, 29, 67, 69–109, 308
G-LOC, 198
Glossolalia, 9
Gluck, Gerald, 261
God, 9–10
Goddard, Trisha, 247
Golden Dawn Society, 218, 221
Golzari, Samad, 55
Good Thinking Society, xxviii
Granqvist, Pehr, 105
Grant, Richard E., 218
Green, Dave, 303–305
Greening, Emma, 96, 99–100
Gregory, Alice, 35, 44, 54
Greyson, Bruce, 193–194, 207, 208
Griffith, Frances, 10–11
Grof, Stanislav, 195
Grom, Jessica, 65
Grossman, Wendy, xxvi
Guided imagery, 157

Haidar, Ramiz, 184
Halawi, Saad, 186, 187
Halifax, Joan, 195
Hallucinations, 39, 40, 42, 46, 47, 52, 56, 57, 64, 80–81, 102–103, 105, 106, 122, 156, 197, 198, 200, 204–205, 211
Halt, Charles, 138=139
Hampton Court Palace, 80, 100, 101
Hand, David, 217, 222
Haque, Usman, 106
Haraldsson, Erlendur, 172
Harden, Mike, 75
Hardy, Thomas, 63
Harr, Amanda, 251
Harris, Melvin, 177
Hasted, John, xiv
Hatzimasouras, Jane, 216
Hearne, Keith, 280–282
Hemingway, Ernest, 63

Hendry, Allan, 141
Heuristics, 243
Hexendrücken, 58
Heydrich, Lukas, 212
Hibous, Mehdi, 185
Hidden causes, 222, 225
Highway hypnosis, 168
Hill, Betty and Barney, 149–152, 166–167, 168
Hmong refugees, 60–61
Hoaxes, 71–77
Hoggart, Simon, 218
Holden, Kate, xxv, xxvi
Holmes, Sherlock, 11, 120
Hopkins, Anthony, 217–218
Hopkins, Budd, 153–154, 165, 168–169
Hot reading, 123
Hough, Peter, 162
Houran, James, 79
Howard, Toby, xxvi
Hufford, David J., 36, 57
Hume, David, 70, 87
Hutchinson, Mike, xvi, 218
Hyde, Deborah, xxvii
Hyman, Ray, xviii–xix, 93, 115, 121, 252
Hynek, J. Allen, 139–140, 142, 147
Hyperactive agency detection device (HADD), 244–245
Hypercarbia, 199
Hypnotic regression, 16, 59, 150–152, 153, 155, 157, 164–167, 174–179, 187, 261, 284
Hypnotic susceptibility, 27, 161, 162, 175, 178
Hypoxemia, 209
Hypoxia, 198–199

I Ching, 12
Icke, David, 14
Ideomotor effect, 125, 258, 285
Imaginary playmates, 29, 30, 161, 188
Imagination inflation, 159, 195

Inattentional blindness, 83–87, 91, 96, 298
Ince, Robin, xxvii
Incubus, 51, 52–53, 57
Infrasound, 105–109
Intelligence quotient (IQ), xxiii
Intruder, 51–52, 53
Intuition, 242–243
Irwin, Harvey, 239, 302
Iverson, Jeffrey, 176–177

Jacobs, David, 169
Jago, Crispian, xxvii
James, William, 316
James Randi Educational Foundation (JREF), 261–262, 263
Jansen, Karl, 198
Jesus, 9–10
Johansen, Mark, 234
Jones, Warren H., 1
Judson, I. R., 197
Juhasz, Panka, 263
Jürgenson, Friedrich, 128

Kahneman, Daniel, 233, 242
Kammann, Richard, 236
Kanashibari, 49, 57
Karma, 173, 181
Katon, Lesley, 275–278
Keogh, Angela, 39–40
Ketamine, 198, 211
King, Brian, 219
Klass, Philip J., 137, 152
Kletti, Roy, 196
Koestler Chair of Parapsychology, xviii, xxiii, xxiv
Koestler Parapsychology Unit, 95
Kokma, 58
Koren helmet (a.k.a. God helmet), 104
Korff, Kal K., 137
Kottmeyer, Martin, 166
Krueger, Freddy, 63
Kurtz, Paul, xix

Lange, Rense, 79
Language, 27
La Plante, Lynda, 248
Law of truly large numbers, 220, 267
Lawrence, Tony, 78
Lawson, Alvin, 167
Ledger, Chris, 181
Legare, Christine, 30
Lenggenhager, Bigna, 213–214
Le Page, Sally, 258
Life after death, 7, 23, 131, 192, 195
Life review, 193
Loch Ness monster, 8
Loftus, Elizabeth, xviii, 98
Louder, John, 56
Lucid dreams, 303–305
Lucky charms, 11
Lund, Frederick, 229
Lutz, George and Kathy, 76–77
Lynn, Steven J., 162
Lytton-Cobbold, Antony, 251
Lytton-Cobbold, Henry, 251
Lytton-Cobbold, John, 251

Mack, John E., 154, 165
MacKinnon, Carla, 41, 43, 63
Magical thinking, 29–31
Magnusson, Magnus, 176
Magritte, René, 62
Majdalani, Tima Khalil, 181
Malton, Jackie, 248, 250
Mancuso, Frank, 76
Mandell, David, 268–174
Marcel, Jesse, 136
Marks, David, 236, 301–302, 314–316
Marshall, Michael, xxviii, 264
Martin, Susan, 56
McCormick, James, 286
McDermott, Kathleen B., 159
Mckears, Stephen, 78
McNally, Richard, 155–156
McWhirter brothers, 223

Media, 28
Mediums, 7, 15, 21, 93, 111–114, 117
Melerb, Melhem, 185
Melville, Herbert, 63
Memory, 26, 27
Memory conformity, 97–100
Merseyside Skeptics Society, xxviii, 264
Meyersburg, Cynthia, 179
Minchin, Tim, xxvii, 14
Miracles, 70
Mirren, Helen, 248
Missing time, 151–152, 168, 170
Monism, 70
Monsters, 29
Moody, Raymond, 192
Morgan, Sally, 112, 264
Morris, Robert ('Bob'), xviii, xxiii–xxiv, 76, 95, 316
Morse, Melvyn, 198
Mossbridge, Julia, 303–305
Mozart, Wolfgang Amadeus, 173
Munroe, James, 263
Murphy, Bridey, 175–176

Naloxone, 198
Narcolepsy, 47, 51
Near-death experience (NDE), 9, 27, 191–214, 274
Nees, Michael, 129
Nelson, Leif, 292–294
Neuropsychology, 26, 27
Night terrors (*pavor nocturnis*), 47
Nightmare, 46, 55
Nonconscious processing, 240–242
Nonverbal communication, 120
Nosek, Brian, 294
Noyes Jr, Russell, 196

Ogilvie, Derek, 258–261
O'Keeffe, Ciarán, 123
Old hag, 38, 57
Opinion polls, 18, 21, 135, 141

Optional stopping, 292, 296
Osman, Magda, 234
Otgaar, Henry, 167
Ouija board, 124–125, 285, 287
Out-of-body experience (OBE), xviii, 27, 53–54, 55, 161, 162, 191, 192, 198, 199–207, 209–214

Påhlsson, Lena, 217
Palm reading, 15, 21
Panic attack, 46
Panic disorder, 54, 65
Paranormal, 2–14
Parapsychology, xiv, xx, xxiv, xxvi, 15–17
Pareidolia, 81, 244
Parnell, June O., 156
Parnia, Sam, 205, 208
Pascal, Blaise, 173–174
Past-life memories, 16, 284
Past-life therapy, 175
Patternicity, 244
Paulos, John Allen, 224
Pauwels, Louis, 218
Peach, Bernard, 56
Pecararo, Ralph J., 76
Penfield, Wilder, 211
Penniston, Jim, 138
Perception, 27, 88, 141–142
Persinger, Michael, 102–105
Personality, 26
Petkova, Valeria, 214
Phillips, Charlotte, 129
Piaget, Jean, 28–30
Plimmer, Martin, 219
Poltergeists, 7, 44, 73–77
Population stereotypes, 226–229, 298
Possession, 9
Post-event misinformation, 98
Post-traumatic stress disorder (PTSD), 54, 155, 156, 194
Prayer, 9, 29
Precognition, 6, 9, 13, 23, 288–291, 314

Precognitive dreams, 224–225, 267–282, 303–305
Preregistration, 294
Presence, Sense of, 36, 37, 46, 52, 55, 64, 71, 80, 105, 107, 168–169, 238–239
Pressure on the chest, 37, 57
Probabilistic reasoning, 231, 298
Probability, 220
Project Blue Book, 140
Project Mogul, 137
Project Sign, 140
Proprioception, 53, 210, 211
Pseudoscience, xx
Psi, xix, xxi, xiv, xxiv, 19, 241, 288, 301, 314, 316
Psychic healing, 13, 18
Psychobiology, 27
Psychokinesis, 6, 9, 23, 75, 95, 96–97, 314
Psychology, 26–31
Psychopathology, xxiii, 27, 56, 156, 178, 240
Psychosis, 81
Publication bias, 290
Puthoff, Harold, 235–236
Putt, Patricia, 261–266

Questionable research practices (QRPs), 292–297

Radford, Ben, xxvii
Radin, Dean, 301, 314
Ramey, Roger, 136
Randi, James, xvi–xvii, xix, xxvii, 75, 259, 261, 280–281, 286
Randolph, Marcy, 278
Randolph, Phil, 279
Randomness, 231, 232, 233, 244
Rapid eye movement (REM) sleep, 50
Rationality, 242–243, 299–300
Raudive, Konstantin, 128
Reasoning, 27
Reincarnation, 7, 147, 171–189, 285, 308
Reliability, 23

Religious belief, 23
Remote viewing, 235–237
Rendlesham, 137–139, 144
Repetition avoidance, 232
Replication crisis, 291–298
Representativeness heuristic, 233
Repression, 165, 284
Resch, John and Joan, 76
Resch, Tina, 74–76
Retroactive facilitation of recall, 289–291
Revised Paranormal Belief Scale (RPBS), 23, 26
Reynolds, Pam, 202–203
Rhine, Joseph Banks, 241
Rhue, Judith W., 162
Richards, Anne, 84, 89, 90, 130
Ridpath, Ian, 139
Ring, Kenneth, 163, 192
Ritchie, Stuart, 289
Ritualized satanic abuse, 157, 284
Rivers, Andy, 219
Roberts, Maxwell, 239
Robinson, Chris, 274–282
Rodeghier, Mark, 162
Roediger III, Henry L., 159
Rogers, Paul, 162, 233
Roll, William, 75
Rosing, Christopher J., 163
Roswell, 135–137
Rowland, Ian, 115
Rubber hand illusion, 212–213

Sabom, Michael, 202
Sagan, Carl, xix, 195, 297
Salusbury, Matt, 37–38
Santa Claus, 10, 29
Savva, Louie, 270
Schegoleva, Anna, 57–58
Schizophrenia, 244
Schouten, Sybo, 113
Schwartz, Gary, 275

Seager, Paul, 239
Séances, 91, 93–95, 112–113
Shakespeare, William, 111
Shanley, Garrett, 38–39
Shannon, Fred, 75
Sharpless, Brian, 35, 36, 49, 56, 58, 64, 65, 66
Shermer, Michael, 244
Simmons, Joseph, 292–294
Simon, Benjamin, 150, 152
Simons, Daniel, 83, 165
Simonsohn, Uri, 292–294
Sinclair, Mrs, 44–45
Singh, Simon, xxvii, xxviii, 264
Skeptic Trumps, xxvii
Skepticism, xvi, xx, xxii
Slade, Henry, 94
Sleep cycle, 50
Sleep hygiene, 65
Sleep onset REM period (SOREMP), 51
Sleep paralysis, 27, 33–67, 69, 81, 168–170, 211, 239
Smith, Gordon, 264
Smith, Martin J., 258
Social marginality hypothesis, 22, 26
Social psychology, 26, 28
Society for Psychical Research, 112, 255
Source attribution error, 178
South Bridge Vaults, Edinburgh, 100–101
Spanos, Nicholas, 162, 178
Spearpoint, Ken, 206–207
Spielberg, Steven, 139, 140
Spiritualism, 23, 91, 93–94, 111, 122
Sprinkle, R. Leo, 156
Stemmen, Roy, 180
Stevenson, Ian, 179, 182, 186
Stokes, Doris, 112
Stone, Anna, 2
Story, Ronald, xiii
Strieber, Whitley, 152–153, 168
Subjective validation, 235–237, 244, 275

Subliminal priming, 242
Succubus, 57
Sudden unexpected nocturnal death syndrome (SUNDS), 60–61, 63
Suggestibility, 105, 108, 166
Suggestion, power of, 94, 96–97, 100
Sullivan, Roy, 219
Supernatural, 9
Superstition, 11–12, 23, 29
Svartalfar, 58
Syllogistic reasoning, 237–240
System 1 thinking, 242–243, 245
System 2 thinking, 242–243, 245

Table tilting (a.k.a. table-turning, table-tapping and table-tipping), 125, 127, 285
Tandy, Vic, 78, 105
Taqamus, 180
Targ, Russell, 235–236
Tarot cards, 12, 15, 21
Tartarini, Fabio, 263
Taylor, John, xiv
Telepathy, 6, 7, 22, 26, 150, 228, 229, 235, 308, 314
Tellegen, Auke, 160
Temporal lobes, 102–103, 105, 199, 211
Temporoparietal areas, 199–200
Temporoparietal junction, 211–212
Tennison, Jane, 248
Terzaghe, Michele, 51
Thalbourne, Michael A., 23, 172, 240–241
Thurkettle, Vince, 139
Tighe, Virginia, 175–176, 177
Time distortion, 160
Tobacyk, Jerome J., 23
Tompkins, Matthew, 63, 86
Tooth fairy, 29
Top-down processing, 88–89, 142, 210–211, 240
Transcranial magnetic stimulation (TMS), 102

Transliminality, 240–242
Triple-blind, 114
Tunnel, 191, 198, 199
Tversky, Amos, 233
Type I errors (a.k.a. false positives), 243–244, 297
Type II errors (a.k.a. false negatives), 243, 297

Ufology, 8–9
Unlucky 13, 11
Unrealistic optimism (a.k.a. optimism bias), 300
Unusual bodily experiences, 51, 53–54

Validity, 23
Vampires, 10, 18
van Lommel, Pim, 194, 203, 208
Van Praagh, James, 111–112
Vestibular system, 53–54, 211
Virtual reality, 212, 213–214
Visual cortex, 199
von Däniken, Erich, xiii

Wagenmakers, Eric-Jan, 295–296
Wallace, Alfred Russell, 112–113
Watt, Caroline, xxiv, 239–240, 302
Weaver, Matthew, 258
Webb, Mark, 130
Weber, William, 77
Westover, William, 278
Westrum, Ron, 169
Whinnery, James, 198–199
Whitton, Kim, 264, 265
Wierzbicki, Michael, 239
Wikipedia, 2, 82
Wilcox, Maureen, 219
Wilde, Oscar, xxii
William, 41–42
Williams, Mark, 255
Wilson, Ian, 177
Wilson, Krissy, 99, 119, 161, 216, 259–260

Wilson, Sheryl C., 161–162
Wiltshaw, E., 197
Wiseman, Richard, xviii, xxi, xxiv, 80, 94–95, 96, 99–101, 117, 239–240, 262, 263, 280, 289
Witchcraft, 23, 56
Witches, 29
Woerlee, G. M., 202
Woolley, Jacqueline, 29
Wright, Elsie, 10–11

Yapko, Michael, 166
Youens, Tony, 97

Zener, Karl, 229
Zener cards, 229, 241
Zetetic, 314
Zombies, 10
Zusne, Leonard, 1